GEOGRAPHIES OF OBESITY

Geographies of Health

Series Editors
Allison Williams, Associate Professor, School of Geography and Earth Sciences, McMaster University, Canada
Susan Elliott, Dean of the Faculty of Social Sciences, McMaster University, Canada

There is growing interest in the geographies of health and a continued interest in what has more traditionally been labeled medical geography. The traditional focus of 'medical geography' on areas such as disease ecology, health service provision and disease mapping (all of which continue to reflect a mainly quantitative approach to inquiry) has evolved to a focus on a broader, theoretically informed epistemology of health geographies in an expanded international reach. As a result, we now find this subdiscipline characterized by a strongly theoretically-informed research agenda, embracing a range of methods (quantitative; qualitative and the integration of the two) of inquiry concerned with questions of: risk; representation and meaning; inequality and power; culture and difference, among others. Health mapping and modeling, has simultaneously been strengthened by the technical advances made in multilevel modeling, advanced spatial analytic methods and GIS, while further engaging in questions related to health inequalities, population health and environmental degradation.

This series publishes superior quality research monographs and edited collections representing contemporary applications in the field; this encompasses original research as well as advances in methods, techniques and theories. The *Geographies of Health* series will capture the interest of a broad body of scholars, within the social sciences, the health sciences and beyond.

Also in the series

Sense of Place, Health and Quality of Life
Edited by John Eyles and Allison Williams
ISBN 978 0 7546 7332 3

Primary Health Care: People, Practice, Place
Edited by Valorie A. Crooks and Gavin J. Andrews
ISBN 978 0 7546 7247 0

There's No Place Like Home: Place and Care in an Ageing Society
Christine Milligan
ISBN 978 0 7546 7423 8

Geographies of Obesity
Environmental Understandings of the Obesity Epidemic

Edited by
JAMIE PEARCE
University of Edinburgh, UK

KAREN WITTEN
Massey University, New Zealand

ASHGATE

© Jamie Pearce and Karen Witten 2010

All rights reserved. No part of this publication may be reproduced, stored in a retrieval system or transmitted in any form or by any means, electronic, mechanical, photocopying, recording or otherwise without the prior permission of the publisher.

Jamie Pearce and Karen Witten have asserted their rights under the Copyright, Designs and Patents Act, 1988, to be identified as the editors of this work.

Published by
Ashgate Publishing Limited
Wey Court East
Union Road
Farnham
Surrey, GU9 7PT
England

Ashgate Publishing Company
Suite 420
101 Cherry Street
Burlington
VT 05401-4405
USA

www.ashgate.com

British Library Cataloguing in Publication Data
Geographies of obesity : environmental understandings of
  the obesity epidemic. -- (Geographies of health)
  1. Obesity--Etiology. 2. Lifestyles--Health aspects.
  I. Series II. Pearce, Jamie. III. Witten, Karen.
  614.5'9398-dc22

Library of Congress Cataloging-in-Publication Data
Geographies of obesity : environmental understandings of the obesity epidemic / [edited] by Jamie Pearce and Karen Witten.
     p. ; cm. -- (Ashgate's geographies of health series)
  Includes bibliographical references and index.
  ISBN 978-0-7546-7619-5 (hardback) 1. Obesity--Epidemiology. 2. Obesity--Prevention. I. Pearce, Jamie. II. Witten, Karen.
  [DNLM: 1. Obesity--epidemiology. 2. Obesity--prevention & control. 3. Environment. 4. Risk Factors. 5. World Health. WD 210 G352 2009]

RC628.G476 2009
614.5'9398--dc22

2009031068

ISBN 9780754676195 (hbk)
ISBN 9780754699415 (ebk)

Printed and bound in Great Britain by
MPG Books Group, UK

Contents

List of Figures   *vii*
List of Tables   *ix*
Contributors' Biographies   *xi*
Foreword   *xxi*
Acknowledgements   *xxiii*

PART I   INTRODUCTION

1   Introduction: Bringing a Geographical Perspective to Understanding the 'Obesity Epidemic'   3
Jamie Pearce and Karen Witten

2   The Emerging Obesity Epidemic: An Introduction   15
Barry M. Popkin

3   Contextual Determinants of Obesity: An Overview   39
Daniel Kim and Ichiro Kawachi

PART II   FOOD ENVIRONMENT AND OBESITY (ENERGY IN)

4   Changing Food Environment and Obesity: An Overview   57
Janet Hoek and Rachael McLean

5   Understanding the Local Food Environment and Obesity   79
Lukar E. Thornton and Anne M. Kavanagh

6   Childhood Obesity and the Food Environment   111
Mat Walton and Louise Signal

PART III   PHYSICAL ACTIVITY, ENVIRONMENT AND OBESITY (ENERGY OUT)

7	The Role of the Changing Built Environment in Shaping Our Shape *Billie Giles-Corti, Jennifer Robertson-Wilson,* *Lisa Wood and Ryan Falconer*	131
8	Understanding the Local Physical Activity Environment and Obesity *Gavin Turrell*	151
9	Childhood Obesity, Physical Activity and the Physical Environment *Melody Oliver and Grant Schofield*	175

PART IV   OBESOGENIC ENVIRONMENTS AND POLICY RESPONSES

10	Policy Responses and Obesogenic Food Environments *Katrina Giskes*	207
11	Policy Responses and the Physical Environment *Mylène Riva and Sarah Curtis*	227

PART V   FUTURE RESEARCH CHALLENGES

12	Residential Environments and Obesity – Estimating Causal Effects *Graham Moon*	251
13	Measuring Obesogenic Environments – Representing Place in Studies of Obesity *Dianna Smith, Kim Edwards, Graham Clarke and Kirk Harland*	277
14	Recourse to Discourse: Talk and Text as Avenues to Understand Environments of Obesity *Robin Kearns*	297

PART VI   CONCLUSIONS

15	Conclusions: Common Themes and Emerging Questions *Jamie Pearce and Karen Witten*	313

Index   *323*

List of Figures

1.1	Examples of environmental characteristics at a range of spatial scales that potentially influence i) food consumption (energy in) and ii) physical activity (energy out)	5
2.1	Global overweight and obesity	17
2.2	Obesity patterns across the world	18
2.3a	Overweight and underweight prevalence in women 20-49y in 36 developing countries ranked by gross national income (gni) per capita – Urban women	19
2.3b	Overweight and underweight prevalence in women 20-49y in 36 developing countries ranked by gross national income (gni) per capita – Rural women	20
2.4	Annual absolute change in the prevalence of overweight and obesity in seven countries from 1985/1995 to 1995/2006	22
2.5	Relationship between the percentage of energy from fat and GNP per capita, 1962 and 1990	25
2.6	Relationship between levels of gross national product per capita and caloric sweetener	27
3.1	International Obesity Task Force Working Group conceptual model	44
5.1	Conceptual framework for the study of nutrition environments	81
6.1	Children's food environment	114
9.1	Percentage of youth meeting guidelines for moderate-to-vigorous physical activity	176
9.2	Conceptual model of environmental factors that may influence children's physical activity	177
10.1	Components of the food supply chain	210
11.1	Policy levels, environments and settings influencing physical activity	229

13.1	Provision scores for access to food outlets in Cardiff, 2002	280
13.2	Fruit and vegetable consumption (>5 portions/day) by OA	287
13.3	Childhood obesity across Leeds (proportion of measured children, in each area, who are obese)	287
13.4	Output areas (the dark shading) where key local determinants of childhood obesity differ from the key global determinants	288

List of Tables

3.1	The ANGELO conceptual framework	42
5.1	A brief overview of studies investigating objective measures of the food environment with individual dietary behaviours and health outcomes	85
8.1	Physical and social characteristics of the neighbourhood environment that have been examined in relation to obesity	154
9.1	Environmental associates of activity in children: Studies reviewed and key findings	179
10.1	Policy instruments under the jurisdiction of different levels of governance that may address the food environment	212
12.1	The Bradford Hill criteria for causation	253
12.2	The hierarchy of epidemiological evidence	255
13.1	Estimating variations in factors associated with obesity	286
14.1	Summary of complementary case studies	298

Contributors' Biographies

Editors

Jamie Pearce (Institute of Geography, School of Geosciences, University of Edinburgh, UK)
Jamie Pearce is a Reader in Human Geography at the University of Edinburgh. He is also an Adjunct Associate Professor at the Department of Geography, University of Canterbury, New Zealand and co-Director of the *GeoHealth Laboratory*. Jamie's work considers processes operating at a range of geographical scales that drive and perpetuate social and spatial inequalities in health. Recent work has focused on the role of context/local neighbourhood in shaping health outcomes and health-related behaviours, including nutritional status, physical activity and obesity. Recently published work has concentrated on the polarisation of health inequalities in New Zealand during the 1980s and 1990s as well as developing new methods to understand the role of neighbourhood as a mediator in the relationship between poverty and health inequalities.

Karen Witten (SHORE, Massey University, New Zealand)
Karen Witten is Associate Professor at the Centre for Social and Health Outcomes Research and Evaluation, Massey University, New Zealand. Karen's research interests centre on interactions between the physical characteristics of neighbourhoods and cities and the social relationships, transport choices and well being of the people living in them. Recent work has included studies on the impacts of school closure of the social relations of place, spatial inequities in access to services and amenities and the impact of differential access on health outcomes, and the relationship between neighbourhood walkability, transport opportunities and choices and physical activity. She has also coordinated the evaluation of a number of locality based community development projects directed at improving the health and social outcomes of residents.

Contributors

Graham Clarke (School of Geography, University of Leeds, UK)
Graham Clarke is Professor of Business Geography at the University of Leeds. His academic career has been driven by a desire to apply spatial modelling methodologies to solve real problems in both the public and private sector. His primary area of work has been in GIS and applied spatial modelling (spatial

interaction and microsimulation models) where he has worked with colleagues on many applications within retailing and health in particular. Research funding from Joseph Rowntree and British Telecom has been used to examine future trajectories of urban welfare (by dynamically simulating urban and regional populations in Britain – ultimately building up to a national SimBritain model). Recent health applications of spatial microsimulation include the study of obesity, type 2 diabetes and the location of community services for anti-smoking and maternity care. A major contribution to research in retail location planning has been to rethink the research agenda in relation to the growth strategies of retail organizations. One such approach has been to focus on the geography of mergers and acquisitions from the point of view of optimising strategic planning. Another important research stream has been work on retail provision, saturation and internationalisation.

Sarah Curtis (Department of Geography, University of Durham, UK)
Sarah Curtis is Professor of Health and Risk at Durham University Department of Geography. She is an internationally renowned specialist in the geography of health focusing on the geographical dimensions of inequalities of health and health care. Her scholarship elucidates how, and why varying geographical settings relate to human health inequalities. Her work has strong applied as well as theoretical aspects, contributing to health policy development and evaluation of health services in the UK, France, Russia, Poland, Canada and the US. Recent research includes: health impact assessment of urban regeneration schemes (for the Department of Health, and other agencies) and development of healthy public policy (with agencies in Canada and UK). Her other research includes studies of effects of the social and physical environment on well-being, resilience and health of adults and children; studies of therapeutic design of psychiatric health care settings; international collaborative work on migration, health and wellbeing in France and Britain; comparative research on geographical variation in psychiatric service use in New York State, USA. Sarah Curtis is author of *Health and Inequality: Geographical Perspectives* (2004, Sage). She has supported academic research and learning in health geography, especially through contributions to the Royal Geographical Society/Institute of British Geographers and its Geography of Health Research Group, and as Senior Editor (Medical Geography) for *Social Science and Medicine*.

Kim Edwards (Department of Epidemiology, University of Leeds, UK)
Kim Edwards is a lecturer in Epidemiology at the University of Leeds. Following a first degree at the University of Sheffield in 'Human Nutrition' she came to the University of Leeds with an ESRC/MRC scholarship in 2004. Working with Graham Clarke, Janet Cade and Joan Ransley she examined the relationship between childhood obesity and obesogenic environment and behaviours (e.g. social capital, deprivation, access to facilities, diet, physical activity, sedentary behaviour). At the heart of this analysis was a spatial microsimulation model called SimObesity.

Following her PhD Kim is now also working on the relationship between fast food (location and consumption) and obesity and the relationship between green spaces, physical activity and obesity. She is also a member of LIGHT, the Leeds Institute of Genetics, Health and Therapeutics (LIGHT). This multi-disciplinary unit performs translational research into complex chronic disorders (including cardiovascular disease, diabetes, cancer and neurodegenerative diseases) to improve the delivery of patient care.

Ryan Falconer (Centre for the Built Environment and Health, School of Population Health, University of Western Australia, Australia)
Ryan Falconer is a Transport and Land Use Planner with experience in policy analysis, strategic planning, TOD, integrated transport planning and transport decision-making. Since April 2008, Ryan has worked for Sinclair Knight Merz as a Transport Planner. During this time he has made significant contributions to the preparation of structure plans for major TOD projects in Perth, the formulation of a Transport Strategy for The University of Western Australia, and planning for the redevelopment of the Fremantle Public Transport Interchange. Ryan has a Doctorate in Sustainable Transport, which he completed in Murdoch University's Institute for Sustainability and Technology Policy. His research was undertaken as part of the RESIDential Environments Project (RESIDE) and was part-funded by the Heart Foundation and the Australian Research Council. He has previously worked for the Christchurch City Council (New Zealand) as an Urban Planner and the Canterbury Regional Council as a Travel Behavior Analyst. He has presented numerous transport-related papers both nationally and internationally. In June 2007, Ryan was the transport track chair and organizer at the 13th Annual International Sustainable Development Research Conference.

Billie Giles-Corti (Centre for the Built Environment and Health, School of Population Health, University of Western Australia, Australia)
Billie Giles-Corti is Director of the Centre for the Built Environment and Health at the School of Population Health, The University of Western Australia (UWA) and a National Health and Medical Research Council (NHMRC) Senior Research Fellow. For more than a decade, she and a multi-disciplinary team of researchers and postgraduate research students at UWA have been studying the impact of the built environment on health, social and health behaviour outcomes including walking, cycling, public transport use, overweight and obesity, social capital and dog walking. A leading health promotion researcher in Australia and recognized internationally for her research on the built form, Professor Giles-Corti serves on numerous international, national and state committees and boards. In 2007, she was awarded a Fulbright Senior Scholar Award that enabled her to spend four months at Stanford University, USA focused on establishing research on the built environment and older adults.

Katrina Giskes (School of Public Health, Queensland University of Technology, Australia)
Katrina Giskes is a Research Fellow in the School of Public Health/Institute of Health and Biomedical Innovation at Queensland University of Technology and from 2004-2008 was an Australian NHMRC Sidney Sax (International) Postdoctoral Research Fellow. She holds a Bachelor of Health Science (Nutrition and Dietetics) with first-class Honours, and a PhD in Nutritional Epidemiology. Her principal research area is socioeconomic inequalities in health and health-related behaviours, and the determinants of these inequalities. Her work includes examining the health impact of social and economic disadvantage at the individual, household and area levels. She has also worked on policy research examining and translating the findings of her own (and other's) research into strategies to reduce health inequalities through public policy, health policy and health promotion interventions.

Kirk Harland (School of Geography, University of Leeds, UK)
Kirk Harland is a Research Fellow at the University of Leeds. Both his undergraduate degree and PhD were devoted to the study of computational geography with a particular focus on modelling social systems and the application of advanced spatial analysis techniques. The technical demands of computational geography have prompted Kirk to develop a core set of technical competencies in Geographical Information System (GIS), information management, database design and to become a Sun Certified Java Programmer. He has over ten years experience of analysing and modelling spatial phenomena ranging from change detection in national parks to developing decision support systems for major international corporations. Over the last four years his research interests have been concentrated on planning, assessing and managing the impacts of change in education and health services.

Janet Hoek (Department of Marketing, University of Otago, New Zealand)
Janet Hoek has a PhD in marketing and a Masterate in early medieval poetry. She is a Professor of Marketing at the University of Otago where she teaches branding and explores the marketing and public policy nexus. Her research has explored tobacco control initiatives, including acting as an expert witness in litigation taken against the tobacco industry, and food marketing and obesity where, along with colleagues from the Centre for Translational Research into Chronic Disease at the University of Otago, she is evaluating the government's Healthy Eating Healthy Action (HEHA) strategy. She was appointed as an expert advisor to the New Zealand Health Select Committee enquiry into obesity and type 2 diabetes and has served on advisory groups developing social marketing campaigns designed to promote healthier eating habits. Janet is a member of the Health Research Council funded teams examining the longer term effects of alcohol marketing on young people, and noise induced hearing loss. She has acted as an expert witness in several intellectual property disputes, including providing evidence in the High Court of New Zealand.

Anne M. Kavanagh (Key Centre for Women's Health in Society, University of Melbourne, Australia)
Anne M. Kavanagh is a Professor of Epidemiology at the University of Melbourne, Australia. She has a prominent research record in social epidemiology with a focus on neighbourhood determinants of health. She has led an extensive research programme into the influence of the local environment on obesity- and diet-related health outcomes. Anne has a particular interest in the role of the local food and physical activity environments in shaping health outcomes, an area in which she has published extensively in leading international public health journals.

Ichiro Kawachi (Department of Society, Human Development and Health, Harvard School of Public Health, USA)
Ichiro Kawachi, is Professor of Social Epidemiology and Chair of the Department of Society, Human Development and Health at the Harvard School of Public Health. Kawachi received both his medical degree and PhD (in epidemiology) from the University of Otago, New Zealand. Since 1992, he has taught at the Harvard School of Public Health. Kawachi has published over 300 articles on the social and economic determinants of population health. He was the co-editor (with Lisa Berkman) of the first textbook on *Social Epidemiology*, published by Oxford University Press in 2000. His other books include *The Health of Nations* with Bruce Kennedy (The New Press, 2002), *Neighborhoods and Health* with Lisa Berkman (Oxford University Press, 2003), *Globalization and Health* with Sarah Wamala of the Swedish National Institute of Public Health (Oxford University Press, 2006), and *Social Capital and Health* (Springer, 2008) co-edited with S.V. Subramanian and Daniel Kim. Kawachi currently serves as the Senior Editor (Social Epidemiology) of the international journal *Social Science and Medicine*.

Robin Kearns (School of Geography, Geology and Environmental Science, University of Auckland, New Zealand)
Robin Kearns is a health geographer and Professor of Geography at the University of Auckland. His research focuses on the role of place in shaping health experience and the place of (particularly health care) services in people's lives. Specifically, he has investigated the effects of social and economic restructuring on community well-being, changing geographies of mental health care, and children's experience of neighbourhoods. Robin's current funded research includes: neighbourhoods and physical activity; activism in the voluntary sector, home maintenance while aging in place; and changing meanings of coastal environments. He obtained his PhD at McMaster University (Canada) in 1987 supported by a Commonwealth Scholarship. He is author of various refereed papers and book chapters as well as two books with Wilbert Gesler, the most recent being *Culture/Place/Health* (Routledge, 2002). He is Associate Editor of *Health and Social Care in the Community* and *Heath and Place* and is a ministerial appointment on New Zealand's National Health Committee.

Daniel Kim (Department of Society, Human Development at Health, Harvard School of Public Health, USA)
Daniel Kim is Research Associate in the Department of Society, Human Development, and Health at the Harvard School of Public Health. He received his medical degree from the University of Toronto and his PhD from the Harvard School of Public Health. Kim's published research has focused on the linkages between neighbourhood socioeconomic as well as other contextual factors (such as social capital) and individual health outcomes including obesity. He co-edited the book *Social Capital and Health* (Springer, 2008) with Ichiro Kawachi and S.V. Subramanian.

Rachael McLean (Edgar National Centre for Diabetes Research, University of Otago, New Zealand)
Rachael McLean works as a Research Fellow at the Edgar National Centre for Diabetes Research, and Lecturer in Epidemiology at the University of Otago. She is medically trained, and worked as a General Practitioner before undertaking her specialist training in Public Health Medicine. She has an interest in obesity and chronic disease prevention, and interventions to promote healthy lifestyles.

Graham Moon (School of Geography, University of Southampton, UK)
Graham Moon is Professor of Spatial Analysis in Human Geography and Co-Director of the Centre for Geographical Health Research at the University of Southampton. His current research focuses on geographical aspects of health-related behaviours and draws particularly on the analysis of secondary datasets using multilevel modelling. He also conducts research on psychiatric and primary health provision. He has authored over 100 academic papers and book chapters, and edited or authored six books. Support for Professor Moon's research has come from the ESRC, UK and international health agencies and the private sector. He is founding editor of the journal *Health and Place* and has been an advisor to UK and overseas governments.

Melody Oliver (Centre for Physical Activity and Nutrition Research, AUT University, Auckland, New Zealand)
Melody Oliver is a National Heart Foundation post-doctoral Research Fellow at AUT University, in Auckland, New Zealand, and is involved in multiple cross-sectional and longitudinal research projects investigating associates of objectively-assessed physical activity in children and adults. Melody has a broad interest in physical activity and health at a population and individual level, including the objective measurement of physical activity dimensions (habitual, transport-related, activity intensity, sedentary behaviours), and the integration of objective activity measurement and associated tools (e.g. pedometry, accelerometry, GPS, and GIS). She is particularly interested in physical activity and sedentary habits formed in childhood and how these can be influenced by children's social and built environments.

Barry M. Popkin (Interdisciplinary Center for Obesity, University of Northern Carolina at Chapel Hill, USA)
Barry Popkin is the The Carla Smith Chamblee Distinguished Professor of Global Nutrition, at UNC-CH where he directs the UNC-CH's Interdisciplinary Center for Obesity. He has a PhD in economics. He developed the concept of the Nutrition Transition, the study of the dynamic shifts in dietary intake and physical activity patterns and trends in obesity and other nutrition-related noncommunicable diseases. His research programme focuses globally on understanding the shifts in stages of the transition and programmes and policies to improve the population health linked with this transition (see www.nutrans.org). His research is primarily funded by a large number of NIH R01's. He has published 290 refereed journal articles and is the author of a new book entitled the WORLD IS FAT (January 2009, Penguin Publishers).

Mylène Riva (Department of Geography, University of Durham, UK)
Mylène Riva is a Research Associate at the Institute of Hazard and Risk Research, Department of Geography, Durham University and Fellow of the Wolfson Research Institute, Durham University Queen's Campus, UK. She holds a Bachelor in Geography and a PhD in Public Health and Health Promotion from the University of Montreal, Canada. Mylène's research focuses on social inequalities in health and well-being in urban and rural areas in Canada and the UK, and on the social and structural determinants of these inequalities. Mylène's doctoral thesis examined the links between the built and socioeconomic context of urban areas and physical activity, especially walking behaviours, from both public health and geographical perspectives. Her current research activities investigate changes through time in the socioeconomic context of urban and rural areas and how these relate to longitudinal changes in population health and well-being. Mylène's approach to examining links between place and health involves secondary analyses of large georeferenced datasets containing information on population health and the conditions of local areas, and applications of quantitative methods such as multilevel modelling, longitudinal analysis, spatial analysis and geographical information systems.

Jennifer Robertson-Wilson (Department of Kinesiology and Physical Education, Wilfrid Laurier University, Canada)
Jennifer Robertson-Wilson is an Assistant Professor in the Department of Kinesiology and Physical Education at Wilfrid Laurier University. Her primary research uses a social ecological lens to better understand youth physical activity within the school context. She is particularly interested in the school physical activity environment and school physical activity-related policy. Her research also includes a focus on the relationships between elements of the built environment, obesity and physical activity. She is currently the Chair of the Pedestrian Charter Steering Committee of Waterloo Region, a citizen's group who uses advocacy and education to promote a pedestrian-friendly and walkable community.

Grant Schofield (Centre for Physical Activity and Nutrition Research, AUT University, Auckland, New Zealand)
Grant Schofield is Professor of Public Health and Director of the Centre for Physical Activity and Nutrition Research at AUT University, in Auckland, New Zealand. Grant brings a psychology background to his current research and teaching in physical activity, nutrition, and health. His primary interests lie in physical activity, lifestyle habits (including nutrition), and obesity. His research is centred firmly around these interests with several large projects including a strong focus on preschool, child and adolescent physical activity measurement, surveillance and intervention; workplace physical activity, health and productivity measurement and intervention; built environment influences on walkability, car dependency, and sustainability; and whole-of-community physical activity health promotion. He has a well-proven research and publication history, having been a lead or named investigator in research projects totalling over NZ$8 million, and having published over 200 peer-reviewed journal publications, book chapters, and conference presentations.

Louise Signal (Department of Public Health, University of Otago, New Zealand)
Louise Signal is a Senior Lecturer at the Department of Public Health. Louise is a social scientist with a PhD in Community Health from the University of Toronto. She has worked and done research in the field of health promotion for 20 years in a range of roles, including Senior Advisor (Health Promotion) for the New Zealand Ministry of Health. Her research interests include tackling inequalities in health, health impact assessment (HIA), tobacco control, and healthy eating and healthy action. Louise is an Executive Director of the Health Promotion and Policy Research Unit and HIA Research Unit. Currently, Louise co-leads a project for the Ministry of Health to assist in reducing inequalities in health. Louise has a number of community service roles including being a member of the Health Promotion Committee of the New Zealand Cancer Society. She teaches undergraduate and postgraduate courses in health promotion, and co-convenes the undergraduate, distance-taught Certificate in Health Promotion.

Dianna Smith (Department of Geography, University of London, UK)
Dianna Smith holds a Medical Research Council (MRC) Population Health Scientist Fellowship in the Department of Geography at Queen Mary, University of London. Her educational background is broadly in geography, with a BA in environmental management (Geography, Texas), MS on food access (Geography, Oregon State) and PhD (Geography, Leeds). Her PhD thesis explored the potential impacts of retail food access on adult obesity and diabetes prevalence in regions of England. Dianna has written on methodological advancements in population simulation for public health policy for *Environment and Planning A* and *Studies in Regional Science*. This work has been further utilized in local branches of the National Health Service (NHS). Her post-doctoral experience has included collaboration on several projects related to obesity, local environments. Dianna's

current work with the MRC explores the impacts of proposed environmental interventions on health outcomes by using simulation methodologies in the UK and abroad.

Lukar E. Thornton (Centre for Physical Activity and Nutrition Research, Deakin University, Australia)
Lukar E. Thornton is a Post-Doctoral Research Fellow at the Centre for Physical Activity and Nutrition Research at Deakin University. His research interests include environmental determinants of dietary behaviours and the development of more sophisticated measures of food access, both through improved conceptualisation and the utilisation of GIS. Lukar completed his PhD at the University of Melbourne in 2008 and during his candidature was involved in a broader research programme investigating neighbourhood determinants of health behaviours. He has also assisted with the translation of research findings to key stake-holders including Local Governments.

Gavin Turrell (School of Public Health, Queensland University of Technology, Australia)
Gavin Turrell is an Associate Professor in the School of Public Health at Queensland University of Technology and is supported by a NHMRC Senior Research Fellowship (2006-2010). His primary research interests are in social epidemiology, with a particular focus on the social determinants of health and health inequalities. His research is mainly population-based, examining how social and economic factors (measured at the individual, group, and area levels) influence health (mortality and morbidity) and health-related behaviours (diet, smoking and physical activity). His work is increasingly focusing on ways to reduce health inequalities through public policy, health policy, health promotion, and other intervention strategies.

Mat Walton (Department of Public Health, University of Otago, New Zealand)
Mat Walton is a doctoral candidate with the Health Promotion and Policy Research Unit, University of Otago, Wellington. Mat's PhD is investigating public policy options to support the role of schools in promoting healthy childhood nutrition. Mat comes from a background of policy analysis with both local and central government, focusing on children and young people, community health and equality in social policy delivery.

Lisa Wood (Centre for the Built Environment and Health, School of Population Health, University of Western Australia, Australia)
Lisa Wood is a postdoctoral fellow on a NHMRC funded Capacity Building Grant. She also holds a Bachelor of Commerce degree (UWA) majoring in marketing and management (First Class Honours) and a Postgraduate Diploma in Health Promotion. Dr Lisa Wood is a post doctoral research fellow at the University of Western Australia and a public health consultant, with 20 years experience in the

arena of health promotion. Her PhD research examined the relationship between neighbourhood environments, social capital and health. Current research interests include: social capital and sense of community; urban design/built environment and health; social determinants of health; healthy communities; tobacco; life-course approaches to health; Aboriginal health; and the translation of research into policy and practice. She also is involved in projects that seek to apply public health models to the issues of domestic violence and child abuse prevention. Lisa has worked with both government and non-government organizations and consulted within Australia and overseas. She has a strong interest and commitment to research that is useful and relevant to the 'real world' and seeks to integrate her health promotion policy and programme experience and collaborative networks into her approach to research.

Foreword

As we all know by now, there have been, and are projected to be, steep and steady rises in overweight and obesity all over the world – in low and middle income countries as well as in high income countries. These increases are predicted to have serious consequences for the health of children and adults, and for many economies. The speed of increase in obesity renders it implausible that this has been caused by a sudden large-scale genetic change, so attention has been increasingly focused on the role of the environment in promoting obesity. Ideas about environmental influences on obesity raise fascinating political and ethical issues in relation to issues such as 'who is to blame', 'the role of the state and the individual', and 'what could and should be done' (Nuffield Council for Bioethics 2007).

This volume provides a comprehensive overview of all the above trends, with contributors from North America, Australasia and the UK covering obesity trends and their environmental drivers in a wide range of societies, including Africa, China and India, and in an evolutionary perspective. Equal attention is given to diet and physical activity. The editors are to be congratulated for assembling such an excellent cast of contributors, and providing both breadth and depth.

The chapters, both individually and collectively, highlight how complex a phenomenon the rise of obesity is, and how difficult it is (and will be) to deal with it effectively because of this complexity (which was well illustrated in the complex 'wiring diagram' in the UK's Foresight report on obesity (Government Office for Science 2007)).

Despite the plausibility in the abstract of there being strong environmental influences on diet and physical health which are driving the obesity epidemic, it is clear from the research summarized in this volume that empirical evidence of specific and universal links between environmental exposures and diet and physical activity behaviours is somewhat equivocal. Detailed empirical studies in specific settings do not all show the expected relationships between environmental exposures and behaviours; observed relationships may be weak; or the relationships observed may vary between different contexts or different social groups. Contributors are admirably frank about this both when reporting their own work and reviewing that of others, and issues of limitations in study design and methods are given full consideration.

The omnipresence and speed of environmental change relevant to obesity may pose limitations for detecting its determinants. As Geoffrey Rose pointed out: 'If everyone in the country had smoked 20 cigarettes a day then clinical, case control and cohort studies alike would have led us to conclude that lung cancer was a genetic disease; and in one sense that would have been true, since if everyone is exposed to the necessary external agent then the distribution of cases becomes wholly determined by individual susceptibility' (Rose 1987, p. 963). If we are all exposed to a more obesogenic environment than we were 50 years ago, paradoxically it may be difficult to detect environmental influences on diet and physical activity since there is no longer sufficient variation in exposures within a population. This poses a challenge to researchers and policymakers concerned with unpicking the relative importance of environmental and individual characteristics driving the obesity epidemic, and seeking effective levers for reversing current trends. A conclusion of this book is that much of the research on environmental drivers is 'in its infancy'. It is to be hoped that future research can build on the excellent foundations described here in order to further our understanding of the complex web of influences on obesity.

Government Office for Science. 2007. *Tackling Obesities: Future Choices*. London, UK: Foresight.
Nuffield Council for Bioethics. 2007. *Public Health: Ethical issues*. London: Nuffield Council for Bioethics.
Rose, G. 1987. Environmental factors and disease: The man made environment. *British Medical Journal*, 294, 963-965.

Sally Macintyre
MRC Social and Public Health Sciences Unit,
University of Glasgow, Scotland

Acknowledgements

Production of this book has involved coordinating the efforts of many dedicated individuals. We wish to thank the chapter authors for their enthusiasm for the project and their timely contributions. Without their collaboration the book would not have been possible. A special thank you goes to Lisa Morice and Jan Sheeran (both at SHORE, Massey University) for their dedicated assistance in preparing the manuscript for publication.

We would also like to acknowledge the Health Research Council of New Zealand as it was through their funding of the Neighbourhoods and Health project that our interest and research in the geographies of obesity developed.

PART I
Introduction

Chapter 1
Introduction:
Bringing a Geographical Perspective to Understanding the 'Obesity Epidemic'

Jamie Pearce and Karen Witten

The Global Context

Globally, the sharp rise in the rates of obesity among both children and adults is a significant public health concern. Since 1980, in some areas such as parts of North America, the United Kingdom, Eastern Europe, the Middle East, the Pacific Islands, Australasia and China obesity rates have risen three-fold or more (World Health Organization 2003). There are more than one billion overweight adults and at least 300 million of them are obese (World Health Organization 2003). By the year 2000 in the United States, over 64 per cent of the population were overweight or obese (Flegal et al. 2002), accounting for somewhere between 100,000 to 300,000 deaths each year (Flegal et al. 2005, Allison et al. 1999). Similarly, rates of childhood obesity have increased over the same period of time (Wang and Lobstein 2006). In the United States for example since the mid-1970s obesity prevalence among children aged 6 to 11 has increased from 6.5 to 17.0 per cent, and for those aged between 12 and 19 from 5.0 to 17.6 per cent (Centers for Disease Control and Prevention 2006). The rise in obesity rates is not limited to industrialized countries as this increase is often more profound in developing countries than in the developed world. The increasing global prevalence of obesity has led to what has been termed the 'obesity epidemic' (Hill and Peters 1998).

A number of negative physical, social and mental health consequences are associated with obesity (Flegal et al. 2007), which now rivals smoking as a public health issue. As a major risk factor for type 2 diabetes, various cancers, cardiovascular disease and hypertension, excess body weight is recognized as a significant health burden for individuals and health systems in many countries (Must et al. 1999). At the same time, the distribution of overweight and obesity is not evenly shared between all social and ethnic groups, or across geographical areas (McLaren 2007, Wang and Beydoun 2007, King et al. 2006). In most developed countries, rates of overweight and obesity tend to be higher among more socially disadvantaged groups. For example, in England obesity rates (obese or morbidly obese) are 68 per cent higher among women in the lowest income quintile of households compared to the highest income quintile (although this trend is less strong among men) (National Centre for Social Research 2008). There are also inequalities in obesity prevalence between geographical areas. In New Zealand for example, rates of obesity are almost twice as high in the most deprived quintile of neighbourhoods compared to the least deprived quintile (Ministry of Health 2008). Similarly, recent trends in developing countries show a shift in obesity prevalence from higher to lower socioeconomic groups. Obesity-related behaviours and health outcomes are likely to be on the causal pathway between social disadvantage and health and therefore contribute to the sharp increase in health inequalities noted in many countries in recent years.

Geographical Perspectives

Although the increasing prevalence of obesity is well documented, explanations for the emerging epidemic, as well as its socio-spatial distribution have proven more elusive. At a fundamental level, obesity arises from an imbalance between the quantity of energy consumed and the amount expended (the energy balance equation). Whilst genetics are a risk factor in explaining an individual's susceptibility to weight gain, it is implausible that they account for the sharp increase in the global prevalence of obesity over such a short period of time. Rather, the rising epidemic reflects profound changes in society over the past two to three decades that have been accompanied by rapid and widespread changes to the environment. These environmental modifications have created a climate that is conducive to increased energy consumption and reduced energy expenditure at a population-level. Therefore, to understand the worldwide rise in obesity prevalence it is necessary to consider a whole host of environmental factors.

Researchers are beginning to pinpoint a suite of environmental features at a range of spatial scales that are integral to influencing energy consumption and reducing energy expenditure. Adequate theorising and empirical testing of the pathways through which environmental factors contribute to obesity is an essential step to understanding the 'obesogenic' characteristics of the environments in which people live, work and play. This wider perspective is consistent with the resurgence of interest in the public health and epidemiology literature on 'place-based' determinants of health and health inequalities. Researchers are increasingly recognizing that the identification of mechanisms linking the features of places to a range of health outcomes (e.g. mortality and morbidity) and health-related behaviours (e.g. smoking) has significant potential for developing successful policy initiatives. Adapting the environment in which people live with a view to improving health outcomes underpins policy frameworks such as the WHO Healthy Cities programme and various urban renewal strategies (Blackman 2006, Davies and Kelly 1993).

The assertion that the environment is integral to understanding obesity raises a number of questions. What is meant by the 'environment'? What are the key environmental features then affect energy intake and energy expenditure? At which geographical scales do these mechanisms operate? In this context, the environment is defined broadly and refers to all factors that are external to the individual including the social, political, economic, built or biophysical spheres. There are numerous theoretical pathways through which features of the environment can influence obesity. The environment can exert a putative influence on obesity at different spatial scales ranging from the macro (e.g. global or national) to the micro (e.g. in schools or workplaces). Figure 1.1 demonstrates a selection of environmental characteristics at assorted scales that potentially shape individual-level food consumption (energy in) and physical activity (energy out). For instance, food consumption may be influenced by macro-level forces such as global food production or international trade agreements. These forces can alter the availability and cost of foods of different nutritional value. At the national-level, the regulation or codes of conduct for advertising food, particularly to children, varies between countries. Different degrees of exposure to advertising and other marketing strategies between countries may partially explain international variations in obesity levels. Finally, at the local level, promotional offers on energy dense food in a neighbourhood supermarket may influence the purchasing decisions, and ultimately the nutritional intake, of its customers.

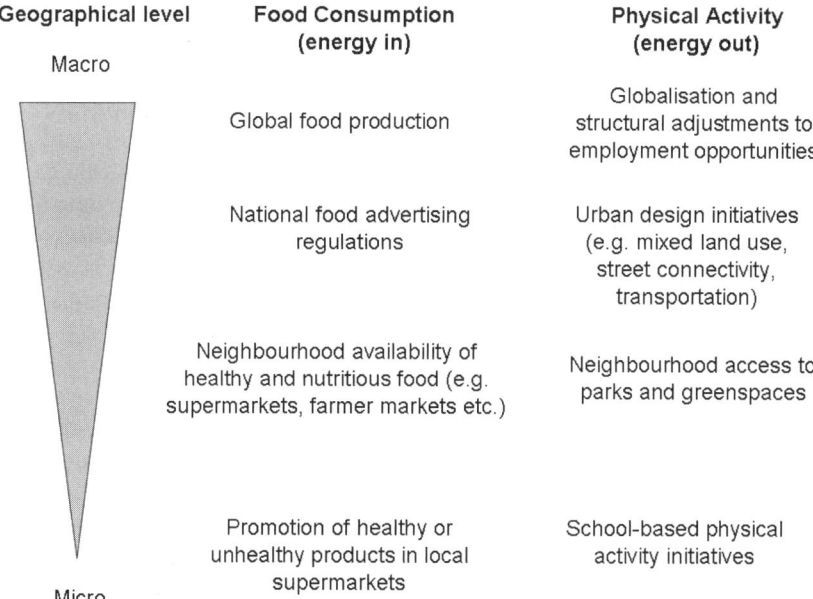

Figure 1.1 Examples of environmental characteristics at a range of spatial scales that potentially influence i) food consumption (energy in) and ii) physical activity (energy out)

With regards to energy expenditure, levels of physical activity may be influenced by patterns of trade associated with globalisation and urbanisation which in turn shape the types of employment opportunities available in different places. In many countries these structural adjustments have been associated with a shift from manual (and often physically active) occupations to more sedentary employment opportunities. At an intermediate level, the urban design strategies of central and local governments are likely to influence the 'walkability' of urban areas. For instance, mixed land use and the variety of community resources accessible locally, such as places of work, parks and retail outlets, increase opportunities for local residents to walk or cycle around their neighbourhood, which in turn is likely to increase physical activity at the population-level. At a micro-level, the strategies of local schools to integrate regular physical activity into the school day can vary significantly. Similarly, characteristics of other settings such as workplaces may assist or hinder employees in remaining physically active.

Policy Perspectives

Given the sharp rise in obesity rates worldwide and the associated health, economic and social costs, it is perhaps unsurprising that weight and obesity has become a policy priority for many national governments. Key policy documents in most countries tend to use staunch language such as 'fighting', 'tackling' or 'battling' the obesity epidemic, implying that firm evidence-based action is being taken. Although the fundamentals of building healthy public policy and creating supportive environments have been central pillars of health promotion theory and practice since they were formally incorporated into the WHO's Ottawa Charter in 1986, to date, strategies to address obesity have been primarily focused on behavioural, educational and medical interventions with a view to encouraging individuals to eat more healthily and exercise more. This individualized perspective is consistent with the neoliberal political agenda implemented over the past 20 to 30 years in developed countries including the United Kingdom, United States and Australia. Health promotion policies in these countries have been consistent with this philosophical direction and tended to devolve responsibility for health from the state to the individual.

Although individualistic approaches to improving health outcomes have been the international vogue for some years, they can be heavily criticized. Overlooking the complex social contexts in which health-related 'decisions' are made is likely to be a major impediment to successful policy development. Health-related behavioural decisions such as smoking and alcohol intake as well as food consumption, nutritional intake and physical activity of course are not taken in a vacuum, but rather are shaped by the political, social, cultural and physical contexts in which they are made. As Caballero (2007) notes '…political leaders still tend to regard obesity as a disorder of individual behaviour, rather than highly conditioned by the socioeconomic environment' (p. 4). Policy approaches that predominantly focus on changing behaviour through educational programmes are unlikely to be sufficiently potent to reduce the prevalence of obesity. Most people will struggle to alter their behaviour in environments that promote high energy intake and physical inactivity. It is therefore unsurprising that policies to address the sharp increase in obesity prevalence have had limited success.

The notion that the environmental context is important for understanding health outcomes is supported by the recommendations from the recent report from the World Health Organization's Commission on Social Determinants of Health (2008) entitled *Closing the Gap in a Generation: Health Equity Through Action*. The report emphasizes the wider conditions of daily life that are integral to understanding and improving health outcomes (see Pearce and Dorling 2009 for an overview). Conditions include providing healthy places to live, work and play in order to lead a healthy life. For example, the report advocates urban planning initiatives that promote healthy behaviours, including ensuring the availability of fresh and nutritious food as well as providing safe places to walk and exercise and a mix of land uses. However, the various policy approaches to addressing the obesity epidemic are often at odds with the more holistic recommendations of the Commission.

Given the persistent rise in the prevalence of obesity and overweight, policy responses that target amending individual behaviour change would seem to be inadequate. It is increasingly being appreciated by researchers and policy makers that successful interventions to address the obesity epidemic require concerted efforts at a range of scales, but with a particular emphasis on modifying the environments in which individual-level decisions are made. In short, to achieve successful behavioural outcomes it is important to recognize that the social, physical and cultural context matters. Policy initiatives that successfully address the high prevalence of obesity will require strategies that alter environments to enable healthy eating and encourage physical activity.

Structure and Content of the Book

This edited book draws together the international evidence for environmental explanations of rising obesity rates. It examines the numerous ways in which the contexts in which people live their lives promote an imbalance between energy intake and energy expenditure. Whilst this book will focus on the environmental or geographical explanations of the 'obesity epidemic', it will also demonstrate the necessity of multidisciplinary thinking in addressing the obesity epidemic. This edited collection draws on the expertise of researchers from a range of disciplines engaged in obesity-related research including geographers, nutritionists, epidemiologists, sociologists and public health researchers. We hope that this book will be of interest to researchers and policy makers in these fields and beyond.

Structural Overview

This volume comprises 15 chapters that are divided into six sections. The first section (Part I) provides an introduction and overview of the obesity issue, including a discussion of why a geographical approach to examining this public health issue is important. Parts II and III then consider the environmental aspects of the two sides of the energy balance equation: energy in and energy out. Various environmental issues from the global to local scales relevant to the emerging obesity epidemic will be discussed. Part IV provides a policy perspective by evaluating potential environmental interventions that could be targeted at addressing the rise in obesity prevalence. Once again this section is structured around the energy balance equation. In Part V, consideration is given to the research challenges for those working in the environmental determinants of obesity field. Discussion of the methodological and conceptual developments and concerns, approaches to measuring obesity-related characteristics of small areas, as well as an important critique of the predominantly quantitative work examining environmental determinants of obesity is provided. The final section (Part VI) draws all of this work together and suggests some future directions for geographical work on obesity.

Part I: Introduction

The first section provides the historical and theoretical context for the book. In Chapter 2, Popkin sets the scene by providing an historical perspective on the emerging obesity epidemic. Drawing on a wealth of personal experience, the chapter is framed in the context of the 'nutrition transition' whereby there has been a global shift from under-nutrition and hunger to obesity over the past three to four decades. Whilst the tendency in developed countries to consume more energy-dense foods (particularly fats, sugars, and processed foods), coupled with significant reductions in energy expenditure is well established, recently the same trends have been noted in developing countries such as India and China. As well as a detailed documentation of this transition, Popkin discusses the implications for the global health burden, particularly with regard to health outcomes such as diabetes, cardiovascular disease, asthma and cancer. Finally, the author provides a synthesis of the global drivers of obesity such as the increased availability of energy dense and sweetened food, as well as the shifts in physical activity at work and for travel and leisure. The implications of these macro-level changes on the health burden of obesity are considered.

In Chapter 3, Kim and Kawachi provide a theoretical overview of the literature on environmental determinants (or contextual effects) of obesity. In particular, the authors note that work has tended to focus on the 'obesogenic' (obesity promoting) *or* 'leptogenic' (encouraging healthy eating and physical activity) properties of local environments. Key theoretical frameworks for examining the environmental determinants of obesity, most notably the well established ANGELO (Analysis Grid for Environments Linked to Obesity) and the International Obesity Task Force

(IOTF) Work Group conceptual framework are outlined. Finally, the scene is set for subsequent chapters by providing a broad overview of the key environmental themes that are integral to considering obesity-related health outcomes. These themes include neighbourhood socioeconomic status, social capital, features of the built environment, as well as school-based characteristics.

Part II: Food Environment and Obesity

International data on food consumption have shown a major shift towards the consumption of high energy food. The environmental factors at the broadest level that have shaped dietary intake in many countries are outlined in Chapter 4 by Hoek and McLean. The authors outline the political, social and economic factors that operate at a national- or international-level to influence dietary intake. Key issues that are discussed include: changes to global food production including its economic imperatives; the implications of targeted government subsidies for food production; the marketing of food that is high in fat, salt and sugar; and the political influence exercised by the food industry. Thornton and Kavanagh (Chapter 5) then shift the discussion from the international scale to factors at a more local level that are critical to the understanding of obesity prevalence. These authors provide a theoretical overview of the local food environment and the potential implications for obesity prevalence. The discussion includes a critical evaluation of neighbourhood (micro-level) factors such as neighbourhood access to food outlets, supermarkets, fast food outlets and advertisements, and the possible links to individual health and nutrition-related outcomes. In the final chapter of this section (Chapter 6), Walton and Signal focus on the critical issue of childhood obesity. They evaluate the environmental factors at a range of geographical scales that are pertinent to understanding the epidemic in childhood obesity. The chapter includes discussion of the trends in childhood obesity over time, the contemporary and future health significance of these changes, the role of the physical environments around schools, availability and quality of school meals, and children's food marketing budgets.

Part III: Physical Activity, Environment and Obesity

In Part III of this edited collection, discussion switches from the food consumption (energy in) side of the energy balance equation to a consideration of the key environmental factors associated with physical activity (energy out). In the first section of the chapter (Chapter 7) Giles-Corti and colleagues present an overview of the changes to the physical environment at the broadest level, which in recent years that have had profound implications for physical activity and related health outcomes. This chapter critically discusses some of the key macro-level drivers of the worldwide reduction in physical activity levels. The themes that emerge include globalization, industrialization and mechanization and urban planning strategies.

Following on from the discussion of the broad macro-level determinants, Turrell considers key features of the local environment that influence physical activity-related health outcomes (Chapter 8). The author presents a systematic review of research evaluating the association between the physical activity environment and various health outcomes with a biologically plausible link. Consideration is given to environmental factors relating to the physical infrastructure (street connectivity, land use mix, residential density, walkability, opportunity structures, transportation, and aesthetics) as well key aspects of the neighbourhood social environment (crime and safety, incivilities, and social capital). Finally in this section, Oliver and Schofield examine aspects of the physical activity environment that specifically influence children's physical activity levels and obesity-related health outcomes (Chapter 9). Neighbourhood characteristics such as a local environment that is conducive to active transport and local access to recreational facilities are discussed. The important role that non-residential settings such as schools (e.g. through the curriculum or the physical design of buildings and layout of school grounds that encourage/discourage play) have in influencing physical activity levels among children is discussed.

Part IV: Obesogenic Environments and the Policy Responses

Previous sections of this book have described the emergence of the obesity epidemic and raised some important explanations at various scales from the global to the local for this international trend. In Part IV a range of potential policy responses are presented. In particular detailed critiques of a range of environmental interventions at various scales that could potentially influence firstly food consumption and secondly physical activity levels are provided. In Chapter 10, Giskes considers potential policy interventions to alter aspects of the food environment at different geographical levels to improve nutritional intake and address obesity levels. At the broad (international and national) level, policy interventions could include adjustments to agricultural subsidies and taxes which may influence the production and pricing of energy dense foods. Other approaches that warrant consideration include adopting food standard regulations which are aligned with public health nutrition regulations (e.g. maximum fat or sugar contents); targeting the labelling of food to clearly indicate nutritional content; and the subsidized distribution of food, particularly to geographically isolated populations. At a more local level, restrictions on food marketing (especially to children) and interventions to alter locational access to healthy or unhealthy food outlets may have efficacy.

The subsequent chapter evaluates environmental policy approaches designed to improve physical activity levels (Chapter 11). Riva and Curtis begin by establishing a clear need for action to modify environments so that they are more conducive to being physically active. After reviewing some key policy frameworks, the authors provide a review of the environmental interventions that are relevant to raising physical activity levels including active transportation and safe commuting as well as land use planning and design. Settings such as schools and workplaces, as well as socioeconomic and informational environments are also reviewed. The chapter concludes by evaluating some of the important barriers to successful policy intervention at the environmental-level in the area of physical activity.

Part V: Future Research Challenges

In the penultimate section of this book, attention turns to the key methodological and theoretical challenges faced by researchers with an interest in the environmental determinants of physical activity. The focus of this section is to outline important methodological developments from a range of disciplines that have considerable potential to advance this field of research. In Chapter 12, Moon details the technical and methodological challenges faced by researchers examining the influence of place-based factors on obesity-related health outcomes. Using a range of case studies, the author covers key methodological issues that can influence the establishment of causality in geographical studies of obesity such as self selection, collinearity and mobility. The chapter outlines some methodological and conceptual opportunities to researchers that have to date been under utilized in obesity-related research such as experimental studies, multilevel analysis and genomics. The following chapter also provides a range of cutting edge methods for advancing the obesity research agenda (Chapter 13). The focus of this chapter by Clarke and colleagues is on methods for measuring and representing obesity-related characteristics of small areas (often neighbourhoods). As previous chapters note, identifying and measuring components of the local food and physical activity environments is a key challenge for researchers. This chapter details a range of approaches to measuring features of neighbourhoods including locational access to resources such as food stores and greenspace, mixed land uses, walkability indices, and exposure to advertising. Innovative approaches for measuring these features will be outlined (e.g. Geographical Information Systems (GIS), Global Positioning Systems (GPS) and various audits/observational techniques). Technical concerns such as MAUP and spatial autocorrelation are briefly reviewed. The authors also present a novel approach to estimating the prevalence of obesity for small areas – microsimulation – an advance that has considerable theoretical and policy potential.

In the final chapter of this section (Chapter 14) Kearns provides a thought provoking overview of the obesity research agenda from a sociocultural perspective including the 'rules' governing obesogenic environments. Issues such as gender, age, ethnicity, traditions, and religion are considered. The chapter includes some international perspectives such as a case study of Pacific peoples in New Zealand. The chapter provides an important critique of the predominantly quantitative and reductionist work that has dominated the literature on the geography of obesity and provides some critical reflection for researchers in this field. There is a commentary on qualitative approaches to understanding these issues, including some suggested priorities for future lines of enquiry.

Part VI: Conclusions

The final section (Chapter 15) draws the book together. A conclusion is provided that synthesizes some of the key environmental features that are pertinent to understanding and tackling the obesity epidemic. We finish by suggesting some research priorities for the future.

Conclusion

The worldwide growth in obesity rates over the past few decades and the ensuing health, economic and social burden has firmly placed weight close to the top of the health agenda of most developed countries. Despite the actions of international agencies such as the WHO, a notable omission from the policy approaches in most of these countries is an explicit recognition that many characteristics of the environment are fundamentally important to understanding and addressing the obesity epidemic. Obesity-related policy initiatives continue to focus on altering individual behaviour, even though they are almost doomed to failure. The chapters in this volume review the evidence on the contributory role of environmental factors in the global increases in obesity rates, an in so doing highlight the inadequacy of policies and intervention strategies that do not extend beyond individual-level biological, behavioural, social and psychological factors. A wide range of factors at a multitude of geographical scales are examined including rapid globalisation and urbanisation, certain aspects of the 'built environment' such as the pervasive presence of fast food outlets selling energy-dense foods, as well as urban planning measures that promote car use and limit opportunities for walking and recreation. The evidence supports the conclusion that in combination these environmental changes have contributed to an increase in energy intake while at the same time sharply reducing the amount of energy we require for basic everyday activities.

This edited collection highlights some of the key ways in which environmental features are accounting for the rapid rise in global obesity prevalence. The book draws attention to the characteristics of the environment that are critical in determining the social distribution of obesity and its related health outcomes. Understanding which characteristics of the 'toxic' environment are pertinent has considerable potential for developing the theoretical understanding of obesity, as well as delivering successful policy interventions.

References

Allison, D.B., Fontaine, K.R., Manson, J.E., Stevens, J. and Vanitallie, T.B. 1999. Annual deaths attributable to obesity in the United States. *JAMA*, 282, 1530-8.

Blackman, T. 2006. *Placing Health: Neighbourhood Renewal, Health Improvement and Complexity.* Bristol: Policy Press.

Caballero, B. 2007. The global epidemic of obesity: an overview. *Epidemiologic Reviews,* 29, 1-5.

Centers for Disease Control and Prevention 2006. *Prevalence of Overweight Among Children and Adolescents: United States, 2003-2004.* Available at: http://www.cdc.gov/nchs/products/pubs/pubd/hestats/overweight/overwght_child_03.htm [accessed 10 December 2008].

Davies, J.K. and Kelly, M.P. 1993. *Healthy Cities: Research and Practice.* London: Routledge.

Flegal, K.M., Carroll, M.D., Ogden, C.L. and Johnson, C.L. 2002. Prevalence and trends in obesity among US adults, 1999-2000. *JAMA,* 288, 1723-7.

Flegal, K.M., Graubard, B.I., Williamson, D.F. and Gail, M.H. 2005. Excess deaths associated with underweight, overweight, and obesity. *JAMA,* 293, 1861-7.

Flegal, K.M., Graubard, B.I., Williamson, D.F. and Gail, M.H. 2007. Cause-specific excess deaths associated with underweight, overweight, and obesity. *JAMA,* 298, 2028-37.

Hill, J.O. and Peters, J.C. 1998. Environmental contributions to the obesity epidemic. *Science,* 280, 1371-4.

King, T., Kavanagh, A.M., Jolley, D., Turrell, G. and Crawford, D. 2006. Weight and place: a multilevel cross-sectional survey of area-level social disadvantage and overweight/obesity in Australia, *International Journal of Obesity,* 30, 281-7.

Mclaren, L. 2007. Socioeconomic status and obesity. *Epidemiologic Reviews,* 29, 29-48.

Ministry of Health. 2008. *A Portrait of Health: Key Results of the 2006/07 New Zealand Health Survey.* Wellington: Ministry of Health.

Must, A., Spadano, J., Coakley, E.H., Field, A.E., Colditz, G. and Dietz, W.H. 1999. The disease burden associated with overweight and obesity. *JAMA,* 282, 1523-9.

National Centre for Social Research. 2008. *Health Survey for England: Cardiovascular Disease and Risk Factors: 2006 16th Annual Health Survey for England.* London: The Information Centre.

Pearce, J. and Dorling, D. 2009. Commentary: Tackling global health inequalities – closing the health gap in a generation. *Environment and Planning A,* 41, 1-6.

Wang, Y. and Beydoun, M.A. 2007. The obesity epidemic in the United States – gender, age, socioeconomic, racial/ethnic, and geographic characteristics: a systematic review and meta-regression analysis. *Epidemiologic Reviews,* 29, 6-28.

Wang, Y. and Lobstein, T. 2006. Worldwide trends in childhood overweight and obesity. *International Journal of Pediatric Obesity,* 1, 11-25.

World Health Organization. 2003. *Information Sheet on Obesity and Overweight.* Geneva.

World Health Organization Commission on Social Determinants of Health. 2008. *Closing the Gap in a Generation: Health Equity Through Action on the Social Determinants of Health. Final Report of the Commission on Social Determinants of Health.* Geneva: World Health Organization.

Chapter 2
The Emerging Obesity Epidemic: An Introduction

Barry M. Popkin

The Context

During 1965-1966 I lived in the slum areas of Old Delhi, India (called Jhuggi Jhompris – translated as jumble of houses made of odds and ends or called 'squatter settlements') and in villages in the north. I saw hunger, starvation, and unemployment everywhere. In the 1969-1971 period, working in urban slums and poor areas of the northern sections of the United States, I saw a different kind of poverty – more related to inequality than absolute deprivation. But I also saw hungry people who were very heavy. Hunger was a common theme in both countries, though its consequences were much harsher in India because it killed so many children, many of them not yet even preschoolers.

If we think of our vision of the developing world and even poverty in some higher income worlds, it is of hunger, food insecurity, and little more. We think of malnourished children, poor sanitary conditions and backbreaking work. At least once a month while living in India I encountered a family in which a child had died in the past few months.

My experiences were very typical of life in India. There were few overweight people – just a few wealthy landowners – in either the urban or rural areas. Women showed their midriffs when dressed in their saris and I rarely saw rings of fat on their bellies except when attending the wedding of a very wealthy person.

When I returned to India in 2006, the world had changed. In squatter areas, there were many overweight and obese men and women. The same is true of China. When I first visited in 1981 no one was overweight or obese and concerns were focused on hunger and poverty. By 1989 huge dietary shifts had occurred, inactivity was increasing and obesity was emerging. Over the subsequent 17 years close to 30 per cent of Chinese adults would become overweight.

In contrast when we think of the higher income world, by the 1950s and 1960s there was minimal under-nutrition and few starving children. In fact, overweight and obesity emerged slowly in the 1900s until this period. But the shifts were gradual. Only later in the 1980s did we see major upward increases in the annual incidence and prevalence rates. One of the most useful studies looked at children in Denmark over the 1930 to 1983 period (Bua et al. 2007). These scholars went back in the Copenhagen municipality school records to study annual recorded measurements of children aged 7-13 over this period. They found that overweight and obesity increased steadily from 1930 until the 1950s, reached a plateau in the 1950-1960 period and rose again rapidly after that. A Nobel Laureate in economics and others have shown similar evidence for adults over a much longer period (Fogel 2004).

Thus for higher income countries we can focus on the large increases beginning in the 1970s or 1980s. In contrast, the lower income world was a world of poverty, hunger, and malnutrition with very little obesity until the last several decades. That is, the emergence of energy imbalance and obesity in lower and middle income countries was mainly a phenomenon of the 1990s and the new Millenium.

So we ask ourselves, how did this world shift so rapidly to one of obesity from one of hunger in just a few decades (Popkin 2008)? Humankind has faced major shifts in dietary and physical activity patterns and body composition since Paleolithic man emerged on this planet. Human diet and nutritional status have undergone a sequence of major transitions – defined as broad patterns of food use and corresponding nutrition-related disease. Over the last three centuries, the pace of dietary change appears to have accelerated to varying degrees in different regions of the world. The concept of the nutrition transition focuses on large shifts in diet and activity patterns, especially their structure and overall composition. These changes are reflected in nutritional outcomes, such as changes in average stature and body composition. Further, dietary and activity pattern changes are paralleled by major changes in health status, as well as by major demographic and socioeconomic changes. These patterns of diet, activity, and body composition and their related underlying causes comprise what I term the nutrition transition and are discussed elsewhere in more detail (Popkin 1993, 1994, 2002).

The Global Transition to a World of Fat: Adult Overweight Seems to Increase First

The Data: As noted above, overweight and obesity are no longer just problems of the affluent, economically more developed world. In the following discussion, except for China, I focus on nationally representative surveys from 38 of the poorest nations in the world and many higher income ones. The lower income country data were collected in an identical manner using protocols from the Measure Demographic and Health Surveys. These included weighing and measuring the stature of women of child bearing age. The other data come from a range of countries. The countries and references for the data are: *Australia* (Cameron et al. 2003, Dunstan et al. 2002, Magarey et al. 2001, MacMahon et al. 1984); *Brazil* (IBGE 2003, Monteiro et al. 2000); *China* (the author's own China Health and Nutrition Survey) (Popkin et al. 1993); *Indonesia* (Frankenberg 2003); *Russia* (the author's own Russian Longitudinal Monitoring Survey) (Popkin et al. 1997); *United Kingdom* (Stamatakis et al. 2005); *United States National Health and Nutrition Examination Surveys* (Flegal et al. 2002); and *Vietnam* (Thang and Popkin 2004). The full set of surveys is discussed in other publications (Popkin et al. 2006, Mendez et al. 2005).

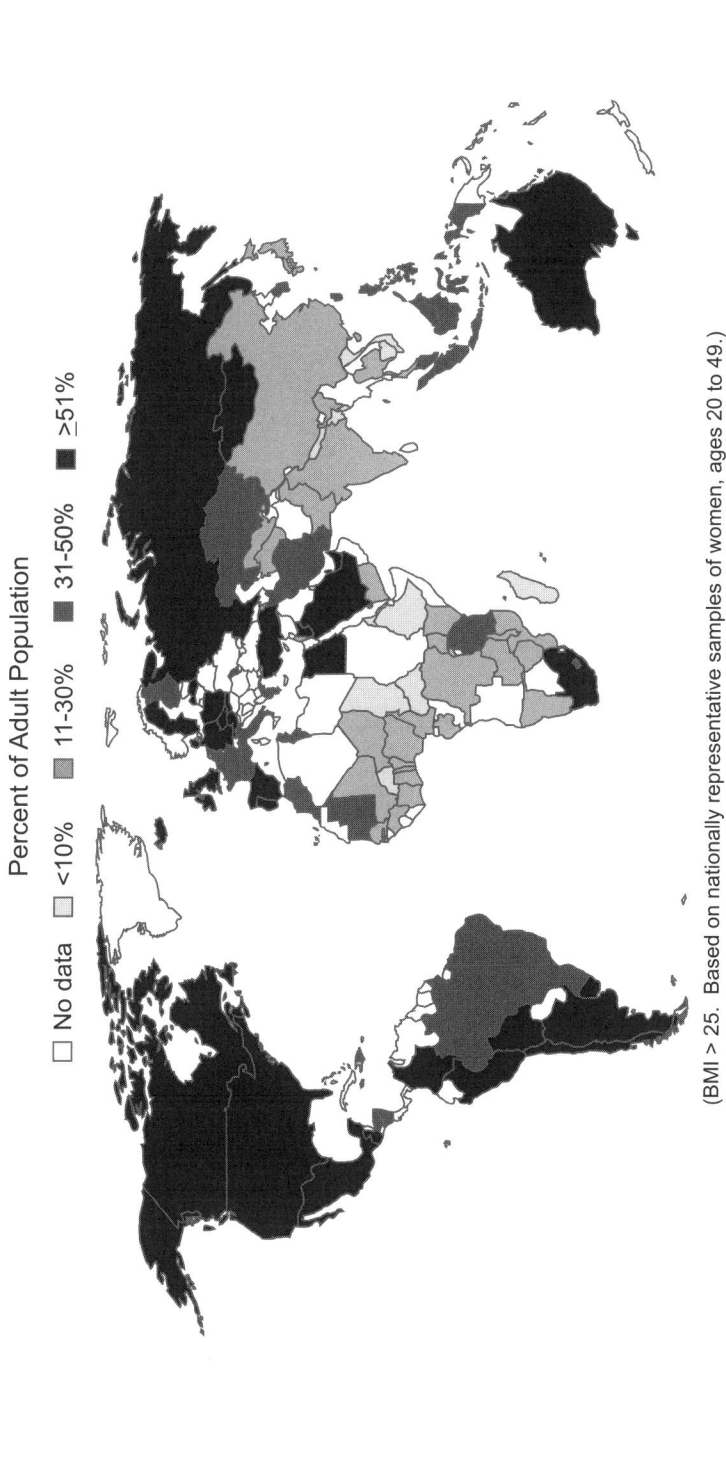

Figure 2.1 Global overweight and obesity

(BMI > 25. Based on nationally representative samples of women, ages 20 to 49.)

Source: The World Is Fat by Barry Popkin. All rights reserved, 2009.

18   *Geographies of Obesity*

Global Prevalence Patterns

Before getting into the numbers for a more limited set of countries, Figure 2.1 presents a map with data from across the globe. We have six countries with close to two-thirds of adults or more overweight and obese. These are Egypt, Mexico, and South Africa among lower income countries and the United States, Australia, and the United Kingdom among higher income ones. But the map also shows a surprisingly large number of other countries where over half of the population are overweight or obese.

Nationally representative data are presented for a large number of countries in Figure 2.2. Only data on Africans (blacks) are presented for South Africa; data from only nine provinces are presented for China, but these data mirror closely national levels and trends (Wang et al. 2006, Ge et al. 1996). Essentially, this figure shows that in an array of high-income countries, overweight and obesity levels are consistently high (but the levels are much lower in other European countries) and several middle- and lower-income countries (e.g. Mexico, Egypt, and South Africa) have equally high levels of obesity among women. Other very large countries, (e.g. China) have rates greater than 20 per cent for women and men.

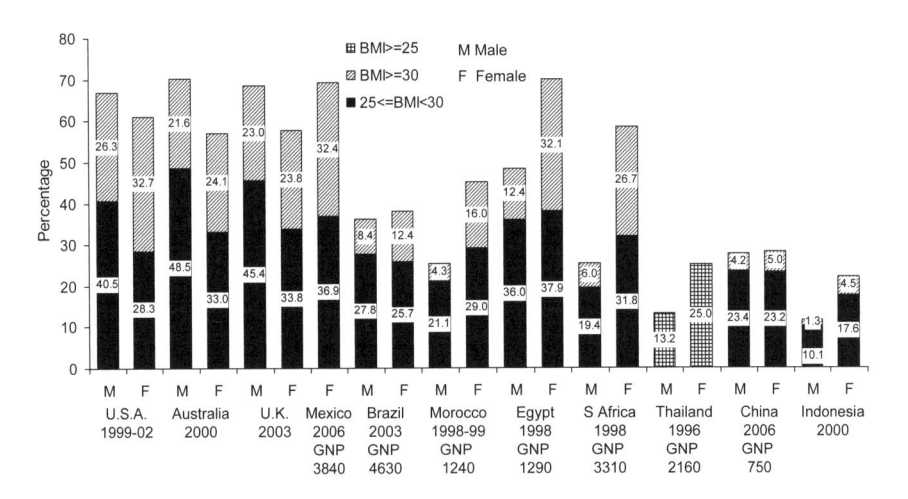

Figure 2.2   Obesity patterns across the world

Obesity is Universally Found in Urban and Rural Areas

Elsewhere we have presented in detail urban vs. rural differences (Mendez et al. 2005). The study presented data for women aged 20-29 years (n=148,579) on both underweight and overweight plus obesity status from 1992-2000 in 36 low-income and transitional countries. The summary shown in Figures 2.3a and 2.3b provide a clear picture of these differential patterns; overweight plus obesity exceeded underweight in the majority of countries. Countries with higher income and urbanization levels not only had higher absolute levels of

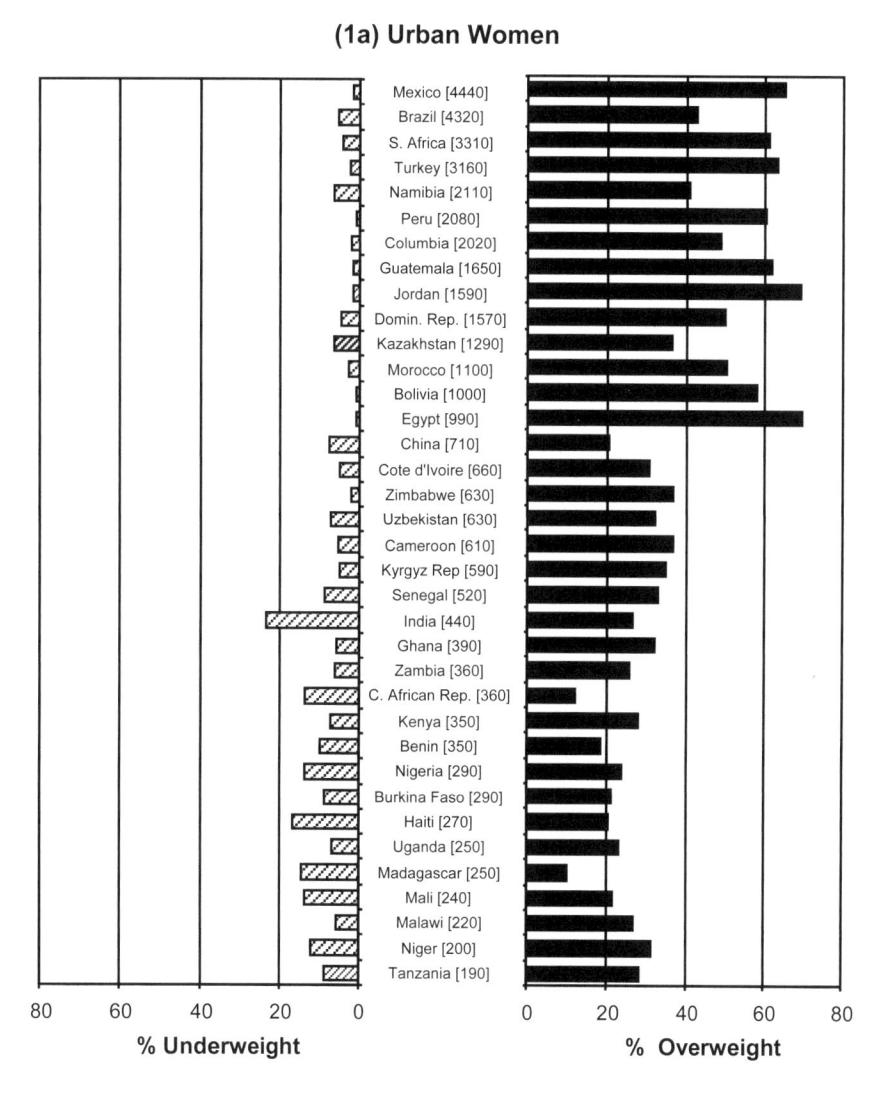

(1a) Urban Women

Figure 2.3a Overweight and underweight prevalence in women 20-49y in 36 developing countries ranked by gross national income (gni) per capita – Urban women

Note: Overweight = BMI ≥25; underweight = BMI <18.5.
Source: Mendez et al. (8): Reprinted with permission of AJCN.

overweight plus obesity, but small urban–rural differences in overweight and very high ratios of overweight plus obesity to underweight. In more developed countries, overweight among low socioeconomic status (SES) women was high

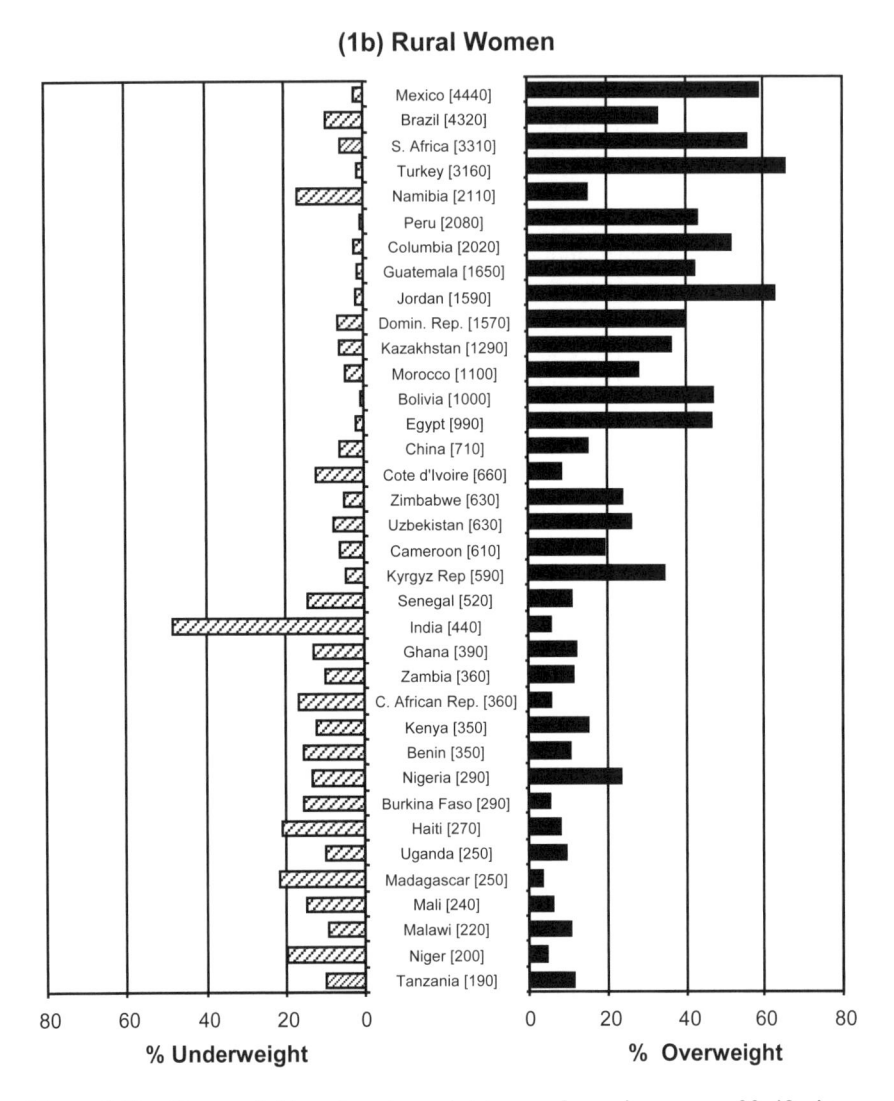

(1b) Rural Women

Figure 2.3b Overweight and underweight prevalence in women 20-49y in 36 developing countries ranked by gross national income (gni) per capita – Rural women

Note: Overweight = BMI ≥25; underweight – BMI <18.5.

Source: Mendez et al. (8): Reproduced with permission of the American Society for Nutrition.

in both rural (38 per cent) and urban (51 per cent) settings. Even many poor countries – where underweight persists as a significant problem – had fairly high levels of rural overweight (Mendez et al. 2005).

Global Trends in the Increase of Overweight and Obesity Prevalence

Absolute rates of increase in overweight plus obesity tend to be higher among adults than they are in children in most countries – much higher in the two low-income countries (i.e. China and Indonesia) – and moderately higher in Brazil and two of the three high-income countries (i.e. the United Kingdom and the United States) (see Figure 2.4). The only country where overweight plus obesity increased more among children than adults was Australia. However, relative rates of increase in overweight indicate faster increases in overweight among children in Brazil and the three high-income countries. As a result, the relative excess of overweight among adults, seen initially in all countries, increased in China, Indonesia, and Russia, but it decreased in Australia, Brazil, the United Kingdom, and the United States.

As noted above, Australia was the only high income country where the rate of increase of the proportion of obesity among children was higher than for adults. However, as I and many other colleagues have observed, child obesity is now emerging in all countries of the world. We have found that 0.3 to 0.6 per cent of all children in each country are becoming overweight each year – a very troubling statistic (Bua et al. 2007, Wang and Lobstein 2006, Stamatakis et al. 2005, Lobstein et al. 2004, Lobstein and Frelut 2003, Ogden et al. 2002, Wang et al. 2002).

Overweight is Becoming a Problem of the Poor

My work with Brazilian colleagues shows that more lower-income and less-educated people are overweight in that country than are the higher-income and better-educated populace – a finding that is replicated globally in all moderate- and higher-income countries (Monteiro et al. 2001, 2004a, 2004b). Even China, with its much lower income, has this same economic class disparity.

This issue of the poor being more overweight than the rich is a phenomena we have lived with for decades in the US and Western Europe. In our countries, obesity, heart disease and diabetes are all problems of the poor much more than of the rich (Wang and Beydoun 2007, Gordon-Larsen et al. 2006, Neumark-Sztainer et al. 2002, Kaufman et al. 1997, Davis et al. 1995, Sobal and Stunkard 1989).

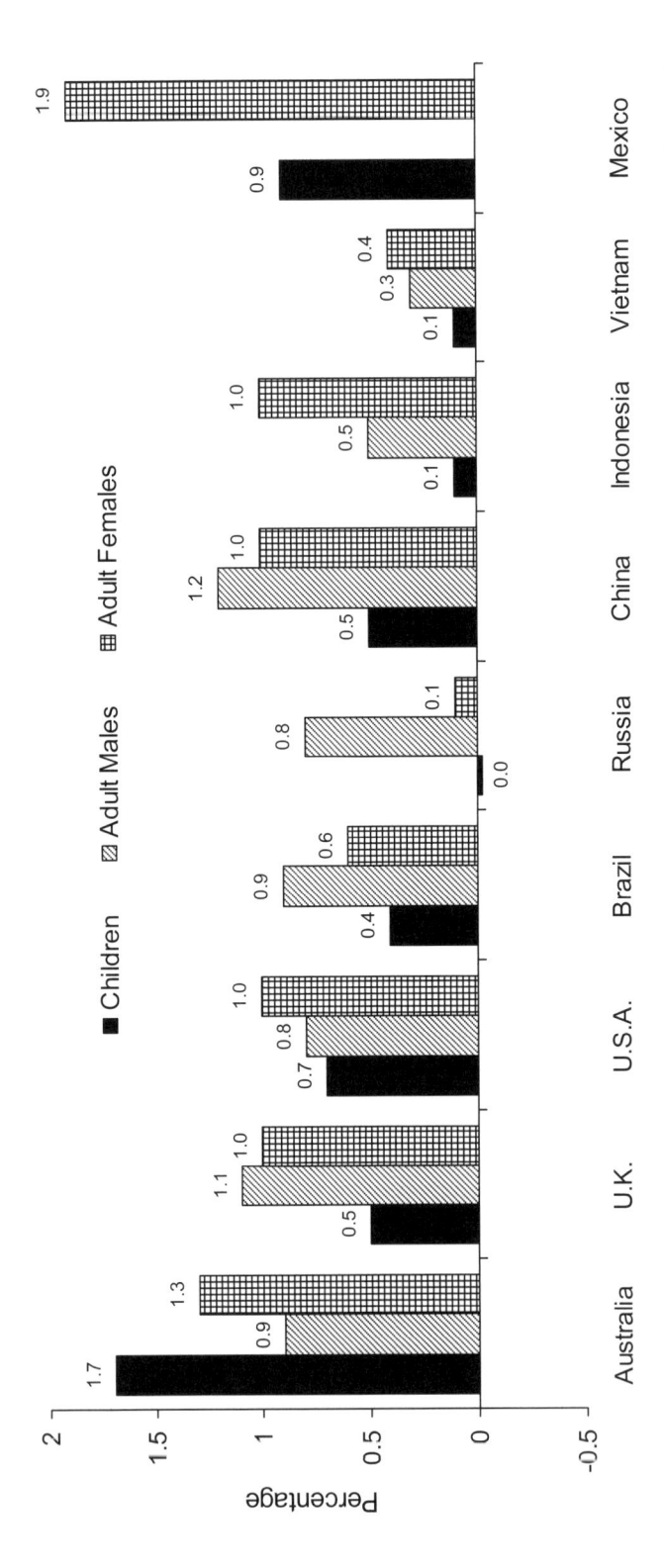

Figure 2.4 Annual absolute change in the prevalence of overweight and obesity in seven countries from 1985/1995 to 1995/2006

Notes: BMI≥25.0 for adults; IOTF equivalent for children.

Are These Changes Slowing Down or Accelerating?

We explore relative rates of change across the globe for adults and children in a new study. We have long-term comparable data for Brazil and the United States (Popkin 2006). Earlier changes in those two countries refer mostly to the mid 1970s, the 1980s, recent changes to the 1990s, and to the early 2000s. In the case of Brazil, time trends indicate a deceleration in the combined overweight and obesity levels for adults; however, the increase in the combined overweight and obesity levels shows acceleration for US adults. Studies of body composition and dietary pattern shifts indicate rates of change are accelerating in China. The prevalence rate of change in overweight plus obese persons accelerated from less than 0.5 per cent in the 1980-1990 period to 1.9 per cent and 0.9 per cent in the 1997-2000 period for men and women, respectively (Wang et al. 2007); preliminary data for 2004 indicates that these changes are still accelerating. In the dietary area, we have documented longitudinally that the income elasticity, or the proportion of food purchases with a 1 per cent increase in income has accelerated at an increasing rate in the past 15 years (Du et al. 2004, Guo et al. 2000).

What Do These Changes Mean for Health Status?

Obesity debilitates but does not kill. Consequently we find that smoking is the number one cause of death in many countries but obesity debilitates. Consider first diabetes. Adult-onset diabetes, the major type of diabetes in the globe, is linked causally to obesity and inactivity (Smyth and Heron 2006, Zimmet et al. 1997, Zimmet 1992).

Diabetes causes many health problems. Uncontrolled blood sugar levels, can lead to circulatory complications and other health problems such as nerve damage in the feet, poor vision, nausea, and weakness can occur much more quickly. More critically, diabetes can lead to heart disease, renal failure, and the loss of eyesight and limbs. People with diabetes are sick for a long time before they die.

One of the first major complications of uncontrolled diabetes is that a person's kidneys stop functioning. We have seen almost a tenfold increase in the number of cases of kidney failure in the United States over the past 30 years (American Diabetes Association position statement 2004), most of it due to uncontrolled high blood sugar. Many diabetics are unable to work. The overall life expectancy of a person with diabetes is reduced to about half that of a normal person.

Further, in the US and globally we are beginning to see young adults and teens with adult onset diabetes. In fact, these two groups represent the major increase in this disease (Pinhas-Hamiel and Zeitler 2007). We know, as well, that these young people do not do well with it. First, they show poor adherence to medical care or to the discipline required to properly manage their condition. Second, there appears to be more rapid development of the sorts of clinical complications we have been talking about, such as the risk of heart attack (Pinhas-Hamiel and Zeitler 2007,

Hillier and Pedula 2003, Kawahara et al. 1994). And, of course, earlier onset of diabetes means development of complications at a younger age as well.

The problem of diabetes is a global one (Wild et al. 2004). In Canada, the same increased rate of obesity is also occurring and is seen in the large increase in diabetes in that country (Lipscombe and Hux 2007). The situation is even more dramatic in India, which has more diabetic patients than any other country. It is predicted that by 2025, India will have more than 60 million diabetic patients (Ramachandran et al. 2001).

Clearly related to the increase in diabetes is an equally marked increase in hypertension and all other forms of heart disease and key cancers also. Global estimates have over 230 million with diabetes, and over 1.5 billion hypertensive; the burden of these nutrition-related non-communicable diseases (NR-NCD's) is very high (Popkin 2008, Smyth and Heron 2006, Kearney et al. 2005, Ford et al. 2005, Kearney et al. 2004, Wild et al. 2004). The role of obesity in the etiology of the most widespread and common cancers is now established so that in terms of both heart disease and cancer, obesity is the major preventable causal factor (WCRF 2007).

Asthma is another outcome of obesity. Globally, the numbers of adults and children who experience asthma is way up – some of this, to be sure, is caused by polluted air. In the United States, for instance, the proportion of children with asthma close to doubled between 1980 and 1996, to 6.2 per cent of all children (Akinbami and Schoendorf 2002). Reports have shown that nearly 75 per cent of emergency room visits for asthma were made by obese individuals and studies have shown that obesity pre-dates asthma. Increasingly studies are finding that obese children are much more likely to become asthmatic (Story 2007, Thomsen et al. 2007, Eder et al. 2006, Schaub and von Mutius 2005).

Obesity and all the underlying hormonal changes related to it are linked with a number of very important cancers. A large team of eminent cancer scholars undertook a four-year, $6.5 million review of research linking any preventable lifestyle determinant to any cancer (WCRF 2007). Among their findings was a significant correlation of obesity and body fatness to breast, colon, endometrial, gall bladder, kidney, esophageal, and pancreatic cancers. Essentially the major finding of this review was that obesity is one of the two most preventable causes of cancer in the world.

Cheap, Plentiful, Edible Oil Fuelled Major Dietary Changes

Globally, our diet is becoming increasingly energy-dense and sweeter. At the same time, higher fibre foods are being replaced by processed versions. There is enormous variability in eating patterns globally but the broad themes seem to be evident in most countries.

The global shifts in the energy density of diet are difficult to document. One can document large increases in consumption of edible oils and animal source foods for selected countries (Du et al. 2002a, 2004) with the help of well-collected, repeated

24-hour recall measures of dietary intake. However, in general, most research has had to focus on use of food disappearance data from the Food and Agricultural Organization. Food disappearance or food balance data essentially estimate the net food available for consumption in the country. With these data, which are not as accurate in picking up smaller shifts in consumption and wastage (Crane et al. 1992), it is possible to see that the shifts in edible oil intake are universal. In one set of analyses that compared edible oil intake patterns in the 1960s with those in the 1990s, based on food disappearance data, large increases in edible oil available for intake was shown particularly for lower income countries (Guo et al. 2000, Drewnowski and Popkin 1997). The crude picture of the increased intake of edible oils is being seen globally as shown in Figure 2.5. However, when one looks at average intake per day in a country such as China, where we have measured the intake with recall plus direct measures of household consumption, we find much higher levels of intake. In our latest China Health and Nutrition Survey the per capita intake of edible oil was over 375 kcals per day.

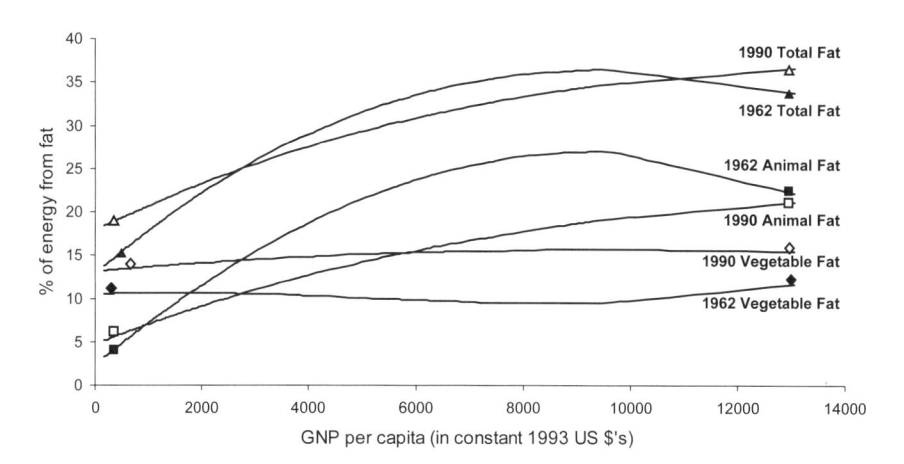

Figure 2.5 **Relationship between the percentage of energy from fat and GNP per capita, 1962 and 1990**

Source: Nonparametric regressions run with food balance data from Food and Agricultural Organization of the United Nations and the GNP data is from the World Bank for 134 countries. Guo (2000): Reprinted with the permission of Econ Dev and Cultural Change.

The Animal Source Foods Revolution Has Hit

A decade ago Delgado, formerly with the International Food Policy Research Institute and now with the World Bank, noted that people in lower and middle income countries consumed milk and meat at levels 25-33 per cent of that consumed by people in higher income countries. He noted also the rapid shift in consumption and termed this the 'Livestock Revolution' (Delgado 2003, Delgado et al. 1999). Higher incomes, urbanization and reduced prices for animal source foods have combined to shift the share and per capita consumption levels of the developing world rapidly. As we have shown in China, the rates of increase for animal source foods and for energy dense foods are far higher than predicted (Du et al. 2004, Guo et al. 2000). China is increasing particularly its intake of pork and poultry while India's increase in milk consumption is having an equally important global effect (Popkin et al. 2001). Animal source foods, including dairy products, beef, pork, poultry, lamb, mutton, fish, and seafood are being increasingly consumed by not only these two countries but other low and moderate income countries with resultant shifts in the demand for the protein-rich grains and oilseeds that provide the food for these items.

The Sweetening of the Global Diet

Sugar is the world's predominant sweetener and this has been the case since approximately the sevententh and eighteenth centuries, as the New World began producing large quantities of sugar at reduced prices (Galloway 2000, Mintz 1986). However, sugar is not the only caloric sweetener and we consider here all the caloric sweeteners from honey to high fructose corn syrup (Duffey and Popkin 2008, Popkin and Nielsen 2003).

Today these caloric sweeteners are found in tens of thousands of foods. Walk down any processed food aisle in any country – rich or poor – and you will see one to three or four caloric sweeteners noted on the labels of the foods. In the United States in 2004 we find that added caloric sweeteners account for over 17 per cent of our total caloric intake and close to a third of all carbohydrate intake (Duffey and Popkin 2008). Furthermore, in this same research we have found that the proportion of caloric sweeteners in food has dropped from two-thirds to one-third as the role of calorically sweetened beverages has grown in the US diet.

There are few countries that measure added caloric sweeteners in all foods so we can not create comparable measures for other countries. In a series of papers we have found in Mexico that caloric beverages doubled in per capita intake in all age groups between 1999 and 2006 and represent over 22 per cent of the energy intake of Mexicans (Barquera et al. 2007, 2008, Rivera et al. 2008).

With food disappearance data we show in Figure 2.6 that across the world, the availability of caloric sweeteners has risen (Popkin and Nielsen 2003). While these intake levels are probably underestimates, they do show a pronounced intake across the low and middle income countries.

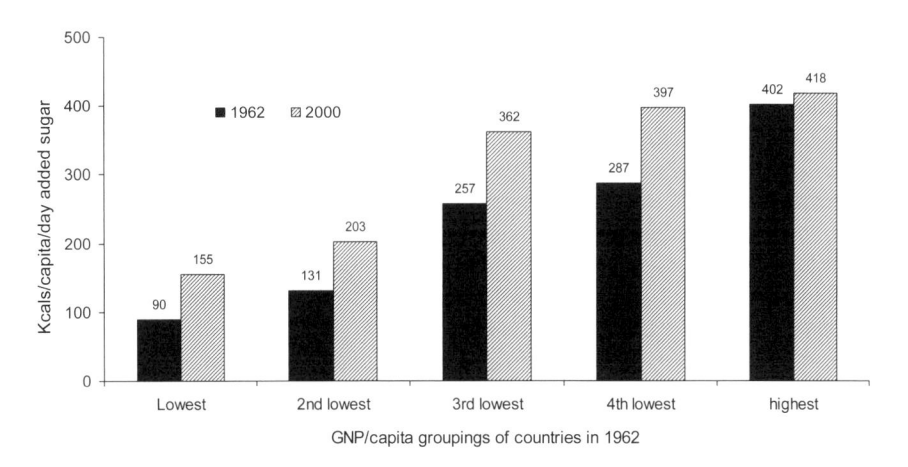

Figure 2.6 Relationship between levels of gross national product per capita and caloric sweetener

The Resultant Shift Toward Increasingly Energy-Dense Diets

With these large increases in animal source foods and edible oils and caloric sweeteners, we do find increased energy density of the food consumed and even that of beverages. Part of this is the result of reductions in fibre intake and fruits and vegetables.

The studies on fibre intake and other changes toward processed foodstuffs are much more incomplete to date. Since the issue of reduced fibre intake in the Western diet was first discussed as a source of major health concerns, there have been few systematic studies of shifts in fibre intake throughout the world. There are, however, important historical case studies that document these shifts for selected population groups and countries (Trowell and Burkett 1981, Hughes and Jones 1979). There is also documentation of specific shifts in diet from coarse grains to refined grains in a few countries (Popkin et al. 1993).

Similarly, studies on fruit and vegetable intake indicate declines in many countries and regions of the world, but again these have not been systematically studied (Popkin et al. 2001, Cavadini et al. 2000, Drewnowski and Popkin 1997). There are also selected countries where fruit and vegetable intake remain very high (e.g. Spain, Greece, and South Korea) (Lee et al. 2002, Moreno et al. 2002, Kim et al. 2000).

Is Westernization of the Food Supply and the Pattern of Eating Occurring?

Many scholars blame the Western soft drink and fast food empires for these rapid shifts in diet so one logical question is: Is the McDonaldization and Cocacolanization of these countries the cause of these shifts? Many scholars have discussed this topic and indeed, the culture of food processing, preparation, and eating away-from-home has changed in some countries. However, our documentation is sparse. We have studies by anthropologists who have studied the way a McDonald's opening in a country might affect the away-from-home food industry (Watson 1997). Indeed one can document some subtle changes caused by the introduction of modern Western ways of preparing food, serving food, and such across many countries. This more subtle effect may be as important as the direct effect.

In one study, we examined comparatively dietary patterns and trends related to child food intake across four countries and could find some countries like the Philippines whose away-from-home, snacking, and modern fast food intakes were comparable to those of the US (Adair and Popkin 2005). But we could also find others like China and Russia where away from home intake was tiny, snacking and modern fast food intake were very low.

Indeed, in other research my colleagues and I have shown that the shift in Chinese diet linked with increased obesity are very much related to increased intake of oils and animal foods. Frying of food is rapidly increasing and has replaced lower calorie boiled and steamed food preparation (Wang et al. 2008). The basic shifts of the Chinese diet are quite independent from the fast food and modern food industry (Du 2002a, 2002b).

At the same time other dimensions, such as the introduction of modern supermarkets have been very important (Reardon et al. 2003). One of the central shifts has occurred in the *global food system* relating to the marketing and sales of food. The fresh (wet or open public) market is disappearing as the major source of supply for food in the developing world. These markets are being replaced by multinational, regional, and local large supermarkets – supermarkets which are usually part of larger chains (e.g. Carrefour or Walmart) – or in other countries such as South Africa and China they are run by local domestic chains patterned to function and look like these global chains. Increasingly, we are finding hypermarkets (very large megastores) as the major force driving shifts in food expenditures in a country or region. For example, in Latin America, supermarkets' share of all retail food sales increased from 15 per cent in 1990 to 60 per cent by 2000 (Balsevich et al. 2003, Reardon and Berdegué 2002). By comparison, 80 per cent of retail food sales in the United States in 2000 occurred in supermarkets. In one decade, the role of supermarkets in Latin America has expanded at a rate equivalent to about a half century of expansion in the United States (Reardon and Berdegué 2002). Supermarket use has spread across both large and small countries, from capital cities to rural villages, and from upper- and middle-class families to the working class. This same process is also occurring at varying rates and different stages in Asia, Eastern Europe, and Africa (Reardon et al. 2003).

In particular, supermarkets tend to provide lower prices, manage the cold chain properly, and provide higher quality food safety (Minten and Reardon 2008). At the same time they provide greater access to processed higher-fat, added-sugar, and salt-laden foods in developing countries. In the context of creating a higher quality, safer food supply with reduced prices and access costs, they are very important vehicles for development.

Shifts in Physical Activity at Work, Travel, and Leisure are Equally Profound!

There are several linked changes in physical activity occurring jointly. One is a shift away from the high energy expenditure activities such as farming, mining and forestry towards the service sector. We have shown this large shift elsewhere (Popkin 1999). Reduced energy expenditures in the same occupation are a second change (Bell et al. 2001). In China a huge shift has occurred in urban areas from physically active to more sedenary patterns of activity, a shift that is significantly related to weight gain (Monda et al. 2007a, 2007b, Popkin 2006, Bell et al. 2001). A third major change relates to mode of transportation and activity patterns during leisure hours (Bell et al. 2002). Adding a car or motor scooter doubled the risk of becoming overweight while adding a bike halved it. In China, 14 per cent of households acquired a motorized vehicle between 1989 and 1997. (In Chapters 6 and 7 changes in urban infrastructure, land use and public transport that follow increasing rates of private vehicle ownership are discussed plus associations between these changes and physcial activity.) A fourth dimension is household production. The introduction of washing machines, electric irons, gas and propane stoves and many other modern conveniences such as indoor plumbing to provide an indoor supply of water and toilets are all major shifts. We have documented the effects of some of these changes on obesity (Monda et al. 2007a, 2007b).

Finally TV ownership and viewer shift has grown exponentially across the globe. For instance, in China when we began to survey TV ownership in 1989, half had a television and 15 per cent had coloured TVs, whereas today over 92 per cent have coloured TV and over 98 per cent have a television that works. The impact of TV viewership on the attitudes and behaviours of individuals in lower income countries has rarely been studied. Dozens of studies have linked TV viewership with increased risk of obesity in higher income countries but only a few studies have been undertaken in lower income countries (Hernandez et al. 1999).

So Is This Really All Genes?

Some scientists have shown that obesity runs systematically in families and hence that it is a genetic disease. There are very few, rare, single gene defects that cause obesity. In most cases, obesity is the result of a complex interaction between many different genes and the environment. Recently, a group of British scientists found that people who inherit one version of a gene they called 'FTO' weighed about seven pounds more than those without the variant (the gene was named FTO for the fused toe trait it caused in mice, but was nicknamed 'fatso' before its association with obesity became known because it is such a large gene) (Frayling et al. 2007). These scholars are careful to note that FTO is not the only gene that influences obesity, and that inheriting a particular variant will not necessarily make anyone fat. Additionally, they note that people with the genetic variant may be slim unless they overeat or don't exercise. In other words, this gene only contributes to a greater risk of being overweight but does not in and of itself cause the obesity.

While there might be a genetic predisposition to obesity and other chronic diseases, genes will not kill us. It is diet and activity and an energy imbalance that do. And these changes can and often do override genetics if people make large enough changes. Nurture does trump nature. But in the current world in which we live it is hard to diet and/or be very physically active and lose weight. The only reason we are fat now is that the food and activity/inactivity patterns are out there to interact with our genes to make us fat. For all of these reasons, then, genetics determinist arguments are palpably naïve. In a sense, we know that it takes thousands of years for genes to change while, in fact, the obesity increases we have seen have taken place only in the past century in the higher income world and in the past one to two decades in countries such as China.

Biology certainly does affect us. Our biology is linked with our evolutionary struggle to reproduce and survive. I think that the shift to calorically-sweetened beverages, larger portion sizes, more eating occasions, and the increased availability of sweeter and fattier foods are caused by vast technological and economic changes. These changes have caused us to eat and move differently. But, these environmental factors do interact with our genetic makeup such that some of us seem more susceptible to these changes, to obesity and even to illnesses when we are obese.

There is another side to the biology of various global subpopulations that I believe might be very important and certainly has implications for many lower and middle income countries as well as for the higher income ones. This might be genetic but there are many other potential explanations for this issue. We find that abdominal obesity, that is, more fat in the central areas of the body, is more common among Asians and Hispanics than among Whites with the same level of BMI. In fact there is a great deal of research to show that even at a low BMI, South Asians from India have more body fat in the abdomen and around critical organs such as the liver and heart, and already have an obesity-linked increased likelihood of becoming diabetic. If you are an American of Mexican origin with a BMI of 22, you are probably much more likely to develop diabetes than an American of Western or Eastern European origin with a BMI of 26 or 27 (Zimmet et al. 1997, Zimmet 1992). As we have said, globally, we use a BMI of 25 to 30 to signify overweight and one above 30 to signify obesity for adults. Children have a BMI standard that will vary with their age, gender and maturational status. But if the adult or child is born of Hispanic or Asian descent, he/she will be much more likely to have abdominal obesity for any BMI than will his/her white counterparts.

Conclusion

The world is indeed fat. Malnutrition continues to be a scourge in many nations of the world, particularly subSaharan Africa and South Asia. At the same time even in these countries overweight and obesity are emerging along with adult-onset type 2 diabetes and other comorbidities of obesity. Very large changes in our food supply, particularly increased intake of energy dense foods and reduced intake of legumes, fibre, fruits, and vegetables and coarse grains are occurring. Concomitant shifts in physical activity are universally seen. These shifts have occurred more slowly in higher income countries while among the lower and middle income countries these changes have typically been seen only in the past two decades.

References

Adair, L.S. and Popkin, B.M. 2005. Are child eating patterns being transformed globally? *Obesity Research,* 13(7), 1281-1299.

Akinbami, L.J. and Schoendorf, K.C. 2002. Trends in childhood asthma: prevalence, health care utilization, and mortality. *Pediatrics,* 110(2), 315-322.

American Diabetes Association position statement. 2004. Nephropathy in Diabetes. *Diabetes Care,* 27(Suppl 1), 79S-83S.

Balsevich, F., Berdegue, J.A., Flores, L., Mainville, D. and Reardon, T. 2003. Supermarkets and produce quality and safety standards in Latin America. *American Journal of Agricultural Economics,* 85(5), 1147-1154.

Barquera, S., Campirano, F., Bonvecchio, A., Hernández, L., Espinosa, J., Rivera, J. and Popkin, B.M. 2007. *Trends and Characteristics of Caloric Beverage Consumption in Mexican Pre-school and School Age Children.* Cuernevaca.

Barquera, S., Hernández, L., Tolentino, M.L., Espinosa, J., Ng, S.W., Rivera, J. and Popkin, B.M. 2008. Energy from beverages is on the rise among Mexican adolescents and adults. *Journal of Nutrition,* 138, 2454-2461.

Bell, A.C., Ge, K. and Popkin, B.M. 2001. Weight gain and its predictors in Chinese adults. *International Journal of Obesity and Related Metabolic Disorders,* 25(7), 1079-1086.

Bell, A.C., Ge, K. and Popkin, B.M. 2002. The road to obesity or the path to prevention: motorized transportation and obesity in China. *Obesity Research,* 10(4), 277-283.

Bua, J., Olsen, L.W. and Sorensen, T.I.A. 2007. Secular trends in childhood obesity in Denmark during 50 years in relation to economic growth. *Obesity,* 15(4), 977-985.

Cameron, A.J., Welborn, T.A., Zimmet, P.Z., Dunstan, D.W., Owen, N., Salmon, J., et al. 2003. Overweight and obesity in Australia: The 1999-2000 Australian Diabetes, Obesity and Lifestyle Study (AusDiab). *Medical Journal of Australia,* 178(9), 427-432.

Cavadini, C., Siega-Riz, A.M. and Popkin, B.M. 2000. US adolescent food intake trends from 1965 to 1996. *Western Journal of Medicine,* 173(6), 378-383.

Crane, N.T., Lewis, C.J. and Yetley, E.A. 1992. Do time trends in food supply levels of macronutrients reflect survey estimates of macronutrient intake? *American Journal of Public Health,* 82(6), 862-866.

Davis, S.K., Winkleby, M.A. and Farquhar, J.W. 1995. Increasing disparity in knowledge of cardiovascular disease risk factors and risk-reduction strategies by socioeconomic status: implications for policymakers. *American Journal of Preventive Medicine,* 11(5), 318-323.

Delgado, C.L. 2003. Rising consumption of meat and milk in developing countries has created a new food revolution. *Journal of Nutrition,* 133(11 Suppl 2), 3907S-3910S.

Delgado, C.L., Rosegrant, M., Steinfield, H., Ehui, S. and Courbois, C. 1999. *Livestock to 2020: The Next Food Revolution.* Food, Agriculture, and the Environment Discussion Paper 28. Washington, DC: International Food Policy Research Institute.

Drewnowski, A. and Popkin, B.M. 1997. The nutrition transition: new trends in the global diet. *Nutrition Reviews,* 55(2), 31-43.

Du, S., Lu, B., Zhai, F. and Popkin, B.M. 2002a. A new stage of the nutrition transition in China. *Public Health Nutrition,* 5(1A), 169-174.

Du, S., Lu, B., Zhai, F. and Popkin, B.M. 2002b. The nutrition transition: a new stage of the Chinese diet, in *The Nutrition Transition: Diet and Disease in the Developing World,* edited by B. Caballero and B. Popkin. London: Academic Press, 205-222.

Du, S., Mroz, T.A., Zhai, F. and Popkin, B.M. 2004. Rapid income growth adversely affects diet quality in China – particularly for the poor! *Social Science and Medicine,* 59(7), 1505-15.

Duffey, K. and Popkin, B.M. 2008. High-fructose corn syrup: Is this what's for dinner? *American Journal of Clinical Nutrition,* 88(suppl), 1722S-1732S.

Dunstan, D.W., Zimmet, P.Z., Welborn, T.A., Cameron, A.J., Shaw, J., de Courten, M., Jolley, D. and McCarty. D.J. 2002. The Australian Diabetes, Obesity and Lifestyle Study (AusDiab) – methods and response rates. *Diabetes Research and Clinical Practice,* 57(2), 119-129.

Eder, W., Ege, M.J. and von Mutius, E. 2006. The asthma epidemic. *New England Journal of Medicine,* 355(21), 2226-2235.

Flegal, K.M., Carroll, M.D., Ogden, C.L. and Johnson, C.L. 2002. Prevalence and trends in obesity among US adults, 1999-2000. *Journal of the American Medical Association,* 288(14), 1723-1727.

Fogel, R.W. 2004. *The Escape from Hunger and Premature Death, 1700-2100 Europe, America, and the Third World.* New York: Cambridge University Press.

Ford, E.S., Mokdad, A.H., Giles, W.H., Galuska, D.A. and Serdula, M.K. 2005. Geographic variation in the prevalence of obesity, diabetes, and obesity-related behaviors. *Obesity Research,* 13(1), 118-122.

Frankenberg, E., Smith, J. and Thomas, D. 2003. Economic shocks, wealth and welfare. *Journal of Human Resources,* 38(2), 280-321.

Frayling, T.M., Timpson, N.J., Weedon, M.N., et al. 2007. A common variant in the FTO gene is associated with body mass index and predisposes to childhood and adult obesity. *Science,* 316(5826), 889-894.

Galloway, J. 2000. Sugar, in *The Cambridge World History of Food,* edited by K. Kiple and K. Ornelas. New York: Cambridge University Press, 437-449.

Ge, K., Zhai, F. and Yan, H. 1996. *The Dietary and Nutritional Status of Chinese Population: 1992 National Nutrition Survey,* Vol 1. Beijing: People's Medical Publishing House.

Gordon-Larsen, P., Nelson, M.C., Page, P. and Popkin, B.M. 2006. Inequality in the built environment underlies key health disparities in physical activity and obesity. *Pediatrics,* 117(2), 417-424.

Guo, X., Mroz, T.A., Popkin, B.M. and Zhai, F. 2000. Structural changes in the impact of income on food consumption in China, 1989-93. *Economic Development and Cultural Change,* 48, 737-760.

Hernandez, B., Gortmaker, S.L., Colditz, G.A., Peterson, K.E., Laird, N.M. and Parra-Cabrera, S. 1999. Association of obesity with physical activity, television programs and other forms of video viewing among children in Mexico City. *International Journal of Obesity and Related Metabolic Disorders,* 23(8), 845-854.

Hillier, T.A. and Pedula, K.L. 2003. Complications in young adults with early-onset type 2 diabetes: losing the relative protection of youth. *Diabetes Care,* 26(11), 2999-3005.

Hughes, R.E. and Jones, E. 1979. A Welsh diet for Britain? *British Medical Journal,* 1(6171), 1145.

IBGE. 2003. *Pesquisa de Orcamentos Familiares 2002-2003. Analise da Disponibilidade Domiciliar de Alimentos e do Estado Nutricional no Brasil.* Rio de Janiero: IBGE.

Kaufman, J.S., Cooper, R.S. and McGee, D.L. 1997. Socioeconomic status and health in blacks and whites: the problem of residual confounding and the resiliency of race. *Epidemiology,* 8(6), 621-628.

Kawahara, R., Amemiya, T., Yoshino, M., Miyamae, M., Sasamoto, K. and Omori, Y. 1994. Dropout of young non-insulin-dependent diabetics from diabetic care. *Diabetes Research and Clinical Practice,* 24(3), 181-185.

Kearney, P.M., Whelton, M., Reynolds, K., Muntner, P., Whelton, P.K. and He, J. 2005. Global burden of hypertension: analysis of worldwide data. *Lancet,* 365(9455), 217-223.

Kearney, P.M., Whelton, M., Reynolds, K., Whelton, P.K. and He, J. 2004. Worldwide prevalence of hypertension: a systematic review. *Journal of Hypertension,* 22(1), 11-19.

Kim, S., Moon, S. and Popkin, B.M. 2000. The nutrition transition in South Korea. *American Journal of Clinical Nutrition,* 71(1), 44-53.

Lee, M.J., Popkin, B.M. and Kim, S. 2002. The unique aspects of the nutrition transition in South Korea: the retention of healthful elements in their traditional diet. *Public Health Nutrition,* 5(1A), 197-203.

Lipscombe, L.L. and Hux, J.E. 2007. Trends in diabetes prevalence, incidence, and mortality in Ontario, Canada 1995-2005: a population-based study. *Lancet,* 369(9563), 750-756.

Lobstein, T., Baur, L. and Uauy, R. 2004. Obesity in children and young people: a crisis in public health. *Obesity Reviews,* 5(Suppl 1), 4-97.

Lobstein, T. and Frelut, M.L. 2003. Prevalence of overweight among children in Europe. *Obesity Reviews,* 4(4), 195-200.

MacMahon, S., Blacket, R., Macdonald, G.J. and Hall, W. 1984. Obesity, alcohol consumption and blood pressure in Australian men and women. The National Heart Foundation of Australia Risk Factor Prevalence Study. *Journal of Hypertension,* 2(1), 85-91.

Magarey, A., Daniels, L.A. and Boulton, T.J. 2001. Prevalence of overweight and obesity in Australian children and adolescents: reassessment of 1985 and 1995 data against new standard international definitions. *Australian Medical Journal,* 174, 561-565.

Mendez, M.A., Monteiro, C.A. and Popkin, B.M. 2005. Overweight exceeds underweight among women in most developing countries. *American Journal of Clinical Nutrition,* 81(3), 714-721.

Minten, B. and Reardon, T. 2008. Food prices, quality, and quality's pricing in supermarkets versus traditional markets in developing countries. *Review of Agricultural Economics*, 30(3), 480-490.

Mintz, S. 1986. *Sweetness and Power: The Place of Sugar in Modern History.* New York City, Penguin.

Monda, K.L., Adair, L.S., Zhai, F. and Popkin, B.M. 2007a. Longitudinal relationships between occupational and domestic physical activity patterns and body weight in China. *European Journal of Clinical Nutrition*, doi: 10.1038/sj.ejcn.1602849.

Monda, K.L., Gordon-Larsen, P., Stevens, J. and Popkin, B.M. 2007b. China's transition: the effect of rapid urbanization on adult occupational physical activity. *Social Science and Medicine,* 64(4), 858-870.

Monteiro, C.A., Conde, W.L., Bing, L. and Popkin, B.M. 2004a. Obesity and inequities in health in the developing world. *International Journal of Obesity Related Metabolic Disorders,* 28(9), 1181-1186.

Monteiro, C.A., Conde, W.L. and Popkin, B.M. 2001. Independent effects of income and education on the risk of obesity in the Brazilian adult population. *Journal of Nutrition,* 131(3), 881S-886S.

Monteiro, C.A., D'A Benicio, M.H., Conde, W.L. and Popkin, B.M. 2000. Shifting obesity trends in Brazil. *European Journal of Clinical Nutrition,* 54(4), 342-346.

Monteiro, C.A., Moura, E.C., Conde, W.L. and Popkin, B.M. 2004b. Socioeconomic status and obesity in adult populations of developing countries: a review. *Bulletin of the World Health Organization,* 82(12), 940-946.

Moreno, L.A., Sarria, A. and Popkin, B.M. 2002. The nutrition transition in Spain: a European Mediterranean country. *European Journal of Clinical Nutrition*, 56(10), 992-1003.

Neumark-Sztainer, D., Croll, J., Story, M., Hannan, P.J., French, S.A. and Perry, C. 2002. Ethnic/racial differences in weight-related concerns and behaviors among adolescent girls and boys: findings from Project EAT. *Journal of Psychosomatic Research*, 53(5), 963-974.

Ogden, C.L., Flegal, K.M., Carroll, M.D. and Johnson, C.L. 2002. Prevalence and trends in overweight among US children and adolescents, 1999-2000. *Journal of the American Medical Association*, 288(14), 1728-1732.

Pinhas-Hamiel, O. and Zeitler, P. 2007. Acute and chronic complications of type 2 diabetes mellitus in children and adolescents. *Lancet*, 369(9575), 1823-1831.

Popkin, B. 1993. Nutritional patterns and transitions. *Population Development Review*, 19(1), 138-157.

Popkin, B. 1994. The nutrition transition in low income countries: An emerging crises. *Nutrition Reviews*, 52(9), 85-98.

Popkin, B.M. 1999. Urbanization, lifestyle changes and the nutrition transition. *World Development*, 27, 1905-1916.

Popkin, B.M. 2002. An overview on the nutrition transition and its health implications: the Bellagio meeting. *Public Health Nutrition*, 5(1A), 93-103.

Popkin, B.M. 2006. Global nutrition dynamics: the world is shifting rapidly toward a diet linked with noncommunicable diseases. *American Journal of Clinical Nutrition*, 84(2), 289-298.

Popkin, B.M. 2008. *The world is fat – The fads, trends, policies, and products that are fattening the human race*. New York: Avery-Penguin Group.

Popkin, B., Baturin, A., Kohlmeier, L. and Zohoori, N. 1997. Russia: Monitoring nutritional change during the Reform Period, in *Implementing dietary guidelines for healthy eating*, edited by V. Wheelock. London: Chapman and Hall.

Popkin, B.M., Conde, W., Hou, N. and Monteiro, C.A. 2006 Is there a lag globally in overweight trends for children as compared to adults? *Obesity*, 14, 1846-1853.

Popkin, B.M., Horton, S., Kim S. 2001. The nutrition transition and prevention of diet-related chronic diseases in Asia and the Pacific. *Food and Nutrition Bulletin*, 22(4 Suppl), 1-58.

Popkin, B.M., Horton, S., Kim, S., Mahal, A. and Shuigao, J. 2001. Trends in diet, nutritional status, and diet-related noncommunicable diseases in China and India: the economic costs of the nutrition transition. *Nutrition Reviews*, 59(12), 379-390.

Popkin, B.M., Keyou, G., Zhai, F., Guo, X., Ma, H. and Zohoori, N. 1993. The nutrition transition in China: a cross-sectional analysis. *European Journal of Clinical Nutrition*, 47(5), 333-346.

Popkin, B.M. and Nielsen, S.J. 2003. The sweetening of the world's diet. *Obesity Research*, 11(11), 1325-1332.

Popkin, B.M., Wolney, C., Hou, N. and Monteiro, C.A. 2006. Is there a lag globally in overweight trends for children as compared to adults? *Obesity,* 14, 1846-1853.

Ramachandran, A., Snehalatha, C., Kapur, A., Vijay, V., Mohan, V., Das, A.K., et al. 2001. High prevalence of diabetes and impaired glucose tolerance in India: National Urban Diabetes Survey. *Diabetologia,* 44(9), 1094-1101.

Reardon, T. and Berdegué, J.A. 2002. The rapid rise of supermarkets in Latin America: Challenges and opportunities for development. *Development Policy Review,* 20, 371-388.

Reardon, T., Timmer, C.P., Barrett, C.B. and Berdegue, J. 2003. The rise of supermarkets in Africa, Asia, and Latin America. *American Journal of Agricultural Economics,* 85, 1140-1146.

Rivera, J.A., Muñoz-Hernández, O., Rosas-Peralta, M., Aguilar-Salinas, C.A., Popkin, B.M. and Willett, W.C. 2008. Consumo de bebidas para una vida saludable: recomendaciones para la población (Beverage consumption for a healthy life: recommendations for the Mexican population). *Salud Publica Mexico,* 50(2), 173-195.

Schaub, B. and von Mutius, E. 2005. Obesity and asthma, what are the links? *Current Opinion in Allergy Clinical Immunology,* 5(2), 185-193.

Smyth, S. and Heron, A. 2006. Diabetes and obesity: the twin epidemics. *Nature Medicine,* 12(1), 75-80.

Sobal, J. and Stunkard, A.J. 1989. Socioeconomic status and obesity: a review of the literature. *Psychological Bulletin,* 105(2), 260-275.

Stamatakis, E., Primatesta, P., Chinn, S., Rona, R. and Falascheti, E. 2005. Overweight and obesity trends from 1974 to 2003 in English children: what is the role of socioeconomic factors? *Archives of Disease in Childhood,* 90(10), 999-1004.

Story, R.E. 2007. Asthma and obesity in children. *Current Opinion in Pediatrics,* 19(6), 680-684.

Thang, N.M. and Popkin, B.M. 2004. Patterns of food consumption in Vietnam: effects on socioeconomic groups during an era of economic growth. *European Journal of Clinical Nutrition,* 58(1), 145-153.

Thomsen, S.F., Ulrik, C.S., Kyvik, K.O., Sørensen, T.I., Posthuma, D., et al. 2007. Association between obesity and asthma in a twin cohort. *Allergy,* 62(10), 1199-1204.

Trowell, H.C. and Burkett, D.P. (eds.) 1981. *Western diseases: Their emergence and prevention.* Cambridge, MA: Harvard University Press.

Wang, H., Du, S., Zhai, F. and Popkin, B.M. 2007 Trends in the distribution of body mass index among Chinese adults aged 20-45 years (1989-2000). *International Journal of Obesity,* 31, 272-278.

Wang, Y. and Beydoun, M.A. 2007. The obesity epidemic in the United States – gender, age, socioeconomic, racial/ethnic, and geographic characteristics: a systematic review and meta-regression analysis. *Epidemiology Review,* 29, 6-28.

Wang, Y. and Lobstein, T. 2006. Worldwide trends in childhood overweight and obesity. *International Journal of Pediatric Obesity,* 1(1), 11-25.

Wang, Y., Mi, J., Shan, X.Y., Wang, Q.J. and Ge, K.Y. 2006. Is China facing an obesity epidemic and the consequences? The trends in obesity and chronic disease in China. *International Journal of Obesity (Lond),* 31, 177-188.

Wang, Y., Monteiro, C., and Popkin, B.M. 2002. Trends of obesity and underweight in older children and adolescents in the United States, Brazil, China, and Russia. *American Journal of Clinical Nutrition,* 75(6), 971-977.

Wang, Z., Zhai, F., Du, S., and Popkin, B.M. 2008. Dynamic shifts in Chinese eating behaviors. *Asia Pacific Journal of Clinical Nutrition,* 17, 123-130.

Watson, J.L. 1997. *Golden arches east: McDonald's in East Asia.* Stanford, CA: Stanford University Press.

WCRF 2007. *Food, Nutrition, Physical Activity and the Prevention of Cancer: a Global Perspective.* Washington DC World Cancer Research Fund in association with the American Institute for Cancer Research.

Wild, S., Roglic, G., Green, A., Sicree, R. and King, H. 2004. Global prevalence of diabetes: estimates for the year 2000 and projections for 2030. *Diabetes Care,* 27(5), 1047-1053.

Zimmet, P.Z. 1992. Kelly West Lecture 1991. Challenges in diabetes epidemiology – from West to the rest. *Diabetes Care,* 15(2), 232-252.

Zimmet, P.Z., McCarty, D.J. and de Courten, M.P. 1997. The global epidemiology of non-insulin-dependent diabetes mellitus and the metabolic syndrome. *Journal of Diabetes and its Complications,* 11(2), 60-68.

Chapter 3
Contextual Determinants of Obesity: An Overview

Daniel Kim and Ichiro Kawachi

Twin studies have established that obesity has a strong hereditary component (Stunkard et al. 1990). However, the rapid increase in the prevalence of obesity throughout the world during the past two decades cannot be attributed to genetic causes. Increasingly, attention is being focused on the environmental contributions to obesity, including the influences of the food environment as well as residential 'built environments'. This chapter aims to provide an overview of the theoretical linkages between so-called contextual factors and obesity. We lay out the theoretical basis for why environmental contexts might matter for obesity, and the characterization of certain environments as 'obesogenic', 'leptogenic', or 'toxic'. Two conceptual frameworks – the ANGELO (ANalysis Grid for Environments Linked to Obesity) framework and the IOTF (International Obesity Task Force) Working Group conceptual model – are discussed and contrasted, with each highlighting environmental factors at the macro and micro-levels. These factors are explained at greater length in Parts II and III of this book. We then provide examples of these environmental contexts and their plausible linkages to obesity-related behaviours and obesity. Overall, these frameworks are useful in conceptualizing and designing interventions and policies (as discussed in Part IV of this book) to promote healthier dietary patterns, physical activity behaviours, and ultimately prevent obesity within the population.

Why do Environmental Contexts Matter for Obesity?

Geoffrey Rose's (1985) theorem concerning the distinction between sick individuals and sick population helps us to understand the population-level drivers of obesity prevalence. According to Rose, the causes of sick individuals are often distinct from the causes of sick populations. For example, the cause of morbid obesity among individuals at the tail of the distribution of body mass index (BMI) is frequently genetic – and that is the case whether we are talking about obese individuals in the United States or in Japan. However, heredity is unlikely to explain either the exceptionally high prevalence of overweight/obesity in the United States compared to Japan, or the rapid increase in the prevalence of overweight/obesity in both societies over time. Rather – as Rose pointed out – the prevalence of obesity in the United States is high because the *average* person in the US is overweight compared to the average person in Japan, i.e. the entire distribution curve of BMI has shifted to the right in America compared to Japan. Similarly, the prevalence of obesity has risen in both societies over time because the entire distributions of BMI in each society have shifted to the right over time – and not because there has been an isolated increase in morbidly obese individuals at the far right-hand tail of the distribution.

Geoffrey Rose's insights strongly imply that population health is driven by forces that drive the location of the distribution of risks. In other words, population variations in risks such as obesity are likely to be driven by contextual forces – such as changes in the food environment or the physical activity environment – operating at the level of the entire population. As discussed in the following sections, contextual forces operating at the level of populations may be conceptualized to encompass not just residential neighbourhoods, schools, and work-places, but also at the policy (e.g. transport policy, urban development policy) and the macro-economy levels (e.g. food production and subsidies).

By 'contextual', we refer to those broad sets of factors which are external to individuals, and produce effects in addition to or in interaction with individual characteristics (Macintyre and Ellaway 2003).

Migrant studies provide insights highlighting the important independent contributions of sociocultural context on diets and body weight, since these differences stem from within members of the same racial/ethnic group. For example, saturated fat intakes of Japanese populations in Japan, Hawaii, and San Francisco have been observed to be 7, 23, and 26 per cent of total energy intakes, respectively (Kato et al. 1973). Serum cholesterol, body weight, and age-adjusted coronary heart disease rates have also been found to be higher respectively with closer distance to the mainland United States (Kato et al. 1973). Likewise, Pima Indian women living in Arizona in the United States have been estimated to have an average dietary fat intake of 41 per cent of total energy intake, body weight of 90 kg, and body mass index (BMI) of 37 kg/m^2, compared to 23 per cent of energy from fat, body weight of 70 kg, and BMI of 25 kg/m^2 in their female relatives remaining in Mexico (Wadden et al. 2002; Ravussin et al. 1994). Because genetic factors are likely to be relatively homogeneous across these respective populations of Japanese/native Indian descent, these patterns implicate effects of *contextual factors* on dietary intakes and obesity.

In keeping with these studies, the US Institute of Medicine (1995: 152) concluded that 'there has been no real change in the gene pool in this period of increasing obesity. The root of the problem, therefore, must lie in the powerful social and cultural forces that promote an energy-rich diet and a sedentary lifestyle.' Given the emergence of the obesity epidemic over the last two decades, and growing knowledge about the influences of place on health behaviours and health, consideration of the role of environmental factors in explaining the obesity epidemic is critical for prevention and the development of policy responses (Wadden et al. 2002, Hill and Peters 1998).

Characterizing Environments

'Obesogenic' or 'leptogenic' are terms that have been coined recently to describe environments that raise or lower the risk of overweight and obesity in populations. 'Obesogenicity' refers to 'the sum of influences the surroundings, opportunities, or conditions of life have on promoting obesity in individuals or populations' (Swinburn et al. 1999: 564). By contrast, the term 'leptogenic' environment corresponds to environments which maintain and promote healthy weights, such as through the encouragement of healthy food choices and engagement in physical activity (Swinburn et al. 1999). Obesogenic factors may be viewed as the barriers and leptogenic factors the enabling and reinforcing factors to maintaining a healthy weight (Swinburn et al. 1999; Green and Kreuter 1990).

Akin to the concept of obesogenicity, Brownell and colleagues (Horgen and Brownell 2002, Wadden et al. 2002, Battle and Brownell 1997) have referred to social and cultural forces promoting energy-rich diets and sedentary lifestyles as constituting 'toxic' environments. By 'toxic', they refer to a myriad of obesity-promoting factors including in children the unprecedented exposure to energy-dense, heavily advertised, inexpensive foods through a growing number of fast food outlets, and school districts signing contracts with soft drink companies, combined with an increasingly sedentary lifestyle due to greater amounts of time spent viewing television and a reduced emphasis on physical education in schools (Wadden et al. 2002).

Hypotheses and Conceptual Models Relating Environments to the Obesity Epidemic

Gene-Environment 'Discordance' Hypothesis

One hypothesis set forth for the emergence of the obesity epidemic (as well of other chronic degenerative disorders, or 'diseases of civilization') is discordance between our human species' genetic makeup and characteristics of the current environmental milieu (Eaton et al. 1988). From an evolutionary perspective, there are stark contrasts between the diets and physical activity of our Paleolithic ancestors and the current western lifestyle in the developed world. Fat intakes of late Paleolithic diets were approximately 10-20 per cent of total energy intakes (vs. over 30 per cent in the current western diet), and present day hunter-gatherers consume lower-fat diets and engage in higher levels of physical activity than members of western civilization, due to labour-saving conveniences in the latter such as motor vehicles and elevators (Poston and Foreyt 1999). Our evolutionary history of hunting and gathering (on a diet low in simple sugars) has created biological constraints that only favour obesity under modern conditions of high caloric and simple sugar intake, and increasingly sedentary behaviours.

The ANGELO Conceptual Framework

The ANGELO (ANalysis Grid for Environments Linked to Obesity) framework was developed to conceptualize obesogenic environments (Egger et al. 2003), and draws upon the environmental emphasis in earlier work in the Host-Agent-Environment epidemiologic triad, as well as the Haddon injury prevention matrix (Egger et al. 2003, Haddon 1980).

The grid (shown in Table 3.1) characterizes environment along two dimensions: size (micro and macro) and type (physical, economic, political, and sociocultural). Microenvironments correspond to places where individuals congregate, including neighbourhoods, workplaces, and schools (Egger et al. 2003). Macroenvironments correspond to entire sectors or industries which can influence population eating and physical activity patterns within microenvironments; macroenvironments are located more diffusely at higher geographical scales, including regions and entire countries. Such macroenvironments include the food manufacturing industry, transportation systems, the media, and local governments (Swinburn et al. 1999).

Table 3.1   The ANGELO conceptual framework

Environmental Size	Environmental Type			
	Physical (Food and PA)	Economic (Food and PA)	Political (Food and PA)	Sociocultural (Food and PA)
Examples of Micro (settings)				
Neighbourhoods	Grocery stores, walking paths	Cost of fitness club memberships		Social norms on healthy eating
Schools	Cafeteria foods, vending machines	Cost of cafeteria foods	Policies on physical education	Role models on physical activity
Examples of Macro (sectors)				
Transportation systems	Availability of public transit	State financing of public transit		
Health regulatory systems			Policies on food product labelling	

Notes: *PA = physical activity. Some examples corresponding to cells in the environmental size by type matrix are shown.

Source: From Swinburn et al. 1999.

In this model, the *physical environment* refers to characteristics of the material environment which can influence diet and physical activity, such as food outlets (e.g. supermarkets, fast-food restaurants) and the availability of green spaces at a microenvironmental level, as well as food manufacturing and transportation systems at a macroenvironmental level (Swinburn et al. 1999). The *economic environment* corresponds to factors such as food prices or the cost of local gym memberships at a micro-level as well as food production and manufacturing costs at a macro-level. The *political environment* refers to the rules and policies which may affect obesity-related behaviours, including school nutrition policies at a micro-level or food labelling regulations at a macro-level. The *sociocultural environment* corresponds to the values, attitudes, and beliefs which may influence diet and physical activity. This may be reflected at the micro-level in the healthy eating-related values promoted by a school, and at the macro-level in mass media messages about physical fitness (Swinburn et al. 1999).

The International Obesity Task Force Working Group Conceptual Framework

An alternative conceptual model (shown in Figure 3.1) was developed by the Working Group of the International Obesity Task Force (IOTF) (Kumanyika et al. 2002). This model approximates a causal web, linking broad distal factors at the international scale to factors at the national, community, and work/school/home levels and on to more proximal factors at the individual-level. Key measurement indicators at the international and national levels include the amount of money spent on marketing foods (corresponding to mass media factors) and obesity prevention initiatives (reflecting economic factors), as well as the presence of integrated nutrition and transportation policies (corresponding to political influences) (Kumanyika et al. 2002).

Several factors are shared with the ANGELO framework (e.g. media advertising, food manufacturing, and transportation at the macro-level, and the residential and school environments at the micro-level). However, one distinct difference and advantage of the IOTF Working Group conceptual model is its explicit linkage of factors at different geographical scales, and the corresponding distinction between 'upstream' and 'downstream' factors (rather than placing macro and micro factors alongside one another). In so doing, it acknowledges both fundamental causes (Link and Phelan 1995) as well as the potentially greater utility of intervening at more upstream points to more effectively bring about change at a population-level. For example, economic growth, urbanization, and globalization of food markets are all upstream societal/environmental factors included in the conceptual model which are thought to underlie the obesity epidemic. All may have contributed to widespread increases in the consumption of energy-dense diets and declines in physical activity (Kumanyika et al. 2002). Individual-level interventions performed in isolation of societal changes may selectively motivate those who already have flexible lifestyle options, and thereby may exacerbate existing health disparities. Furthermore, because of the linkages between factors

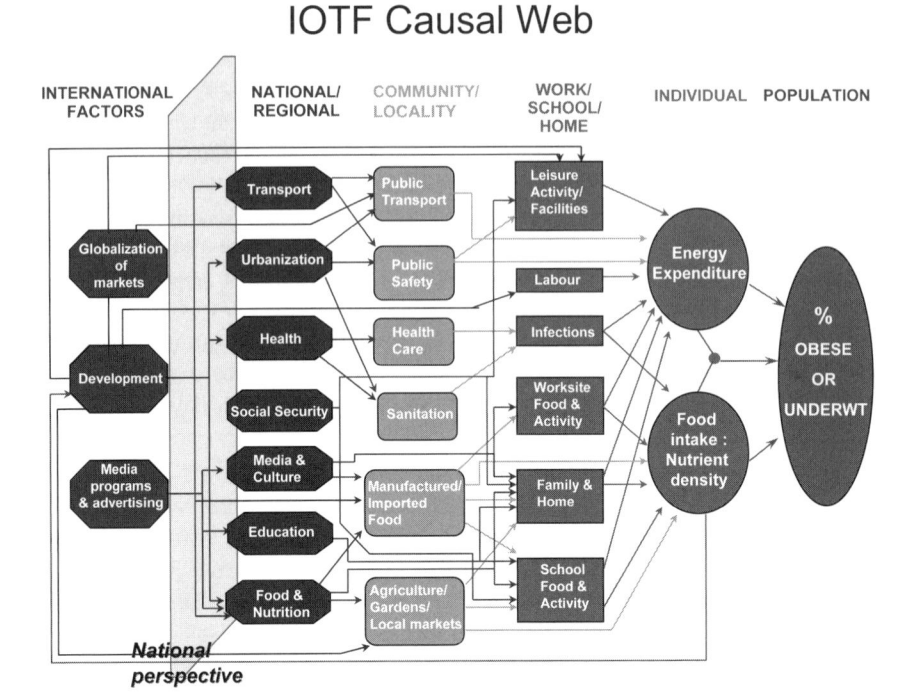

Figure 3.1 International Obesity Task Force Working Group conceptual model
Source: Reprinted with permission from Macmillan Publishers Ltd.

in the causal web, this model highlights that a single factor cannot be addressed without impacting on other factors, a useful perspective for the development of effective interventions and policies (Kumanyika et al. 2002).

A component that would benefit both conceptual models is a temporal dimension, both at the environmental and individual-level. In particular, environmental contexts are dynamic and changing over time, and considerations of worsening trends in obesogenic/toxic environmental factors (or improving trends for leptogenic factors) can help to better explain the occurrence and trajectory of the obesity epidemic. Moreover, such dynamic factors may interact with individual-level characteristics, which in and of themselves may change or may have variable sensitivities for behavioural/obesity effects over the lifecourse of individuals.

Drawing on these two conceptual models, we now highlight several key environments – food manufacturing and marketing at the macroenvironmental-level, and neighbourhoods and schools at the microenvironmental-level – as examples of environments with plausible linkages to obesity-related behaviours and obesity. Instances of changing environments (e.g. advances and trends in food manufacturing) as well as individual-environment interactions (e.g. early life environmental exposures) are further described.

The Food Manufacturing Industry and Marketing (MACRO-Level)

The mass preparation of food has been implicated as a particularly strong driving force for the marked increase in calories consumed in recent decades internationally, and the emergence of the obesity epidemic (Cutler et al. 2003). Since around 1970, technological innovations including vacuum packing, better preservatives, and microwaves, have allowed food manufacturers to cook food centrally, and to get it to consumers more efficiently for consumption in the home and in fast food restaurants. This transition from individual to mass preparation has plausibly led to increases in the quantity and variety of foods consumed.

As Nestle (2003) has noted, the US food industry alone produces food in sufficient quantities that could feed the entire population twice over on a daily basis. This level of efficiency and over-production has led inevitably to a mass marketing imperative on the part of the food industry to get Americans to consume more, for profit maximization. These marketing imperatives include: taste ('make foods sweet, fat, and salty'); cost ('add value but keep costs low'); convenience ('make eating fast'); confusion ('keep the public puzzled'); advertising (with food companies spending more than $11 billion annually on direct media advertising, nearly 70 per cent of which is for convenience foods, candy and snacks, soft drinks, and desserts, and only 2 per cent of which is for fruits, vegetables, grains, and beans); and serving larger portions i.e. 'supersizing' (Nestle 2003). With the conglomeration of food corporations and the trend towards greater economic efficiency, specialization, and food manufacturing in recent years, each of these marketing imperatives has only become stronger (Nestle 2003).

Food advertising as on television has also been increasingly targeted to children in the US in recent years (due to their considerable influence on family purchase decisions), particularly for inexpensive fat- and sugar-loaded processed foods. Between 1992 and 2000, expenditures on marketing to children by US food corporations nearly doubled (Center for Media Education 2000). Such advertising has implications for obesity over the lifecourse into adulthood, through establishing lifelong dietary habits (including 'brand loyalties') and affecting possible biological 'setpoints' in early life.

Neighbourhood Environments (MICRO-Level)

Neighbourhood Socioeconomic Environments

Within the United States, neighbourhood socioeconomic status tends to map closely with physical resources and amenities that directly promote (or constrain) the behaviours of residents such as food intake and physical activity. For example, low-SES neighbourhoods may be characterized by higher crime rates causing residents' concerns about safety. Combined with the lack of recreational spaces, this may make residents less willing to engage in physical activity (Huston et al.

2003, Seefeldt et al. 2002). Such neighbourhoods (along with low-income African-American and other minority neighbourhoods) may also be targeted and saturated with fast food restaurants to capitalize on the lack of food choices of residents, thereby contributing to poor eating habits and high-fat consumption (LaPoint 2003). The availability of inexpensive, low-fat and nutritious food options may be lower in poor neighbourhoods (e.g. Morland et al. 2002a, 2002b, Sooman et al. 1993), which has similar implications for dietary intakes. Furthermore, eating calorie-dense fast foods may serve a functional value for individuals under the conditions of economic and environmental stress, as evidence suggests for smoking (Emmons 2000; Romano et al. 1991). Under such conditions, 'living in the cocoon of obesity is a very comfortable thing to do' (Critser 2003: 112).

One possible consequence of the resultant poor nutrition for pregnant women living in more socioeconomically deprived neighbourhoods is early growth retardation and foetal programming effects. The nutritional exposures during critical periods of foetal development could have latent effects on the development of obesity in later life (Ben-Shlomo and Kuh 2002, Centers for Disease Control 2002, Ong and Dunger 2002, Barker 1992).

Notably, these patterns in US urban settings may not necessarily be generalizable across all societies and populations. For example, contrary to what has been generally suggested, Macintyre et al. (2005) found that fast-food outlets were more likely to be located in more affluent rather than socioeconomically-deprived areas of Glasgow, Scotland. Likewise, contrary to hypothesized directions of associations, Pearce et al. found that in New Zealand, more deprived communities were more likely to have fast food outlets, but also health-promoting resources such as recreational facilities and supermarkets (Pearce et al. 2007a, 2007b). Witten et al. (2008) further found that sedentarism was *lower* and physical activity was *higher* in neighbourhoods with worse access to parks in New Zealand (estimated using GIS methods). Such exceptions (as described later in this book) call attention to the fact that in order to more fully account for associations between neighbourhood environments and obesity across multiple settings and populations, we may need to draw on additional hypotheses/conceptual frameworks, or adapt our existing ones accordingly.

Neighbourhood Social Capital

Apart from physical resources and amenities, levels of neighbourhood social capital may influence obesity-related behaviours and obesity. Social capital has been defined as the features of social organization, including social trust, civic participation, and norms of reciprocity that facilitate cooperation for mutual benefit (Putnam 2000, Kawachi et al. 1997). Some mechanisms by which social capital may lower the risk of obesity include the diffusion of knowledge about healthy promotion, maintenance of healthy behavioural norms through informal social control, promotion of access to local services and amenities, and psychosocial processes which provide affective support and mutual respect (Kawachi and Berkman 2000). For instance, drawing on the diffusion of innovations theory (Rogers 2003), residents of high social capital neighbourhoods in which exercise and healthy eating (e.g. the avoidance of sugar-sweetened beverages) are prevalent may be more likely to adopt these behaviours through diffusion of knowledge about the behaviours. Informal social control may also exert influence over these health behaviours (Sampson 2003). Secondly, based on social cognitive theory (Bandura 1991), the belief in collective agency is linked to the efficacy of a group in meeting its needs. Thus, a neighbourhood high in social capital and trust would be expected to effectively work together in lobbying for local services relevant to dietary intakes and physical activity, such as adequate numbers of supermarkets and green spaces. Thirdly, the social resources theory (Lin 2001) proposes that access to and use of social resources can lead to improved socioeconomic status. In turn, socioeconomic gains in individuals would be anticipated to improve dietary and exercise behaviours. Lastly, psychosocial processes including high levels of social support and trust may buffer the deleterious effects of stress, or may have independent beneficial effects on mental and physical health (Stansfeld 1999). The lack of social resources within one's neighbourhood to buffer stress levels may have the effect of chronically activating the hypothalamic-pituitary-adrenal (HPA) axis, and thereby contributing over the long term to higher cortisol levels, adipose deposition, and obesity (Reaven 1988).

At the same time, it should be kept in mind that social influence and social capital may additionally have adverse impacts on obesity. One recent example is a study of social networks in the well-established Framingham Heart Study, in which Christakis and Fowler (2007) determined that having social ties with obese individuals (e.g. friends, spouses, siblings) increased one's risk of becoming obese. This might be attributed to such social ties changing one's perception of the social norms concerning the acceptability of obesity (Christakis and Fowler 2007).

Finally, plausibly, the experiences of poor nutrition and stress/affective states in mothers during foetal critical periods as a result of living in a neighbourhood low in social capital (as in the case of living in a socioeconomically-deprived neighbourhood) may contribute to the development of obesity in offspring in later life. Some evidence suggests that certain forms of social capital are also low in socioeconomically-deprived neighbourhoods (Kim et al. 2006).

The Built Environment and Urban Sprawl

The built environment encompasses features of the physical environment which are human-made, including urban design (e.g. street design and orientation), land development and land use (e.g. the location, density, and mix of residential and commercial spaces), and transportation systems (e.g. street connectivity, presence of bicycle lanes, and availability of public transit) (Papas et al. 2007, Diez Roux 2003). Each of these features has been hypothesized to provide supportive environments for higher levels of physical activity. Systematic literature reviews have established general linkages between the built environment and physical activity (Transportation Research Board and Institute of Medicine 2005). More recent evidence suggests the need to distinguish between types of physical activity that are conceptually and empirically linked to aspects of the built environment. In a study of adults in Western Australia, McCormack et al. (2008) found that the overall proximity to and mix of local destinations was related to walking for transport, but not recreational walking. Being close to multiple recreational destinations was, however, associated with recreational walking (but not transport-related walking). In the Twin Cities, US, Oakes et al. (2007) found that lower street connectivity was associated with more leisure-related walking (which is contra-positive to the built environment hypothesis), but unrelated to walking for transport. Such empirical findings point to the need to refine our understanding of the obesity-related relevance of features of the built environment, and have important policy implications for designing and building communities (Sallis 2008).

At the metropolitan scale, *urban sprawl*, corresponding to features of urban design including street connectivity and accessibility and residential density, is plausibly associated with higher levels of sedentarism and obesity, over and above the contributions of individual-level demographic factors (Ewing et al. 2003). Urban sprawl has been expanding over time (with the percentage of residents living in metropolitan areas outside of the central city more than doubling from 23 per cent in 1950 to 49 per cent in 1996), and may further contribute to declining levels of social capital. The latter may occur through reductions in time engaged in informal and formal social interactions (due to increased time spent in transit) and rises and disruptions in community bounded-ness, which may lead to lower levels of community investment by residents (Putnam 2000).

School Environments (MICRO-Level)

Children and adolescents spend a considerable amount of time in school environments, which thereby may offer critical contexts for the formation of dietary and exercise habits into adulthood (Dietz and Gortmaker 2001). One source of dietary influence is the National School Lunch Program in the United States, which subsidizes lunches for 28 million children from low-income households in 98,000 public and non-profit private schools. In the past, concerns have been raised by several nutrition experts for the programme's provision of lower quality foods to participants, many of whom already have limited choices as a result of their families' incomes.

In the United States, soft drinks and fruit drinks are sold in school vending machines and at school events, with as much as 56 to 85 per cent of children in school drinking at least one soft drink daily, and 20 per cent of these children consuming four or more servings daily (Gleason and Suitor 2001). The American Academy of Pediatrics Committee on School Health in the United States (2004) has issued a policy statement recommending the reduced consumption and advertising of soft drinks in schools due to the associated nutritional implications. Public health concerns have been raised because soft drink consumption may displace milk consumption (an important source of calcium and other nutrients, at critical periods of peak bone mass formation in adolescence) as well as other key nutrients (Lytle et al. 2000, Johnson et al. 1998). Sweetened drinks have been associated with overweight and obesity (Bellisle and Rolland-Cachera 2001, Tordoff and Alleva 1990), and may in part be driving the ongoing obesity epidemic.

Economic influences in the school environment are exemplified by 'pouring-rights' contracts, which offer large payments to school districts in exchange for exclusive sales of company products at schools. The Committee on School Health has recommended that school districts convene public discussions before any decision to create soft drink contracts (American Academy of Pediatrics Committee on School Health 2004). Social capital (contributing to the neighbourhood social environment) at the local-level could in theory play a role in influencing such contracts, based on the degree of collective resistance to allowing such contracts to take effect. Finally, the cost of foods in school environments (as sold in school vending machines and cafeterias) may plausibly affect dietary consumption. This effect is supported by several experimental studies on the effects of prices on food purchases in schools (French et al. 1997, 2001).

Theoretical Frameworks and Causal Inference

The theoretical models that we have highlighted are useful for framing the way that we conceptualize the causes of the obesity epidemic, from the proximal to the more distal. However, such models are ultimately useful for prevention only to the extent to which causality can be demonstrated in any of the chains that link the various environment characteristics to the individual's risk of obesity. As empirical studies begin to mount, however, the evidence for some of the hypothesized connections has become murkier in some instances. To wit, causal inference remains a challenge. For example, demonstrating an association between residential proximity to fast food outlets and risk of obesity may be insufficient for establishing causality. The reason is because of the well-known problem of endogeneity in the location of fast food outlets, i.e. fast food franchises do not randomly choose to open in certain neighbourhoods. Their decisions – based on an eye for the bottom line – are usually based upon an assessment of latent local consumer demand. In other words, a profit-maximizing fast food outlet owner will tend to open his store in a location where he believes he will make the most profit, i.e. where local residents have the greatest taste for burgers and French fries. If selective sorting of this type is all that is going on, then removing the fast food outlet (for example, via a local ordinance) would do little to affect the obesity prevalence in the neighbourhood, since presumably the residents will obtain their fast food elsewhere. Similarly, residents do not move into different neighbourhoods randomly. If exercise-conscious individuals move into neighbourhoods that are more walkable and blessed with athletic facilities, then observational studies will tend to demonstrate that there is an association between built environments and physical activity. That association, however, is not causal.

Although it is beyond the scope of the present chapter, innovative methods have been proposed by researchers for improving causal inference in observational studies (Subramanian et al. 2007). These methods include leveraging the results of natural experiments that occur in the real world on an ongoing basis (e.g. the opening of a new supermarket in a neighbourhood, or the implementation of a new public transport policy), as well as the related use of instrumental variable estimation to identify the causal effect of environmental influences on obesity. Cluster randomized trials would help too, though life is often too short – or funding too scarce – to wait for the results of an active manipulation of environmental characteristics (e.g. installing new sidewalks) on obesity outcomes. As the detailed chapters in the remainder of this book will also emphasize, robust causal inference is still lacking for a number of hypothesized linkages between aspects of the 'obesogenic' environment and overweight/obesity. The theoretical frameworks are invaluable, however, for pointing out the places where we should turn our gaze.

Conclusions

Drawing on prior hypotheses and conceptual work, this chapter has highlighted the importance of considering different environmental contexts for explaining the occurrence of obesity. The ANGELO framework and IOFT Working Group conceptual framework are particularly useful in detailing and discussing the extra-individual factors and mechanisms which may influence diets, levels of physical activity, and ultimately affect obesity incidence and prevalence in populations. Refinements and modifications in these frameworks, as new theoretical understandings emerge, are essential. Considerations of temporal trends in macro-level factors such as food manufacturing, and interactions at different time points over the lifecourse of individuals are additional helpful perspectives to these models, which may more comprehensively explain the ongoing obesity epidemic. Overall, such theoretical considerations may help to produce a more nuanced scientific inquiry, and form the basis for more effective obesity-related interventions and prevention policies.

References

American Academy of Pediatrics Committee on School Health. 2004. Soft drinks in schools. *Pediatrics*, 113(1 Pt 1), 152-4.
Bandura, A. 1991. Social cognitive theory of moral thought and action, in *Handbook of moral behavior and development*, edited by W.M. Kurtines and J.L. Gerwitz. Hillsdale: Lawrence Erlbaum, 45-103.
Barker, D.J.P. 1992. *Fetal and infant origins of adult disease*. London: British Medical Journal.

Battle, E.K. and Brownell, K.D. 1997. Confronting a rising tide of eating disorders and obesity: Treatment vs. prevention and policy. *Addictive Behaviours,* 21(6), 755-65.

Bellisle, F. and Rolland-Cachera, M.F. 2001. How sugar-containing drinks might increase adiposity in children. *Lancet*, 357(9255), 490-1.

Ben-Shlomo, Y. and Kuh, D. 2002. A life course approach to chronic disease epidemiology: conceptual models, empirical challenges and interdisciplinary perspectives. *International Journal of Epidemiology*, 31(2), 285-93.

Center for Media Education. 2000. *Marketing to children harmful: experts urge candidate to lead nation in setting limits* [Online]. Available at: http://www.cme.org.press/001018pr.html [accessed: 1 November 2004].

Centers for Disease Control. 2002. Infant mortality and low birth weight among black and white infants – United States, 1980-2000. *Morbidity and Mortality Weekly Report,* 51(27), 589-92.

Christakis, N. and Fowler, J.H. 2007. The spread of obesity in a large social network over 32 years. *New England Journal of Medicine*, 357(4), 370-9.

Critser, G. 2003. *Fatland: how Americans became the fattest people in the world.* New York: Houghton Mifflin Co.

Cutler, D.M. Glaeser, E.L. and Shapiro, J.M. 2003. Why have Americans become more obese? *Journal of Economic Perspectives*, 17(3), 93-118.

Dietz, W.H. and Gortmaker, S.L. 2001. Preventing obesity in children and adolescents. *Annual Review of Public Health*, 22, 337-53.

Diez Roux, A.V. 2003. Residential environments and cardiovascular risk. *Journal of Urban Health,* 80(4), 569-89.

Eaton, S.B., Konner, M. and Shostak M. 1988. Stone agers in the fast lane: Chronic degenerative diseases in evolutionary perspective. *American Journal of Medicine,* 84(4), 739-49.

Egger, G., Swinburn, B. and Rossner, S. 2003. Dusting off the epidemiological triad: could it work with obesity? *Obesity Reviews*, 4(2), 115-9.

Emmons, K.M. 2000. Health behaviors in a social context, in *Social Epidemiology*, edited by L. Berkman and I. Kawachi. New York: Oxford University Press, 242-66.

Ewing, R., Schmid, T., Killingsworth, R., Zlot, A. and Raudenbush, S. 2003. Relationship between urban sprawl and physical activity, obesity, and morbidity. *American Journal of Health Promotion*, 18(1), 47-57.

French, S.A., Jeffery, R.W., Story, M., Breitlow, K.K., Baxter, J.S., Hannan, P. and Snyder, M.P. 2001. Pricing and promotion effects on low-fat vending snack purchases: the CHIPS Study. *American Journal of Public Health*, 91(1), 112-7.

French, S.A., Story, M., Jeffery, R.W., Snyder, P., Eisenberg, M., Sidebottom, A. and Murray, D. 1997. Pricing strategy to promote fruit and vegetable purchase in high school cafeterias. *Journal of the American Dietetic Association*, 97(9), 1008-10.

Gleason, P. and Suitor, C. 2001. *Children's Diets in the Mid-1990s: Dietary Intake and Its Relationship with School Meal Participation.* Alexandria, VA: US Department of Agriculture, Food and Nutrition Service, Office of Analysis, Nutrition and Evaluation.

Green, L.W. and Kreuter, M.W. 1990. Health promotion as a public health strategy for the 1990s. *Annual Review of Public Health*, 11, 319-34.

Haddon, W. 1980. Advances in the epidemiology of injuries as a basis for public policy. *Public Health Reports*, 95(5), 411-21.

Hill, J.O. and Peters, J.C. 1998. Environmental contributions to the obesity epidemic. *Science*, 280(5368), 1371-4.

Horgen, K.B. and Brownell, K.D. 2002. Confronting the toxic environment: Environmental, public health actions in a world crisis, in *Treating Addictive Behaviours*, edited by T.A. Wadden and A.J. Stunkard. New York: Guilford Press, 95-106.

Huston, S.L., Evenson, K.R., Bors, P. and Gizlice, Z. 2003. Neighborhood environment, access to places for activity, and leisure-time physical activity in a diverse North Carolina population. *American Journal of Health Promotion*, 18(1), 58-69.

Institute of Medicine. 1995. *Weighing the options: Criteria for evaluating weight management programs*. Washington, DC: National Academy Press.

Johnson, R.K., Panely, C. and Wang, M. 1998. The association between noon beverage consumption and the diet quality of school-age children. *Journal of Child Nutrition and Management,* 22(2), 95-100.

Kato, H., Tillotson, J., Nichaman, M.Z., Rhoads, G.G. and Hamilton, H.B. 1973. Epidemiologic studies of coronary heart disease and stroke in Japanese men living in Japan, Hawaii and California. *American Journal of Epidemiology*, 97(6), 372-85.

Kawachi, I. and Berkman, L.F. 2000. Social cohesion, social capital, and health, in *Social Epidemiology*, edited by L. Berkman and I. Kawachi. New York: Oxford University Press, 174-90.

Kawachi I., Kennedy, B.P., Lochner, K. and Prothrow-Stith, D. 1997. Social capital, income inequality, and mortality. *American Journal of Public Health,* 87(9), 1491-8.

Kim D., Subramanian, S.V. and Kawachi, I. 2006. Bonding versus bridging social capital and their associations with self-rated health: a multilevel analysis of 40 US communities. *Journal of Epidemiology and Community Health*, 60(2), 116-22.

Kumanyika, S., Jeffery, R.W., Morabia, A., Ritenbaugh, C. and Antipatis, V.J. 2002. Obesity prevention: the case for action. *International Journal of Obesity*, 26(3), 425-36.

LaPoint, V. African American perspectives on marketing fast food to children [Online: Campaign for a Commercial-Free Childhood]. Available at: http://www.commercialexploitation.com/articles/african_american_perspectives.htm [accessed: 1 October 2008].

Lin, N. 2001. *Social Capital: A Theory of Social Structure and Action.* New York: Cambridge University Press.

Link, B.G. and Phelan, J. 1995. Social conditions as fundamental causes of disease. *Journal of Health and Social Behavior,* Spec No:80-94.

Lytle, L.A., Seifert, S., Greenstein, J. and McGovern, P. 2000. How do children's eating patterns and food choices change over time? Results from a cohort study. *American Journal of Health Promotion*, 14(4), 222-8.

Macintyre, S. and Ellaway, A. 2003. Neighbourhoods and health: an overview, in *Neighbourhoods and Health*, edited by I. Kawachi and L. Berkman. New York: Oxford University Press, 20-44.

Macintyre, S., McKay, L., Cummins, S. and Burns, C. 2005. Out-of-home food outlets and area deprivation: case study in Glasgow, UK. *International Journal of Behavioral Nutrition and Physical Activity*, 2, 16.

McCormack, G.R., Giles-Corti, B. and Bulsara, M. 2008. The relationship between destination proximity, destination mix and physical activity behaviors. *American Journal of Preventive Medicine*, 46(1), 33-40.

Morland, K., Wing, S., Diez Roux, A. and Poole, C. 2002a. Neighborhood characteristics associated with the location of food stores and food service places. *American Journal of Preventive Medicine*, 22(1), 23-9.

Morland, K., Wing, S. and Diez Roux A. 2002b. The contextual effect of the local food environment on residents' diets: the atherosclerosis risk in communities study, *American Journal of Public Health*, 92(11), 1761-7.

Nestle, M. 2003. *Food politics*. Los Angeles: University of California Press.

Oakes, J.M., Forsyth, A. and Schmitz, K.H. 2007. The effects of neighborhood density and street connectivity on walking behavior: the Twin Cities walking study. *Epidemiologic Perspectives and Innovations*, 4, 16.

Ong, K.K. and Dunger, D.B. 2002. Perinatal growth failure: the road to obesity, insulin resistance and cardiovascular disease in adults. *Best Practices in Research of Clinical Endocrinology and Metabolism*, 16(2), 191-207.

Papas, M.A., Alberg, A.J., Ewing, R., Helzlsouer, K.J., Gary, T.L. and Klassen, A.C. 2007. The built environment and obesity. *Epidemiologic Reviews*, 29, 129-43.

Pearce, J., Blakely, T., Witten, K. and Bartie, P. 2007a. Neighborhood deprivation and access to fast-food retailing: a national study. *American Journal of Preventive Medicine*, 32(5), 375-82.

Pearce, J., Witten, K., Hiscock, R. and Blakely. 2007b. Are socially disadvantaged neighborhoods deprived of health-related community resources? *International Journal of Epidemiology*, 36(2), 348-55.

Poston, W.S.C. and Foreyt, J.P. 1999. Obesity is an environmental issue. *Atherosclerosis*, 146(2), 201-9.

Putnam, R.D. 2000. *Bowling Alone: The Collapse and Revival of American Community*. New York: Simon and Schuster.

Ravussin, E., Valencia, M.E., Esparza, J., Bennett, P.H. and Schulz, L.O. 1994. Effects of a traditional lifestyle on obesity in Pima Indians. *Diabetes Care*, 17(9), 1067-74.

Reaven, G.M. 1988. Role of insulin resistance in human disease. *Diabetes*, 37(12), 1595-607.

Rogers, E.M. 2003. *Diffusion of Innovations*. New York: The Free Press.

Romano, P.S., Bloom, J. and Syme, S.L. 1991. Smoking, social support, and hassles in an urban African-American community. *American Journal of Public Health*, 81(11), 1415-22.

Rose, G. 1985. Sick individuals and sick populations. *International Journal of Epidemiology*, 14(1), 32-8.

Sallis, J. 2008. Angels in the details: Comment on 'The relationship between destination proximity, destination mix and physical activity behaviors'. *American Journal of Preventive Medicine*, 46(1), 6-7.

Sampson, R. 2003. Neighborhood-level context and health: lessons from sociology, in *Neighborhoods and Health,* edited by I. Kawachi and L. Berkman. New York: Oxford University Press.

Sampson, R.J., Raudenbush, S.W. and Earls, F. 1997. Neighborhoods and violent crime: a multilevel study of collective efficacy. *Science,* 277(5328), 918-24.

Seefeldt, V., Malina, R.M. and Clark, M.A. 2002. Factors affecting levels of physical activity in adults. *Sports Medicine*, 32(3), 143-68.

Sooman, A., Macintyre, S. and Anderson, A. 1993. Scotland's health: a more difficult challenge for some? The price and availability of healthy foods in socially contrasting localities in the West of Scotland. *Health Bulletin*, 51(5), 276-84.

Stansfeld, S.A. 1999. Social support and social cohesion, in *Social Determinants of Health*, edited by M. Marmot and R.G. Wilkinson. Oxford: Oxford University Press, 155-78.

Stunkard, A.J., Harris, J.R., Pedersen, N.L. and McClearn, G.E. 1990. The body-mass index of twins who have been reared apart. *New England Journal of Medicine*, 322(21), 1483-7.

Subramanian, S.V., Glymour, M.M. and Kawachi I. 2007. Identifying causal ecologic effects on health: a methodologic assessment, in *Macrosocial Determinants of Population Health*, edited by S. Galea. New York: Springer, 301-32.

Swinburn, B., Egger, G. and Raza, F. 1999. Dissecting obesogenic environments: the development and application of a framework for identifying and prioritizing environmental interventions for obesity. *Preventive Medicine*, 29(6 Pt 1), 563-70.

Tordoff, M.G. and Alleva, A.M. 1990. Effect of drinking soda sweetened with aspartame or high-fructose corn syrup on food intake and body weight. *American Journal of Clinical Nutrition*, 51(6), 963-9.

Transportation Research Board and Institute of Medicine. 2005. *Does the built environment influence physical activity? Examining the evidence.* Washington, DC: National Academies Press.

Wadden, T.A., Brownell, K.D. and Foster, G.D. 2002. Obesity: responding to the global epidemic. *Journal of Consulting and Clinical Psychology*, 70(3), 510-25.

Witten, K., Hiscock, R., Pearce, J. and Blakely, T. 2008. Neighborhood access to open spaces and the physical activity of residents: A national study. *Preventive Medicine*, 47(3), 299-303.

PART II
Food Environment and Obesity (Energy In)

Chapter 4

Changing Food Environment and Obesity: An Overview

Janet Hoek and Rachael McLean

Introduction

There is no dispute that obesity has become a major health problem affecting developed and developing nations alike (Maziak et al. 2007, Morrill and Chinn 2004). However, on-going discussion continues about the factors most likely to have contributed to this situation, and the measures that could most effectively reduce obesity levels. As Alderman et al. (2007) commented, although the cause of obesity – excessive energy intake relative to expenditure – is easily understood, the reasons why this imbalance has increased so rapidly are less apparent (see also Friedman and Schwartz 2008, Kersh and Marone 2005).

Arguments that changes in population genetics and physical activity levels have produced the epidemic proportions of obesity evident internationally are no longer regarded as the most likely explanation for increases in obesity prevalence (Maziak et al. 2007, Bray et al. 2004, Jeffery and Utter 2003, see also Chapter 3). Although a genetic basis for obesity exists and individuals' genetic profile may predispose them to obesity, this profile is strongly influenced by gene–gene and gene–environmental interactions, and the susceptibility of these interactions to increasingly permissive and obesogenic environments (Froguel and Blakemore 2008, Yang et al. 2007, Mutch and Clement 2006). A recent review concluded that while several genes may predispose individuals to obesity, their effect size is not large (Yang et al. 2007); this finding lends further support to the view that obesity is primarily shaped by environmental forces.

Research exploring these forces has focussed on changes in energy intake and energy expenditure, and concluded the former are largely responsible for promoting higher body weight (Westerterp and Speakman 2008, Jeffery and Harnack 2007). For example, Peters et al. (2002) noted: 'The environment encourages overeating through an abundant food supply that is high in fat and energy density, easily available, relatively inexpensive, good tasting and served in large portions' (p. 70). This comment highlights the myriad environmental changes that have occurred and their combined influence on consumption. As well as highlighting researchers' focus on energy intake, Peters et al.'s (2002) comment recognizes the shift in emphasis from individual factors that might lead to obesity to environmental attributes that pre-dispose and reinforce obesity-related behaviours (Jeffery and

Utter 2003, Egger and Swinburn 1997). Of these, changes in food supply and demand, and factors such as food policies and marketing strategies, have received increasing research attention.

This chapter analyses the environmental, political, social and economic factors that affect food supply and influence consumers' choices and consumption practices, focusing primarily on food environments in Western developed countries. We begin by summarizing global trends in food supply and the relationships between these and the growing prevalence of obesity. We then examine the commercial imperatives that determine how food is marketed and compare the promotion of foods high in fat, salt and sugar with initiatives designed to support healthier food choices. Finally, we conclude by suggesting more fundamental changes that may be required if consumers' food environments are to become less obesogenic.

Food Supply

Changes in food supply have occurred at many levels and typically come about in response to political and economic influences; those that have occurred post-World War II have emphasized agricultural production as a means of achieving economic prosperity. Tillotson (2003) suggested that two important events occurred during the twentieth century: industrialization, which increased the range of food available and rapid advances in nutrition knowledge. Industrialization meant food production became more efficient, making food itself an increasingly commercial item capable of delivering economic benefits. As a consequence, incentives to promote increased consumption and enhanced efficiencies emerged. Although food production could have been guided by enhanced nutrition knowledge, particularly in relation to non-communicable diseases, Tillotson argued that little of this newly discovered knowledge was translated into practices that would benefit population level health. Food production thus became driven by economic goals that did not take full cognisance of the potential public health consequences pursuit of a commercial agenda would bring about.

Industrialization has also been supported by government policies, particularly within major producer groups such as the United States and the European Union. These policies influence the costs and profits of food production by supporting the price received for specific crops and thus the attractiveness of particular raw materials to manufacturers. Tillotson (2003) commented on the widespread availability of low-cost commodity products and the plentiful availability of energy dense items manufactured from these. More specifically, Story et al. (2008) suggested the extensive use of high fructose corn syrup (HFCS) is likely to have resulted from the inexpensive raw materials used in its manufacture (see also Morrill and Chinn 2004). Several researchers have commented on the rapid growth in products using HFCS and have linked this to US government subsidies that support corn production, make manufacture of HFCS less expensive, and create incentives for manufacturers to use it in other food products (Schwartz and Brownell 2007).

The effect of these interventions has been marked; Bray et al. (2004) reported that while HFCS represented only 1 per cent of sweeteners available for consumption in the United States in 1970, it grew to account for more than 40 per cent in 2000 (see also Kersh and Marone 2005, Morrill and Chinn 2004). According to Bray et al.'s (2004: 524) estimates, intake of sweeteners has increased rapidly since the 1970s and has paralleled the growth in obesity, leading them to suggest increased consumption of HFCS, particularly in beverages, may be linked to weight gain and obesity (Malik et al. 2006, Bray et al. 2004).

If the HFCS example illustrates how government policy can directly influence food supply, it would seem possible that policy interventions could promote greater production, use and availability of nutrient rich, lower energy foods. However, Story et al. (2008) noted that fresh fruit and vegetables have received very few subsidies; they reported that price increases in this category exceeded those observed in processed food categories. Food subsidies thus represent a powerful incentive that promotes continued production and processing of subsidized commodity products, and directly influences the foods available to consumers.

Improved production efficiencies increase and change food supply, decrease the costs of foods in abundant supply, and foster initiatives that add value to commodity foods via further processing. Winson (2004) argued that unprocessed foods typically returned less profit than products that underwent substantial transformation, and suggested the ability to 'add value' at each production point promoted greater food processing and refinement. Drewnowski's (2007) comparison of the food supply during 1970-1974 with 2000 figures supports this contention; he reported that the availability of flour and cereal products had increased by nearly 50 per cent; added fats and oils increased by 38 per cent, and corn sweeteners by 277 per cent. Against these trends, he found the availability of fruit and fruit juices was 1.4 servings per day while vegetables were 3.8 daily servings, half of which was made up of potato products (French fries, fresh potatoes and potato chips) canned tomatoes and iceberg lettuce. Overall, the twin forces of over-production in developed countries and the higher profits available from processed foods have created compelling economic incentives to process and refine agricultural products so these possess greater value (Chopra et al. 2002).

To ensure food manufactured for consumption is sold, it is priced at a level that makes it attractive to consumers. Rigby et al. (2004) described this situation as a 'policy paradox' (p. 425) and suggested that at the same time as governments recommend the public reduce consumption of high fat, salt and sugar (HFSS) items, their policies actually support the over-production of these items. Nestle (2002) had earlier identified the same problem when she concluded that too much food is produced, thus creating intense competition that promotes over-consumption. Governments have thus been criticized for protecting producers' interests and promoting economic benefits, particularly where growing evidence suggests these often undermine population health (Weiss and Smith 2004: 380).

Tillotson (2003) explored the dilemma that governments face. He noted that the food industry is one of America's largest and most powerful industries; it has experienced considerable growth, attracted large investment, and is regarded as having made an important contribution to the United States' economy. To maintain growth, Tillotson suggested that food manufacturers became increasingly marketing oriented and placed a premium on developing processed foods that would generate consumer demand. He cited earlier work noting that Americans, whom he suggested could be assumed to be broadly representative of Western consumers, were influenced primarily by taste, convenience and price, rather than by nutritional factors. Because food manufacturers could highlight the former attributes, consumers paid less attention to the latter, in which most were less interested anyway. Government policies have simultaneously assisted manufacturers to create and meet consumer demand while also promoting an obesogenic environment.

Yet while some commentators have criticized governments for continuing to subsidize food production despite the chain of events linking it to obesity, Tillotson suggested policy makers were initially unlikely to have realized that production subsidies could lead to serious public health problems. However, even now that obesity has been recognized as having reached pandemic status, governments have been slow to introduce policy changes that could reduce incentives to produce and market HFSS items.

Furthermore, problems arising in the developing world suggest the wider consequences of production and trade policies have not been considered in detail. For example, free trade policies (and imports of food grown in developed countries) have threatened the viability of local farmers and food production, and thus reduced food security. Similarly, the abolition of local food subsidies and introduction of expensive genetically modified crops that require annual seed purchase have undermined traditional agricultural practices and reduced food security (McMichael 2001, Patnaik 1999).

Overall, researchers have identified economic and political factors that have promoted increased food production and stimulated the development of more highly processed foods. As a result, the food supply available to consumers has altered and become both sweeter and more energy dense (Popkin 2005, Rigby et al. 2004, Morrill and Chinn 2004) (although Jeffery and Utter (2003) noted changes in food supply that could both support and militate against increases in overall body weight). At the same time as food supply changed, the food industry itself was also evolving rapidly; the following section explores these developments.

Food Industry Structure and Food Accessibility

In addition to macro political and economic factors affecting food supply, structural changes within the food industry, which have also arisen in response to economic pressures, have affected food accessibility. The consolidation of food companies into large multi-national organizations with considerable marketing power has arguably influenced food supply, purchase and consumption internationally (Maziak et al. 2007). In many countries, the food supply chain has become more concentrated and, where there were once multiple participants, two or three mega-companies now dominate some markets. Winson (2004) suggested the increasing concentration of supermarket chains had fundamentally changed consumers' 'foodscape' and contributed to the increased emphasis on value-added foods. Tillotson (2003) made similar points when he argued that large food manufacturers and retailers alike could exert considerable marketing leverage, and through their size and the resources available to them, could undertake campaigns that prompted, maintained and supported consumption of HFSS foods.

Like food manufacturers, supermarket chains are driven by profit goals; they thus have a greater incentive to stock products that will contribute to these goals, and to support these by allocating them more space and higher visibility facings. Because processing creates new opportunities for 'value' to be added to products, processed energy dense foods typically offer higher profit margins. Winson's (2004) analysis of the space allocated to these products led him to conclude that these were 'colonising...the supermarket foodscape...at an unprecedented level' (p. 307). He also noted that whereas these foods were once traditionally found in supermarkets, they were now appearing in non-traditional sites, such as airports and service stations, and sites that are within easy access of schools. He cited the Coca Cola company executive quoted by Nestle and Jacobsen (2000), who explained the company strategy thus: 'To build pervasiveness of our products, we're putting ice-cold Coca-Cola classic and our other brands within reach, wherever you look: at the supermarket, the video store, the soccer field, the gas station – everywhere' (Winson 2004: 308). Accessibility and brand ubiquity have simplified purchase and promoted increased consumption.

Paradoxically, although supermarkets have increased the range of foods available and the ease with which consumers can access these, not all social groups can purchase healthy food items easily. Morland et al. (2002) reported that supermarkets were more likely to be found in higher socioeconomic neighbourhoods than in lower socioeconomic status SES areas, leading to 'food deserts' where individuals were unable to purchase fruit and vegetables (see also Friedman and Schwartz 2008). Smoyer-Tomic et al.'s (2006) analysis of Edmonton supermarkets confirmed this finding and reported that, while most areas of the city had good access to supermarkets, 'food deserts' (typically older and lower SES neighbourhoods) were evident. They suggested these areas existed because of economic disincentives to provide unprocessed, low-margin foods to lower income neighbourhoods (see also Friedman and Schwartz, 2008). Story et al. (2008) referred to the growing body of international literature documenting differences between low and high income consumers' ability to access and afford healthy food items (see also Ford and Dzewaltowski 2008, Foresight 2007a, 2007b). Inter country variations in access to healthy and unhealthy foods in more and less advantaged neighbourhoods have been observed. In New Zealand Pearce et al. (2007) reported that higher deprivation neighbourhoods have increased access to both fast food outlets and supermarkets. However where lower SES areas have less access to healthy food but greater access to energy dense, nutrient poor (EDNP) foods, they face a double jeopardy situation that is highly likely to exacerbate existing inequities in obesity prevalence.

Recently, initiatives that would reduce the number and size of 'food deserts' have been tested. Pothukuchi (2005) reported on a programme to attract supermarkets to low income neighbourhoods; she suggested that providing transport, on-trolley calculators, and assistance to non-English speaking immigrants, could produce good business outcomes. Although this study was a pilot scale intervention, Pothukuchi's results indicate that making healthful foods accessible to low income families has the potential to promote public health as well as business outcomes.

Changes to food retailing have not been confined to supermarkets. However, whereas these have amalgamated and increased in size and power, other trends are apparent among restaurants, which have proliferated in number and variety. The following section explores changes in food settings.

Food Consumption Outside the Home

At the same time as consumers have been able to access a wider array of food products, social changes affecting their disposable income and the time they have available to prepare meals, have increased their ability to purchase these foods. Whereas two generations ago, meals were largely prepared and consumed at home, a growing proportion of foods today are eaten outside the home. Several social and technological changes are thought to account for this shift in food preparation and consumption patterns. Higher proportions of household adults in the paid workforce mean other activities may now compete for the time formerly spent on food preparation. In addition, larger disposable incomes resulting from greater workforce participation have made eating out of home a more viable option (Popkin 2005, Morrill and Chinn 2004, Rigby et al. 2004, Story et al. 2001). Social trends have focussed on interactions that accompany eating; these are stimulated and reinforced by food marketing, which has fostered a social culture that is oriented around food (Tillotson 2003).

Together, these changes have created an environment where a substantial proportion of the food consumed is not prepared by the individuals who purchase and eat it. Stanton (2006) reported that more than one third of Australian food expenditure is now on foods purchased and consumed outside the home environment. Nestle and Jacobsen (2000) reported that Americans spend half their food budget and obtain a third of their energy requirements from outside the home, while more recent figures cited by Chopra et al. (2002) estimated that only 38 per cent of meals consumed in the US were home made (2002: 953). Rosenkranz and Dzewaltowski (2008) reported that US children obtain a third of their energy intake from food prepared away from home, with 20 per cent of this coming from 'fast foods'. Although food eaten outside the home could have a similar nutritional content to food prepared at home, recent studies reported that expenditure on energy dense foods has increased. For example, Schwartz and Brownell (2007) cited statistics that US consumers' expenditure on 'fast food has increased eighteen-fold since 1970' (p. 79), while Kim and Kawachi (2006) reported large increases in soft drink and salty snack consumption among 2-38 year old Americans during 1977-1996. Tillotson (2003) estimated that, by 2005, 49 per cent of food expenditure would be on food eaten outside the home, a trend he predicted would continue if not curbed by regulatory measures or other interventions.

Other changes in consumption settings have also occurred. Schools are an increasingly important food supplier and the widespread availability of EDNP foods in schools has generated considerable concern, since children are captive audiences whose preferences are likely to be shaped by foods available within school environments. In addition, school neighbourhoods expose children to high levels of food marketing; Maher et al. (2005) found that the majority of advertisements found in a 1 km radius of a sample of New Zealand secondary schools were for food; of these, 70 per cent promoted foods that were classified as 'unhealthy'. The Labour-led government of 1999-2008 introduced guidelines to improve the quality of food sold in schools; however, these were rescinded by the Anne Tolley, the National-led government's Minister of Education, in early 2009, despite evidence that they had led to improved nutrition and other outcomes, and before any formal evaluation could take place.

Irrespective of the source or context, the research evidence points consistently to a striking change in food consumption settings. Nor have these changes in consumption patterns been benign; Ludwig et al.'s (2001) longitudinal study showed that each additional serving of a sugary drink was associated with a 1.6 times greater risk of obesity. This finding held after demographic, dietary and other potential confounding factors had been controlled.

Changes in the energy density of foods consumed reflect patterns in food availability. Story et al. (2001) noted that the number of fast food restaurants in the United States increased by nearly 150 per cent between 1972 and 1995, a finding reflected in the fact that nearly half of all US adults ate at a restaurant on any given day during 1998 (p. 312). Increased out of home food consumption would not contribute to obesity if consumers had access to, and chose, nutritious lower energy foods. However, Story et al. (2008) reported that it was difficult to purchase a healthy meal option from some outlets. This difficulty is due largely to changes in pre-prepared foods, which are reportedly higher in fat and lower in micronutrients (Nestle and Jacobsen 2000). In addition, the perceived costs of pre-prepared foods relative to home-prepared meals also appear to have supported the trend to purchase and consume more meals out of home. As well as eating out more frequently, consumers also have a greater array of food products that can be incorporated into home-made meals. This rapid diversification in the foods available has promoted consumption of more processed foods, many of which have higher fat or sugar levels and thus result in increased energy intake (Bray et al. 2004). As Maziak et al. (2007) noted: 'Supply, variety, price, manipulation of taste, packaging, fast food and aggressive marketing are leading to increasing numbers of people joining the consumption euphoria of various food products' (2007: 36).

While changes in disposable income may explain some of the growth in out of home food consumption, the fact that obesity disproportionately affects lower socioeconomic groups suggests a more complex relationship between income and consumption exists. Story et al. (2001) noted that income was not associated with the total quantity of food consumers ate, but with the types of foods they consumed (2001: 318). Food purchases in higher income markets are less energy dense relative to lower income markets, although Story et al. (2001) concluded that while increased food affordability promoted greater consumption, changes in dietary intake could not be explained solely by variations in food prices.

Meals eaten out of home are often served in larger portions, a factor also held responsible for the growth in obesity (Kersh and Marone 2005, French et al. 2001, Story et al. 2001). Whereas varying portion sizes may once have had little effect on overall obesity rates, there is now increasing evidence that lay consumers and experts alike find it difficult to estimate the energy content of different foods (Stanton 2006). Story et al. (2008) commented on the trend towards larger portion sizes, often two to three times the size of 'normal' servings, and noted that some restaurant entrees alone may contain more than half the energy individuals require each day. In further evidence of 'upsizing', Story et al. (2001) noted that what is currently marketed as a 'child size' drink was described as 'king size' in the 1950s. Overall, strategies promoting larger portions and 'bundled' items are a response to increased competition and consumer demands for 'value for money' (Nestle and Jacobsen 2000).

Popkin (2005) summarized many of the factors that have affected consumers' environments, and noted changes had occurred globally (see also Chapter 2). The above sections have examined factors that shape consumers' food supply, how they access food, and where consumption occurs. However, affordability also directly influences food purchase and the following section examines food pricing.

Food Pricing

Rosenkranz and Dzewaltowski (2008) noted that price was a critical macro-level factor that influenced food purchase and consumption (see also Glanz et al. 1998). At present, however, prices do not promote purchase of low fat, sugar and/or salt (LFSS foods). For example, Drewnowski and Darmon (2005) found that 'lean meats, fish, fresh vegetables, and fruit generally cost more than less healthy alternatives, such as energy-dense foods made from refined grains, sugars, and fats' (2005: 124).

Drewnowski (2004) suggested that as foods with high fat and sugar levels provide the lowest cost energy sources, pricing creates incentives for low income consumers to select energy dense foods because these are less expensive (see also Drewnowski 2007, Schwartz and Brownell 2007, Weiss and Smith 2004). As Story et al. (2008) noted: 'Fresh fruit and vegetables are more expensive on a per calorie basis than are fats and sugar' (2008: 263). This discrepancy between healthfulness and affordability has existed for some time. Brownson et al. (2006) reported that between 1985 and 2000, the cost of fresh fruit and vegetables increased by more than 100 per cent whereas the increase in the cost of EDNP foods was less than half this increase.

Kim et al. (2006) recommended imposing taxes on HFSS items and suggested these would provide an opportunity to generate revenue that could support obesity-prevention initiatives. However, they noted that taxes could have unintended consequences; for example, taxes may affect low income consumers more than high income consumers and may result in one risk behaviour being replaced by another. To test these possibilities, and to estimate the behavioural consequences of HFSS taxes, Kim et al. recommended future work examine the likely population coverage taxation would achieve and how it could be integrated with other initiatives, such as food labelling. Overall, the affordability of low fat, salt, sugar (LFSS) and HFSS foods reflects imbalances in production incentives and food availability. Although food marketing has the potential to address this imbalance, experts suggest it has instead reinforced existing disparities. The following section examines food marketing strategies and their influence on food environments.

Food Marketing

Marketing uses environmental stimuli to shape and reinforce consumers' behaviour. Story et al. (2008) noted that marketers begin applying these stimuli to young children by introducing them to brands, linking these brands to positive and desirable attributes, and maintaining brands' salience. The strategies aim to foster and support purchase behaviour (see also Hayne et al. 2004). While marketing to children has occurred for many decades, Alderman et al. (2007) suggested that it has increased in intensity very recently and cited the proliferation of media designed to appeal to and reach children (2007: 97).

Although public health researchers and advocacy groups argue children should be protected from marketing, advertisers claim they have a right to provide information to children and cite the United Nations Convention for the Rights of the Child (UNROC) to support their arguments (ASA 2006). This argument relies on Article 13, which states that children have the right to receive or impart information of all kinds. However, the New Zealand Advertising Standards Authority's Code for Advertising to Children regards children as 'below the age of 14', which is inconsistent with Article 1 where children are defined as: 'every human being below the age of eighteen years unless under the law applicable to the child, majority is attained earlier' (UNROC 1990). Advertisers' use of the UNROC would have greater credibility if it was consistent with all Articles.

Despite calls for restrictions on marketing to children, self-regulation has promoted liberal marketing environments. Analyses of food marketing suggest it has affected primary as well as secondary demand (Hastings et al. 2003). That is, food marketing's effects went beyond shifting consumers' preference between competing brands in a given product category and increased overall demand for particular food categories. McGinnis et al. (2006) reached similar conclusions; they found that the foods marketed most heavily to children and young people were inconsistent with a healthy diet and cited food marketing as a 'probable' cause of changes in food consumption that have contributed to obesity. Several studies have implicated television advertising as a factor that has contributed to obesity (Salmon et al. 2006, Wilson et al. 2006, Lobstein and Dibb 2005, Hill and Radimer 1997). However, other media, particularly emerging electronic media, have received less attention and require more detailed analysis. Chester and Montgomery (2007), for example, documented widespread use of social networking media to promote HFSS brands and noted the need for regulatory vigilance to extend beyond traditional media.

Logically, food marketing would be expected to contribute to obesity as food companies need to maintain or increase food sales to a population that needs to decrease consumption (Nestle 2002). This logic suggests public health concerns risk becoming subordinate to the profit goals that marketing strategies support. As Nestle and Jacobsen (2000) concluded: 'Promotions, pricing, packaging, and availability all encourage Americans to eat *more* food, not less' (2000: 18). Even attempts to promote alternative consumption habits are unlikely to success since, as Jeffery and Utter (2003) pointed out, the sums spent by the food industry on marketing greatly exceed those spent by government to promote healthy eating messages (see also Hayne et al. 2004). Although supportive of healthy food promotions, Schwartz and Puhl (2003) commented on the discrepancy between advertising expenditure that supports HFSS items and the amount spent on nutrition education. They argued that greater modelling of healthy behaviour would assist children to develop healthy eating habits and then regard these as the norm: 'If the environment naturally provided exposure to foods that are consistent with the food guide pyramid, children would adhere to its principles more easily' (2003: 63).

Food marketers have responded to these concerns by reviewing self-regulatory codes, implementing new advertising review procedures, and reformulating some foods to remove fat, salt and sugar from the overall food supply. However, these initiatives have not satisfied many public health researchers, who regard self-regulation as a flawed model, fatally compromised by self-interest (Simon 2006). Efforts to reformulate food products have also been questioned; Maziak et al. (2007) noted that reformulation typically focusses on one ingredient and products marketed as 'low fat' risk misleading consumers if they remain high in energy (because of sugar levels, for example). Furthermore, as Schwartz and Brownell (2007) pointed out, although changes in product composition may improve individual food items, they may encourage greater consumption of the reformulated product. Describing this as the 'Snackwell phenomenon' (p. 83), they called for more research to examine the public health implications of food reformulation before these initiatives are regarded as having made changes to the 'toxic environment'.

The introduction of 'healthier' product lines has also raised important questions. While new products could improve nutrition, improvements will only occur if consumers reduce their consumption of less health foods and replace these with healthier choices. Further research is required to establish whether these changes have occurred, or whether healthier products are instead appealing to a new group of consumers. Until this work has been conducted, the introduction of 'healthier' foods cannot be assumed to have prompted changes in consumers' behaviour.

In summary, consumers' food environments have undergone substantial and wide-ranging changes. These changes have affected food supply, composition, accessibility, affordability, consumption settings and salience, and they have prompted calls for interventions that would render the environment less obesogenic. Several of those calling for action have drawn specific parallels between food marketing and tobacco marketing, and have suggested the latter should serve as a model to guide change (Maziak 2007). The final section examines these arguments and explores the changes that would be required to address them.

Calls for Change

The strongest calls for intervention have argued that marketing practices should be restricted or banned if what Schwartz and Brownell (2007) described as 'toxic environments' (2007: 79) are to be changed. They argued that 'the key drivers of human over-consumption are flavour, variety, large portions, visibility, and proximity' (2007: 79) and suggested that only public health policy measures could change these environments. Kim and Kawachi (2006) had earlier supported this view, arguing that environmental interventions have an advantage over campaigns directed at individuals as the former do not rely on informed consumers, but instead create conditions where healthier choices are simplified. Friedman and Schwartz (2008) extended this reasoning when they called for policies to create 'optimal defaults' (2008: 718) that made health promoting behaviours the easiest and most salient options for consumers to choose. For example, they suggested fast food restaurants could offer several low fat, salt and sugar options, which could become 'optimal defaults' if they represented better value (i.e. had lower prices or other value incentives), were the only ones linked to premiums or collectibles, or were the most visible images featured on in-store display boards.

However, despite growing political interest in exploring food marketing regulation, food industry members have criticized calls for greater regulation of their practices and argued these would be both 'draconian' (Irwin 2005) and unnecessary. Indeed, researchers, health professionals or politicians who call for regulation of the food environment have been derogated and described as 'nanny statist' or 'food Nazis' (McCully 2008, Irwin 2006, Wilson 2004). Analysis of food industry arguments suggests these have relied on two lines of reasoning: calls for greater individual responsibility and claims there is no evidence that policy changes would reduce obesity.

Individual Responsibility

Kersh and Marone (2005) noted that the food industry had avoided government regulation by shifting attention from macro-environmental interventions, such as regulation, to individual responsibility. Policy intervention, it has been argued, would obviate the need for individuals to assume responsibility for their own actions and the consequences of these and would introduce an unwelcome precedent of the state intervening in people's lives. However, this reasoning assumes that free enterprise and commercial speech should be valued as highly as public health and implies that state intervention could only be negative. These arguments parallel those formerly advanced by the tobacco industry, which advocated that government intervention is 'a cure far worse than the disease' (Kersh and Marone 2005: 847; see also Maziak et al. 2007).

Arguments that individuals need to exert greater responsibility for the foods they consume and the quantities in which they consume these relocate the locus of control for obesity to individuals and deflect attention from marketing practices that complicate and undermine individuals' control (Schwartz and Puhl 2003). Obesity is thus a consequence of individuals having exerted insufficient self-control rather than a function of the environment that shapes their choices.

Grotz (2006) suggested that obesity was initially cast as a cosmetic, rather than a public health problem; a status that reinforced industry members' view that individuals could exert control over the factors shaping their behaviour (and appearance). However, Schwartz and Brownell (2007) concluded, there is no 'evidence documenting an epidemic of decreased personal responsibility over the last thirty years' (2007: 81). Furthermore, the UK Foresight Programme concluded:

> People in the UK today don't have less willpower and are not more gluttonous than previous generations. Nor is their biology significantly different to that of their forefathers. Society, however, has radically altered over the past five decades, with major changes in work patterns, transport, food production and food sales. These changes have exposed an underlying biological tendency, possessed by many people, to both put on weight and retain it (Foresight 2007a).

Because macro-environmental factors are beyond individuals' control, Friedman and Schwartz (2008) argued there is little logic in calling on individuals to assume greater responsibility. Winson (2004) had earlier supported this view when arguing that the 'structural constraints to healthy eating' (2004: 301) required more detailed consideration. Alderman et al. (2007) used a similar approach when criticizing interventions that target individual behaviour instead of the social context in which behaviour occurs (see also Müller and Danielzik 2006). They suggested legal interventions could introduce changes to individuals' social context and, in so doing, 'address the production of risk factors at the societal level and break the causal chain connected to the individual' (2006: 92).

Furthermore, as Alderman et al. (2007) have argued: 'The moral tone of "personal responsibility" rhetoric is one of pure negative freedom…[where] the only function of the state is non-interference…This ignores the illusory nature of many of the choices consumers supposedly make' (2007: 102). Alderman et al.'s logic highlights a further policy paradox: individuals may only be able to take full responsibility for their decision-making when governments have redirected the macro-level forces that shape purchase and consumption behaviours. That is, industry arguments overlook the fact that individuals in the midst of what Linn (2004) described as a 'marketing maelstrom' may be unable to assert responsibility, since their environment militates against healthy behaviour. Identical reasoning led Story et al. (2008) to conclude that: 'Individual behaviour to make healthy choices can occur only in a supportive environment with accessible and affordable healthy food choices' (2008: 254) and reinforces concerns about deep-seated structural inequalities that mean these choices are limited (Marmot 2008).

Alderman et al. (2007) anticipated and pre-empted the argument that environments result from consumer, rather than corporate, desires when they asserted: '[Businesses] convince consumers that the environment is a result of their choices rather than a reflection of corporate desires; industry encourages consumers to be wary of government regulation of their private lives to draw attention away from their own power in creating and defining existing social conditions' (2007: 102). This logic highlights the value businesses place on liberal regulatory environments, the influence they are able to exert in these contexts, and their desire to maintain relaxed policy climates.

Public health practitioners and researchers do not dispute that individuals make decisions about their actions; however, they argue strongly that few can assume full responsibility for the consequences of these in obesogenic environments. This reasoning is well-established in health promotion; the 1986 Ottawa Charter for Health Promotion emphasizes 'a secure foundation in a supportive environment, access to information, life skills and opportunities for making healthy choices' (World Health Organization 1986: 1). Decision making thus follows, rather than precedes, environmental factors. Calls for regulation that would reduce the influence of food marketing, increase the affordability, accessibility and salience of healthier foods, and promote a healthier food supply reflect the need to create environments that promote individual responsibility. Although proponents of free markets have not resiled from their argument that individuals should assume more responsibility for their behaviours, irrespective of environmental factors that militate against this, they have advanced a second line of reasoning; this claims the evidence base supporting regulatory intervention is inadequate.

Arguments that regulation of food supply or demand should be only be implemented when the effects of proposed measures can be demonstrated requires a level of proof inconsistent with social science research (Michaels and Monforton 2005). Glass and McAtee (2006) noted that causal effects could not be identified in observational studies because it was either 'unethical or impossible' (2006: 1657) to design experimental studies that could test multiple factors. They suggested that rather than struggle to identify the causal role played by different environmental features, researchers should instead examine 'powerful levers of behaviour change' (2006: 1658), the macro-environmental forces that determine behaviour. Furthermore, they suggested efforts to develop research that might obtain elusive proof of causality would risk deferring action that could bring about more rapid change. Swinburn et al. (2005) recognized the same difficulty when they noted the lack of conclusive scientific evidence on obesity prevention was partly due to the rapid rise in obesity around the world, and the time required to show population level differences. However, they argued that this difficulty should not deter policy makers from implementing courses of action to combat obesity 'using the "best evidence available" as distinct from the "best evidence possible"' (Swinburn et al. 2005: 24). Furthermore, they noted that traditional criteria for evidence-based medicine such as Randomized Controlled Trials, long considered the 'gold standard' for evidence-based medicine, are too narrow in the context of public health interventions.

Although there has been no rapprochement between food industry members and public health advocates, governments have attempted to develop collaborative solutions that bring together multiple stakeholders to work jointly on measures that could reduce obesity. However, while politically palatable, health researchers have questioned the likely effectiveness of this approach. Some have questioned the food industry's commitment to 'being part of the solution' when they continue to spend billions of dollars promoting HFSS items to children (Friedman and Schwartz 2008), while others argue that food companies face an 'irreconcilable conflict' (Ludwig and Nestle 2008: 1809), given their imperative to maximize profit. This requirement leads to an increased emphasis on processing foods, even though, as Ludwig and Nestle noted, this emphasis is often inversely related to the final nutritional value or satiating properties of the food. Given these conflicts, Ludwig and Nestle concluded: 'this collaborative approach seems better suited to the interests of industry than to those of the public' (2008: 1809) and warned that governments risked subordinating a public health agenda to a profit oriented mandate.

Instances where food companies or suppliers have developed alliances with sporting codes or with social and community groups have also created some suspicion. Although publically described as a means of promoting healthful behaviours, particularly among children, these actions replicate the extensive sponsorship networks developed by tobacco companies, which successfully linked unhealthy products and behaviours with sporting heroes (Hoek et al. 2007). These on-going concerns about the broader food environment have prompted consideration of differing interventions; the following section examines these and their likely effects.

Proposed Solutions

Rothschild (1999) suggested that environmental interventions could involve education, social marketing or regulatory initiatives. Many food industry participants have developed education campaigns, in the belief that promulgating nutrition information will fill 'knowledge gaps' and promote more healthful behaviours. However, Swinburn (2007) argued that obesity was not a 'knowledge deficit problem', and there is ample evidence that education on its own does not change behaviour for the vast majority of the population (Egger and Swinburn 1997).

Education strategies assume consumers' behaviour will be influenced more by information than by the incentives and rewards offered in marketing programmes. Evidence from tobacco control suggests education has been unsuccessful in reducing smoking prevalence. In addition, some commentators have suggested that education campaigns enable food industry members to demonstrate concern for obesity-related problems while protecting their ability to promote the desirability of free choices made by informed individuals (Kersh and Marone 2005: 848).

Even mandated education has had variable success. Tillotson (2003) noted that although the Nutrition Education and Labelling Act 1990 was seen as a statute that would promote changed food purchase and consumption behaviours, consumers did not use labels very effectively and funding that might have supported label use was inadequate. To date, the initiatives designed to inform consumers, such as more detailed on-pack nutrition labelling, have taken little cognisance of consumers' own preferences, had little or no effect on their ability to differentiate between products' nutrition profiles, and done little to change behaviour (Maubach and Hoek 2008). These findings link logically to other studies that have documented the failure of education programmes to address serious public health problems (Chopra et al. 2002: 954, Glass and McAtee 2006) and led to some cynicism about industry-led initiatives, which Alderman et al. (2007: 98) described as 'more talk than action'.

Tillotson urged some caution before other policies, whose dynamics may not also be fully understood, are implemented. He argued that obesity followed a predictable cycle where government initially took a permissive view of entrepreneurial behaviour, since this was expected to generate economic benefits. However, if these benefits did not accrue, or if they were also associated with costs, government's initial laissez faire approach was replaced by one where 'undesirable societal side effects…[were ameliorated] by the introduction of remedial public policies' (2003: 639). If this pattern is correct, Tillotson's logic implies policies designed to ameliorate activities that undermine health objectives are now urgently needed. This, in turn, implies that stronger interventions are required.

Social marketing campaigns, particularly those that offer a benefit in return for behaviour change, can be more successful in introducing new behaviours. For example, offering subsidized nicotine replacements has promoted cessation beyond the levels achieved by education campaigns. However, the most successful tobacco control measures have undoubtedly been increases in taxation, restrictions on marketing (including promotion, distribution, product formulation and pricing) and supply restrictions, all of which have combined to reduce the visibility, salience and accessibility of tobacco products (Brownson et al. 2006). Yet despite the documented success of these measures, they are entirely dependent on governments having the political will to impose regulation (Schwartz and Brownell 2007). Regulators wishing to reduce the prevalence of obesity could learn a great deal from the experiences of tobacco control researchers. In particular, they could note the speed with which regulation can bring about environmental change.

Policy is critically important to bring about rapid change in what is variously described as a toxic and obesogenic environment. As Hayne et al. (2004: 392) noted regulation 'can transform the entire environment in one moment'; they suggested it could be used to promote an environment that supported healthier food choices. Furthermore, researchers have argued that other 'softer' approaches have been tried and found wanting. For example, Rigby et al. (2004) argued that the '"personal responsibility" approach has been tested over many decades and as a public health policy has clearly failed' (2004: 429). They called for societal controls, again citing tobacco as an example, to avoid the problems they attributed to a 'free market' approach. Brownson et al. (2006) also concluded that environmental and policy changes are more likely to promote sustainable behaviour change and will create a context that supports subsequent, more targeted, interventions. This reasoning suggests the logic that dominates many regulators' approach needs to be reversed; that is, it should focus on policy change to create environments that support healthy behaviours. Reliance on individual behaviour change to create healthier macro-environments is charming in its optimism, but lacks any logical or empirical basis (Hoek and Gendall 2006).

Conclusions

Changes in the food environment have created conditions that have contributed to obesity and have not been adequately addressed by policy responses. Instead, policies continue to support increased food production and create incentives for manufacturers to use the most efficient raw materials, even when these lack nutritional merit. While developments in food production have greatly increased the choices available to consumers, many of these choices are highly processed. Furthermore, changes in food supply have benefitted higher socioeconomic groups and appear to have exacerbated existing inequalities. At the same time as food supply and availability has changed, promotions to stimulate consumer demand for food products have increased. In particular, marketing of EDNP foods, especially to children, has increased.

Policy makers have been reluctant to implement measures that would reduce the adverse consequences of the changed food environment. In particular, they have relied on the illogical 'individual responsibility' argument and have failed to recognize that environmental change is required to support healthy food choices. While acceptance of this deeply flawed reasoning continues, the outlook for a reduction in obesity levels is grim.

References

Alderman, J. Smith, J.A. Fried, E.J. and Daynard, R.A. 2007. Application of law to the childhood obesity epidemic. *Journal of Law, Medicine and Ethics*, 35(1), 90-112.

ASA. 2006. Advertising Standards Authority Code for Advertising to Children. [Online]. Available at: http://www.asa.co.nz/code_children.php [accessed: 12 November 2008].

Bray, G.A. 2008. Fructose – How worried should we be? *Medscape Journal of Medicine*. [Online]. Available at: http://www.pubmedcentral.nih.gov/articlerender.fcgi?artid=2525467 [accessed: 26 November 2008].

Bray, G.A. Nielsen, S.J. and Popkin, B.M. 2004. Consumption of high-fructose corn syrup in beverages may play a role in the epidemic of obesity. *American Journal of Clinical Nutrition*, 79, 537-43.

Brownson, R.C., Haire-Joshu, D. and Luke, D. 2006. Shaping the context of health: A review of environmental and policy approaches in the prevention of chronic diseases. *Annual Review of Public Health,* 27, 341-70.

Chester, J. and Montgomery, K. 2007. Interactive food and beverage marketing: Targeting youth in a digital age. Center for Digital Democracy. [Online]. Available at: http://digitalads.org/documents/digiMarketingFull.pdf [accessed: 12 October 2007].

Chopra, M. Galbraith, S. and Darnton-Hill, I. 2002. A global response to a global problem: the epidemic of overnutrition. *Bulletin of the World Health Organization*, 80(12), 952-8.

Drewnowski, A. 2004. Obesity and the food environment: Dietary energy density and diet costs. *American Journal of Preventive Medicine*, 27(3), 154-62.

Drewnowski, A. and Darmon, N. 2005. Food choices and diet costs: an economic analysis. *Journal of Nutrition,* 135, 900-4.

Drewnowski, A. 2007. The real contribution of added sugars and fats to obesity. *Epidemiologic Reviews Advance Access* published 24 June 2007. [Online]. Available at: http://epirev.oxfordjournals.org/cgi/reprint/mxm011v1 [accessed: 26 November 2008].

Egger, G. and Swinburn, B. 1997. An ecological approach to the obesity pandemic. *British Medical Journal*, 315, 477-80.

Elliott, S.S., Keim, N.L., Stern, J.S., et al. 2002. Fructose, weight gain, and the insulin resistance syndrome. *American Journal of Clinical Nutrition*, 2002, 911-22.

Ford, P.B. and Dzewaltowski, D.A. 2008. Disparities in obesity prevalence due to variation in the retail food environment: three testable hypotheses. *Nutrition Reviews*, 664, 216-28.

Foresight. 2007a. *Tackling Obesities: Future Choices – Project Report: 2nd Edition*, Government Office for Science.

Foresight. 2007b. *Trends and Drivers of Obesity: A Literature Review for the Foresight Project on Obesity*. London: Government Office for Science.

French, S., Story, M. and Jeffery, R. 2001. Environmental influences on eating and physical activity. *Annual Review of Public Health,* 22, 309-35.

Friedman, R. and Schwartz, M. 2008. Public policy to prevent childhood obesity, and the role of pediatric endocrinologists. *Journal of Pediatric Endocrinology and Metabolism*, 21, 717-25.

Froguel, P. and Blakemore, A.I.F. 2008. The power of the extreme in elucidating obesity. *New England Journal of Medicine*, 3599, 891-3.

Glanz, K., Basil, M., Maibach, E., Goldberg, J. and Snyder, D. 1998. Why Americans eat what they do: Taste, nutrition, cost, convenience, and weight control concerns as influences on food consumption. *Journal of the American Dietetic Association*, 98(10), 1118-26.

Glass, T. and McAtee, M. 2006. Behavioral science at the crossroads in public health: Extending horizons, envisioning the future. *Social Science and Medicine*, 62(7), 1650-71.

Grotz, V.L. 2006. A look at food industry responses to the rising prevalence of overweight. *Nutrition Reviews*, 64(1), S48 – S52.

Hastings, G., Stead, M. and McDermott, L. 2003. *A review of research on the effects of food promotion to children*. Report prepared for the Food Standards Authority by the Centre for Social Marketing, Strathclyde University. [Online]. Available at: http://www.ism.stir.ac.uk/pdf_docs/final_report_19_9.pdf [accessed: 30 September 2003].

Hayne, C., Moran, P. and Ford, M. 2004. Regulating environments to reduce obesity. *Journal of Public Health Policy*, 25(3/4), 391-407.

Hill, J. and Radimer, K. 1997. A content analysis of food advertisements in television for Australian children. *Australian Journal of Nutrition and Dietetics*, 54, 174-81.

Hoek, J. and Gendall, P. 2006. Advertising and obesity: a behavioral perspective. *Journal of Health Communication*, 11, 409-23.

Hoek, J., Mansfield, P., Maubach, N. and Schofield, G. 2007. *CSR or spin?: A logical analysis of the Food Industry Accord*. Paper presented at the Australian and New Zealand Marketing Academy Conference, Dunedin, New Zealand, 1-3 December.

Irwin, J. 2005. Presentation to National Nutrition and Physical Activity Public Health Conference, 28 May, Christchurch, New Zealand. [Online]. Available at: http://www.anza.co.nz/docs/Speech%20Notes%20National%20Nutrition%20and%20Physical%20Activity%20Public%20Health%20Conference%2025May05.doc [accessed: 6 June 2005].

Jackson, M., Hastings, G., Wheeler, C., Eadie, D. and MacKintosh, A. 2000. Marketing alcohol to young people: implications for industry regulation and research policy. *Addiction,* 95(Suppl 4), S597-S608.

Jeffery, R. and Harnack, L.J. 2007. Evidence Implicating Eating as a Primary Driver for the Obesity Epidemic. *Diabetes*, 5611, 2673-76.

Jeffery, R. and Utter, J. 2003. The changing environment and population obesity in the United States. *Obesity Research*, 11, S12-S22.

Kersh, R. and Marone, J. 2005. Obesity, courts, and the new politics of public health. *Journal of Health and Political Policy Law*, 30(5), 839-68.

Kim, D. and Kawachi, I. 2006. Food taxation and pricing strategies to 'thin out' the obesity epidemic. *American Journal of Preventive Medicine*, 30(5), 430-7.

Linn, S. 2004. *Consuming Kids: The Hostile Takeover of Childhood*. New York: New Press.

Linn, S. and Novostat, C. 2008. Calories for sale: Food marketing to children in the twenty-first century. *Annals of the American Academy of Political and Social Science*, 615(1), 133-55.

Lobstein, T. and Dibb, S. 2005. Evidence of a possible link between obesogenic food advertising and child overweight. *Obesity Reviews,* 6(6), 203-8.

Ludwig, D. and Nestle, M. 2008. Can the food industry play a constructive role in the obesity epidemic? *Journal of the American Medical Association,* 300(15), 1808-11.

Ludwig, D., Peterson, K. and Gortmaker, S. 2001. Relation between consumption of sugar-sweetened drinks and childhood obesity: a prospective, observational analysis. *Lancet*, 357 (9255), 505-8.

Maher, A., Wilson, N. and Signal, L. 2005. Advertising and availability of 'obesogenic' foods around New Zealand secondary schools: a pilot study. *New Zealand Medical Journal*, 118, 1218.

Malik, V.S., Schulze, M.B. and Hu, F.B. 2006. Intake of sugar-sweetened beverages and weight gain: a systematic review. *American Journal of Clinical Nutrition,* 84, 274-88.

Marmot, M. 2008. Health in a just society. *Lancet*, 372, 881-2.

Maubach, N. and Hoek, J. (2008). *Alternative Nutrition Information Disclosure Formats: Using the Elaboration Likelihood Model to investigate consumers' attitudinal responses*. Paper presented at the Australian and New Zealand Marketing Academy Conference, University of Western Sydney, December.

Maziak, W., Ward, K.D. and Stockton, M.B. 2007. Childhood obesity: are we missing the big picture? *Obesity Reviews*, 9(1), 35-42.

McCully, M. 2008. A Weekly Report from the Keyboard of Murray McCully. Available at: http://www.scoop.co.nz/stories/PA0803/S00494.htm [accessed: 28 March 2008].

McGinnis, J., Gootman, J. and Kraak, V. (eds) 2006. *Food Marketing to Children and Youth: Threat or Opportunity?* Washington, D.C.: National Academies Press.

McMichael, P. 2001. The impact of globalisation, free trade and technology on food and nutrition in the new millennium. *Proceedings of the Nutrition Society*, 60, 215-22.

Michaels, D. and Monforton, C. 2005. Manufacturing Uncertainty: Contested Science and the Protection of the Public's Health and Environment. *American Journal of Public Health*, Supplement 1, July.

Morland, K., Wing, S., Diez Roux, A. and Poole, C. 2002. Neighborhood characteristics associated with the location of food stores and food service places. *American Journal of Preventive Medicine*, 22, 23-29.

Morrill, A. and Chinn, C. 2004. The obesity epidemic in the United States. *Journal of Public Health Policy*, 25(3/4), 353-66.

Müller, M. and Danielzik, S. (2006) Childhood overweight: is there need for a new societal approach to the obesity epidemic? *Obesity Reviews*, 8 (1), 87-90.

Mutch, D.M. and Clement, K. 2006. Genetics of human obesity. *Best Practice and Research Clinical Endocrinology and Metabolism*, 204, 647-64.

Nestle, M. 2002. *Food Politics: How the Food Industry Influences Nutrition and Health*. University of California Press.

Nestle, M. and Jacobsen, M. 2000. Halting the obesity epidemic: A public health policy approach. *Public Health Reports*, 115(1), 12-24.

Patnaik, U. 1999. The Cost of Free Trade: The WTO regime and the Indian economy. *Social Scientist,* 27(11-12), 3-26.

Pearce, J., Blakely, T., Witten, K. and Bartie, P. 2007. Neighbourhood deprivation and access to fast-food retailing. A national study. *American Journal of Preventive Medicine,* 32(5), 375-82.

Peters, J., Wyatt, H., Donahoo, W. and Hill, J. 2002. From instinct to intellect: the challenge of maintaining healthy weight in the modern world. *Obesity Reviews*, 3(2), 69-74.

Popkin, B.M. 2005. Global Trends in Obesity, in *Food, Diet and Obesity*, edited by D. Mela. Cambridge: Woodhead Publishing Ltd, 1-14.

Pothukuchi, K. 2005. Attracting supermarkets to inner-city neighborhoods: Economic development outside the box. *Economic Development Quarterly*, 19(3), 232-44.

Powell, L.M., Slater, S., Mirtcheva, D., Bao, Y. and Chaloupka, F. 2007. Food store availability and neighborhood characteristics in the United States. *Preventive Medicine,* 44, 189-195.

Rigby, N., Kumanyika, S. and James, P. 2004. Confronting the epidemic: The need for global solutions. *Journal of Public Health Policy*, 25(3/4), 418-34.

Rosenkranz, R. and Dzewaltowski, D. 2008. Model of the home food environment pertaining to childhood obesity. *Nutrition Reviews,* 66(3), 123-40.

Rothschild, M. 1999. Carrots, sticks and promises: A conceptual framework for the behavior management of public health and social issues. *Journal of Marketing*, 63, 24-37.

Salmon, J., Campbell, K. and Crawford, D. 2006. Television viewing habits associated with obesity risk factors: A survey of Melbourne schoolchildren. *Medical Journal of Australia*, 184(2), 64-7.

Schwartz, M. and Brownell, K. 2007. Actions necessary to prevent childhood obesity: Creating the climate for change. *Journal of Law, Medicine and Ethics*, 35(1), 78-89.

Schwartz, M. and Puhl, R. 2003. Childhood obesity: A societal problem to solve. *Obesity Reviews*, 4, 1-15.

Simon, M. 2006. Can food companies be trusted to self-regulate? An analysis of corporate lobbying and deception to undermine children's health. *Loyola of Los Angeles Law Review*, 39(1), 169-236.

Smoyer-Tomic, K., Spence, J. and Amrhein, C. 2006. Food deserts in the prairies? Supermarket accessibility and neighborhood need in Edmonton, Canada. *Professional Geographer*, 58 (3), 307-26.

Stanton, R. 2006. Nutrition problems in an obesogenic environment. *Medical Journal of Australia*, 184(2), 76-9.

Story, M., Kaphingst, K., Robinson-O'Brien, R. and Glanz, K. 2008. Creating healthy food and eating environments: Policy and environmental approaches. *Annual Review of Public Health,* 29, 253-72.

Story, M., Stevens, J., Evans, M., et al. 2001. Weight loss attempts and attitudes toward body size, eating, and physical activity in American Indian children: Relationship to weight status and gender. *Obesity Research,* 9(6), 356-63.

Swinburn, B. 2007. Expert presentation to the Health Select Committee Enquiry into Obesity and Type 2 Diabetes. Cited in FOE report available at: http://foe.org.nz/2007/03/28/nz-foe-media-release-new-report-highlights-divisions-over-obesity-prevention/.

Swinburn, B., Gill, T. and Kumanyika, S. 2005. Obesity prevention: A proposed framework for translating evidence into action. *Obesity Reviews*, 6(1), 23-33.

Tillotson, J. 2003. Pandemic obesity: Agriculture's cheap food policy is a bad bargain. *Nutrition Today*, 38(5), 186-190.

UNROC, 1990. *Convention on the Rights of the Child*. Available at: http://www.unhchr.ch/html/menu3/b/k2crc.htm [accessed: 10 January 2008].

Weiss, R. and Smith, J. 2004. Legislative approaches to the obesity epidemic. *Journal of Public Health Policy*, 25(3/4), 379-90.

Westerterp, K.R. and Speakman, J.R. 2008. Physical activity energy expenditure has not declined since the 1980s and matches energy expenditures of wild mammals. *International Journal of Obesity,* 32, 1256-63.

Wilson, N., Signal, L., Nicholls, S. and Thomson, G. 2006. Marketing fat and sugar to children on New Zealand television. *Preventive Medicine,* 42(2), 96-101.

Wilson, S. 2003. *Fast food advertising ban more nanny statism* (Press release). Libertarianz Party, 5 March.

Wilson, S. 2004. Don't ban alcohol biscuits, just don't eat them! Press release Libertarianz Party, 5 February. Available at: http://www.scoop.co.nz/stories/PO0402/S00040.htm.

Winson, A. 2004. Bringing political economy into the debate on the obesity epidemic. *Agriculture and Human Values,* 21(4), 299-312.

World Health Organization. 1986. *Ottawa Charter for Health Promotion*. First International Conference on Health Promotion, Ottawa, 21 November. Available at: http://www.who.int/hpr/NPH/docs/ottawa_charter_hp.pdf [accessed: 10 January 2009].

Yang, W., Kelly, T. and He, J. 2007. Genetic Epidemiology of Obesity. *Epidemiologic Reviews*, 29, 49-61.

Chapter 5
Understanding the Local Food Environment and Obesity

Lukar E. Thornton and Anne M. Kavanagh

Introduction

Micro-environments, such as neighbourhoods, have received more attention in recent years due to increasing evidence of their potential importance as a determinant of health behaviours, particularly in relation to area-level disadvantage (King et al. 2006, Turrell et al. 2006, Kavanagh et al. 2005, Giles-Corti and Donovan 2002, van Lenthe and Mackenbach 2002, Sundquist et al. 1999, Diez Roux et al. 1997). Specifically, small area (neighbourhood) differences in rates of obesity have been reported in Australia (King et al. 2006), Europe (van Lenthe and Mackenbach 2002, Ellaway et al. 1997) and North America (Do et al. 2007, Robert and Reither 2004) with most of these studies showing positive associations between rates of obesity or Body Mass Index (BMI) and increasing levels of area disadvantage. Additionally, area-level differences in obesity related disease have also been detected for cardiovascular disease (Diez Roux et al. 2000) and diabetes (Ford et al. 2005).

Dietary behaviours also vary between areas and again this is patterned by area disadvantage (Turrell et al. 2009, Kavanagh et al. 2007, Diez-Roux et al. 1999). As a result of these area-level differences in diet, researchers have turned their attention to environmental factors that may contribute to this patterning. Neighbourhoods with different degrees of disadvantage may vary with respect to access to both healthy and unhealthy foods across areas (contextual effects) (Pearce et al. 2008a, Macdonald et al. 2007, Pearce et al. 2007a, 2007b, Block et al. 2004, Reidpath et al. 2002). Contextual effects may also occur in other settings such as workplaces (for example access to vending machines that stock energy dense foods).

In addition, the micro-environmental setting of supermarkets may also influence dietary behaviours through within-store factors such as the availability, price and quality of food items, presence of end-aisle displays, special promotions, and snacks at checkouts (Moore et al. 2008a, 2008b, Glanz et al. 2005, 2007, Ball et al. 2006b, Dibb 2004).

Conversely, area-level clustering may also be attributable to compositional (individual) effects related to the people living in the areas (for example a preference for less healthy foods because of taste considerations). Of course, health behaviours and outcomes are most likely influenced by a combination of both compositional and contextual factors.

Further, contextual settings, or micro-environments (for example homes, workplaces, neighbourhoods, supermarkets) are likely to be influenced by multiple macro-environmental factors (for example governments, industry, marketing), although these are not necessarily aiming to achieve the same goals (Swinburn et al. 1999). For instance, although government may aim to improve health outcomes with policies towards ensuring people meet dietary guidelines, industry and marketing forces are profit driven and may seek to increase to sales of unhealthy food products as that is where the larger profits are to be made (see Chapter 4).

This chapter sets out to explore variations in the characteristics of food environments at the micro-environmental level (between neighbourhoods and within stores) and whether these factors influence dietary behaviours and associated health outcomes. Key methodological shortfalls and inconsistencies will also be highlighted.

Contextual Framework

Glanz and colleagues have previously proposed a framework to explore some key micro-environmental settings that influence dietary behaviours (Glanz et al. 2005) (Figure 5.1). This model draws on the research from multiple fields including public health, health and consumer psychology, and urban planning and shows hypothesized relationships between dietary behaviours at the macro–policy environmental, micro-environmental and individual-levels.

Within this framework, Glanz et al. (2005) identified three sets of environmental variables that can potentially affect dietary behaviours: the community, consumer and organizational nutrition environments (Figure 5.1).

Community Nutrition Environment

The community nutrition environment refers to characteristics such as the number, type, location and accessibility of food stores. This has been the environment most explored in public health research, probably because it is simpler to measure. When researchers have examined the community nutrition environment, it has usually been in the context of residential neighbourhoods.

Consumer Nutrition Environment

The food environment consumers experience within-stores (or at the point of purchase) is described as the consumer nutrition environment. The information that needs to be collected in order to measure the consumer nutrition environment includes the availability, quality and cost of food choices (Glanz et al. 2005, 2007, Saelens et al. 2007). Other factors may also include portion sizes, the availability of nutrition information, in-store promotions, availability of competing stores, displays within the stores, promotions and the strategic location of snack foods near counters (Glanz et al. 2005, 2007, Saelens et al. 2007, Ball et al. 2006b, Dibb 2004, Winson 2004).

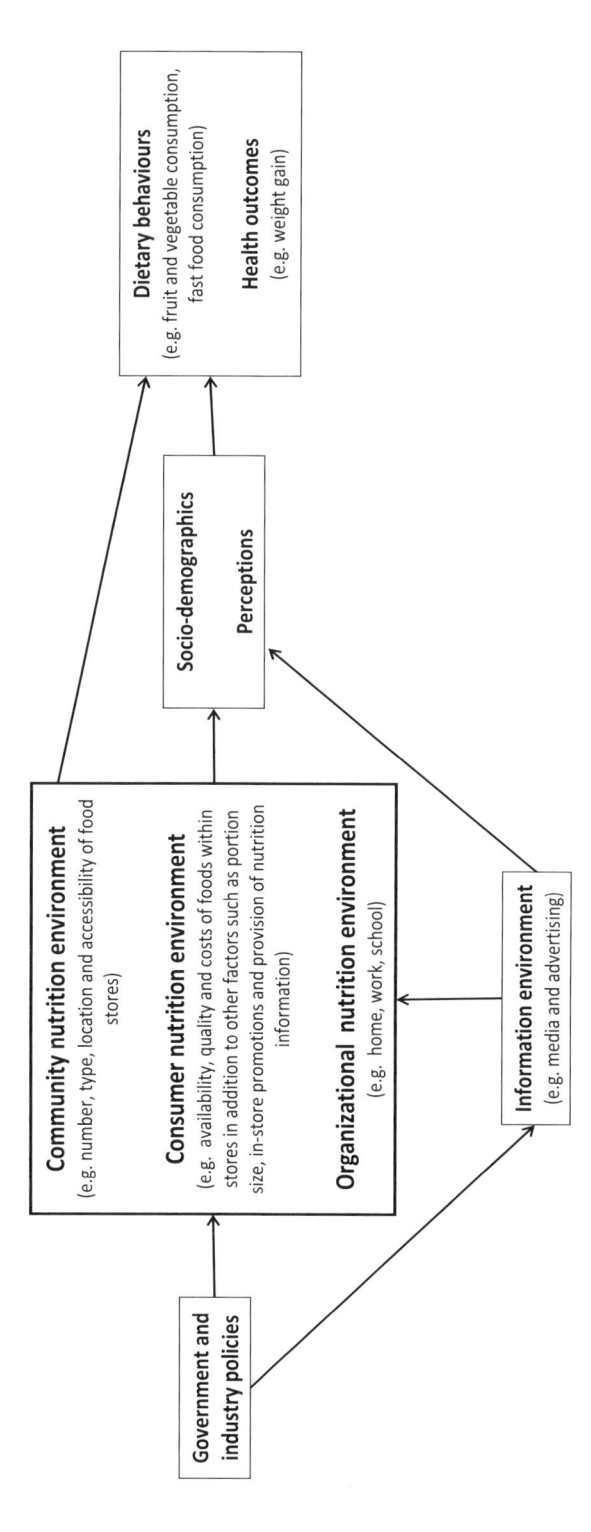

Figure 5.1 Conceptual framework for the study of nutrition environments

Source: Adapted from Glanz et al. (2005: 331).

The measurement of consumer nutrition environments need not be limited to supermarkets and other grocery stores and could be collected for a variety of store types including takeaway food stores, restaurants and vending machines (Saelens et al. 2007). Up until recently, these environments have rarely been explored in academic research, which is possibly the result of the cost and time pressures associated with collecting such data.

Organizational Nutrition Environment

People are part of multiple contexts in the course of their day-to-day life including home, work, neighbourhood, social and school. Exposure to these environments varies across the lifecourse because people move to different residential and work environments and from educational to work environments. Each of these could be seen as micro-environments that can potentially expose individuals to different food environments. As Ball and colleagues have suggested, these multiple contexts make it challenging to accurately measure the nutrition environments that individuals are exposed to (Ball et al. 2006b). This is also the case when people have a particular attachment to areas in which they prefer to shop in (because they grew up there or because they seek ethnic specific foods) but may not live in (Cummins et al. 2007).

Other Determinants

Other determinants of eating behaviours proposed in the nutrition environments framework include government and industry policies, the information environment (for example advertising) and individual variables including socio-demographic characteristics and perceptions of the environment. Issues around policy and advertising have previously been discussed elsewhere (Nestle 2002).

Using the Glanz et al. (2005) framework, we provide an overview of the evidence regarding how community and consumer nutrition environments vary according to the socioeconomic characteristics of micro-environments. We provide a detailed review of research that has investigated how characteristics of the community and consumer nutrition environments influence health behaviours and outcomes.

Evidence Regarding Community Nutrition Environments

Although there is an awareness of the multiple contexts within which people live their daily lives, studies of food environments tend to be limited to the examination of a single organizational environment, most frequently residential environments. In this context, the community nutrition environment is commonly explored.

Access to food stores within neighbourhoods is often determined through crude measures to assess the presence and density of food stores. Studies have tended to examine either a single food outlet type or multiple outlets, which commonly include chain brand supermarkets, midsize independent grocery stores, smaller convenience stores, greengrocers, fast food outlets, and restaurants. In these studies, chain supermarkets and greengrocers are commonly used as proxies for healthy foods while convenience stores and fast food outlets are used as proxies for unhealthy foods. A more detailed discussion on the way foods stores are defined is undertaken later in this chapter.

We review studies that have investigated the distribution of these stores over small areas in relation to variations in area-level characteristics and assess the implication of their presence or absence for dietary behaviours and health outcomes.

Area Disadvantage and the Community Nutrition Environment

Areas that have little or no food access have been termed food deserts however their existence in deprived areas is not always apparent and this varies both between and within countries (Larsen and Gilliland 2008, Apparicio et al. 2007, Smoyer-Tomic et al. 2006, Wrigley et al. 2003, Cummins and Macintyre 2002a).

Further inconsistencies exist with regard to whether residents of disadvantaged areas have reduced access to larger supermarkets with some studies suggesting that this is the case (Burns and Inglis 2007, Powell et al. 2007c, Block and Kouba 2006). In addition, evidence from the US suggests that neighbourhoods with a higher proportion of African American residents have reduced access to larger supermarkets but have increased access to smaller independent food stores (Powell et al. 2007c, Zenk et al. 2005, Morland et al. 2002b). Others outside of the US find that access to supermarkets and greengrocers is equally distributed across areas or in fact that areas with higher levels of socioeconomic disadvantage have better access (Pearce et al. 2008a, Smoyer-Tomic et al. 2008, Apparicio et al. 2007, Kavanagh et al. 2007, Pearce et al. 2007b). In conjunction with planning laws, a possible explanation for the differences between US findings and those reported elsewhere relate to the lower levels of residential segregation found between cities in the US and those located in places like New Zealand, Australia and parts of Europe including the UK (Pearce et al. 2009, Cummins and Macintyre 2006).

In relation to fast food restaurants, the majority of the international evidence has shown that areas with higher levels of disadvantage are more likely to have a higher exposure to fast food restaurants in their local environment (Pearce et al. 2008a, Smoyer-Tomic et al. 2008, Burns and Inglis 2007; Macdonald et al. 2007, Pearce et al. 2007a, Powell et al. 2007b, Cummins et al. 2005b, Block et al. 2004, Reidpath et al. 2002). The racial composition of neighbourhoods has been associated with the prevalence of fast food restaurants, although the direction of the effect is mixed and dependent upon the study location (Powell et al. 2007b, Block et al. 2004, Morland et al. 2002a, 2002b).

The role of outdoor advertising has recently been investigated within four cities in the US (Los Angeles, Austin, New York, and Philadelphia) in relation to obesity (Yancey et al. 2009). Of interest were features including (but not limited to) billboards and bus shelter advertisements, particularly those that promoted unhealthy (high calorie/low nutrient) food products. The density of these advertisements was patterned by neighbourhood level variables related to racial composition and income with poorer and minority communities (Latinos) most exposed. To date, research evidence on neighbourhood level advertising is sparse and further work is required, particularly with regard to how the influence of advertising may interact with the built environment.

Community Nutrition Environment as a Determinant of Diet and Health

We reviewed studies that explored associations between the food environment and dietary behaviours and the individual health outcomes of local residents; these are summarized in Table 5.1. Articles were sourced through searches of the main public health databases and journals, and scanning the references lists of articles collected. Studies are grouped by geographic region because some of the issues related to food store location, dietary behaviours and health outcomes may be context dependent at the national level as a result of cultural influences or government policies.

The majority of the studies reviewed were conducted in the US which possibly reflects the greater availability of both individual data collected through health surveys and spatial datasets with information on food stores. Within the US most studies were interested in the effect of food environments on weight status (Mehta and Chang 2008, Wang et al. 2007, 2008, Powell et al. 2007a, Inagami et al. 2006, Jeffery et al. 2006, Morland et al. 2006, Sturm and Datar 2005, Burdette and Whitaker 2004, Maddock 2004) while in Australia and New Zealand the key focus has been on dietary behaviours (Thornton et al. 2009, Pearce et al. 2009, Pearce et al. 2008b, Timperio et al. 2008, Turrell and Giskes 2008, Giskes et al. 2007, Ball et al. 2006a).

Supermarkets and greengrocers Studies exploring the associations between access to supermarkets and greengrocers have not been consistently linked to health outcomes and dietary behaviours. For example, some studies linked better supermarket access to a decrease in weight status (Powell et al. 2007a, Morland et al. 2006) while others have reported no association (Sturm and Datar 2005). In fact, one study linked improved supermarket access to a higher weight status amongst women (Wang et al. 2007).

Fruit and vegetable intake was found to be unrelated to supermarket access in Australia (Ball et al. 2006a), New Zealand (Pearce et al. 2008b) and the UK (Pearson et al. 2005). In the US, one study reported reduced supermarket access was linked to a less healthy diet (Moore et al. 2008b) while another found that better access increased intake of both healthy and unhealthy foods (Morland et al. 2002a).

Table 5.1 A brief overview of studies investigating objective measures of the food environment with individual dietary behaviours and health outcomes

Author (year)	Country	Sample/study design	Predictor	Outcome	Findings
North America					
Burdette and Whitaker (2004)	US	Cross-sectional analysis of 7,020 children (36 to 59 months of age) from low income households in Cincinnati.	Distance to the nearest fast food restaurant along a road network from each participant's household location.	A measure of overweight defined as being equal to or above the 95th percentile of BMI for the child's age and sex.	Distance to the nearest fast food restaurant was unrelated to weight status.
Inagami et al. (2006)	US	Multilevel cross-sectional analysis of 2,144 individuals from 65 census tracts in Los Angeles.	Respondents provided details of which grocery store they shopped in and an area-level disadvantage score for that neighbourhood was used as a proxy for grocery store quality; this was also compared to the area-level disadvantage score from the neighbourhood where respondents reside. Distance to grocery store was estimated from geographic centroid of respondents' neighbourhood to centroid of neighbourhood containing the grocery store where they shopped.	Self-reported height and weight used to calculate BMI.	BMI was found to be 0.24 units higher for individuals who shopped in grocery stores located in areas with higher levels of disadvantaged than the area they reside in. Those who travelled further than 1.76 miles to a grocery store had a BMI of 0.78 units higher.
Jeffery et al. (2006)	US	Cross-sectional analysis of 1,033 adults in Minnesota.	Number of fast food and non-fast food restaurants within 0.5, 1 and 2 miles (Euclidean distance) from participants' home and work addresses.	Frequency of fast food consumption and BMI calculated from self-reported height and weight.	Density of 'non-fast food' restaurants near household location was positively associated with consumption (β 0.001; $p<0.05$), however no significant associations were found for 'fast food' restaurants. Density of 'fast food' (β -0.029; $p<0.01$) and 'non-fast food' restaurants (β -0.022; $p<0.05$) around work address was inversely associated with BMI in men only.

Table 5.1 continued   A brief overview of studies investigating objective measures of the food environment with individual dietary behaviours and health outcomes

Author (year)	Country	Sample/study design	Predictor	Outcome	Findings
Maddock (2004)	US	Cross-sectional ecological analysis of individual level outcome data aggregated to the State level with 50 States included in the final analysis.	Number of fast food restaurants (per square mile and per population) within each State.	State-level obesity rates calculated from individual level self-reported height and weight.	States with more fast food restaurants per population had higher rates of obesity ($r = -0.53$; $p<0.001$). Fast food restaurants per square mile and per population accounted for 6% of the variance in State obesity rates.
Mehta and Chang (2008)	US	Multilevel cross-sectional analysis of 714,054 individuals from 544 counties.	Number of fast food restaurants per 10,000 population in each county.	Self-reported height and weight used to calculate BMI (continuous outcome) and obesity status (dichotomous outcome).	Fast food restaurant density (comparison between the 25th and 75th percentile) was associated with increased BMI (β 0.09; $p<0.05$) and likelihood of obesity (OR 1.05; $p<0.01$) while full service restaurant density was linked to a decrease in these outcomes (BMI β -0.32; $p<0.001$; Obesity OR 0.86; $p<0.001$). BMI (β 0.20; $p<0.001$) and obesity rates (OR 1.08; $p<0.001$) were highest in areas with more fast food restaurants than full service restaurants.

Table 5.1 continued   A brief overview of studies investigating objective measures of the food environment with individual dietary behaviours and health outcomes

Author (year)	Country	Sample/study design	Predictor	Outcome	Findings
Moore et al. (2008b)	US	Multilevel cross-sectional analysis of 2,384 individual from three study sites (Maryland, North Carolina, New York). Analysed without random effects because they had no influence on fixed effect estimates.	Density of supermarkets within one mile of participants' homes as determined by kernel density analysis. A further two measures of food access were based on perceptions (not discussed in findings).	Two measures of dietary quality (Alternate Healthy Eating Index and 'fats and processed meats' dietary pattern) derived from a 120-item food frequency questionnaire.	The objective measure of the environment revealed that having no supermarkets near household location (compared to those with the greatest access) was associated with a lower likelihood of having a healthy diet as measured by an Alternate Healthy Eating Index (relative probability (RP) 0.75; 95% CI 0.59-0.95) and 'Fats and processed meats' dietary pattern (RP 0.54; 95% CI 0.42-0.70).
Morland et al. (2002a)	US	Multilevel cross-sectional analysis of 10,623 individuals from 208 census tracts located in Mississippi, North Carolina, Maryland and Minnesota.	Number of supermarkets, grocery stores, full-service restaurants and fast food restaurants within census tracts.	Self-reported dietary intake.	For African Americans, increased supermarket access was associated with an increase in fruit and vegetable intake (relative risk (RR) 1.54; 95% CI 1.11-2.12) but also with a higher fat and saturated fat intake (RR 1.30; 95% CI 1.07-1.56). For White Americans a higher total fat intake (RR 1.09; 95% CI 1.01-1.18) was likely with an increase in supermarket density. Outcomes were largely unrelated to other features of the food environment.

Table 5.1 continued   A brief overview of studies investigating objective measures of the food environment with individual dietary behaviours and health outcomes

Author (year)	Country	Sample/study design	Predictor	Outcome	Findings
Morland et al. (2006)	US	Multilevel cross-sectional analysis of 10,763 individuals from 207 census tracts located in Mississippi, North Carolina, Maryland and Minnesota.	The presence of at least one supermarket, grocery store, convenience store, full-service restaurant, franchised fast food restaurant and limited service restaurant within census tracts.	Clinically measured weight status (overweight or obese), and presence of diabetes, hypertension and hypercholesterolemia.	The presence of grocery stores was associated with an increased prevalence of being overweight (prevalence ratio (PR) 1.03; 95% CI 1.00-1.07) or having hypertension (PR 1.08; 95% CI 1.00-1.17) while the presence of convenience stores was linked to an increase in both overweight (PR 1.06; 95% CI 1.02-1.10) and obesity (PR 1.16; 95% CI 1.05-1.27). Conversely, having supermarkets present within census tracts was associated with a lower prevalence of residents that were overweight (PR 0.94; 95% CI 0.90-0.98) or obese (PR 0.83; 95% CI 0.75-0.92).
Powell et al. (2007a)	US	Multilevel cross-sectional study of 73,079 adolescents from a nationally representative sample analysed as a single level regression.	Density (number per 10,000 people) of chain supermarkets, non-chain supermarkets, grocery stores, and convenience stores within the children's school ZIP code.	Self-reported heights and weight used to calculate BMI and a measure of overweight defined as being equal to or above the 95th percentile of BMI for the adolescents' age and sex.	Adolescent BMI was lower in areas with more chain supermarkets (β -0.1093; p<0.01) and higher in areas with more convenience stores (β 0.0295; p<0.05).

Table 5.1 continued   A brief overview of studies investigating objective measures of the food environment with individual dietary behaviours and health outcomes

Author (year)	Country	Sample/study design	Predictor	Outcome	Findings
Sturm and Datar (2005)	US	Multilevel longitudinal analysis of over 7,600 children from 724 schools from when they were in kindergarten to grade three.	Specific food items prices (meat, fruit and vegetables, dairy and fast food) and density (number per 10,000 people) of grocery stores, convenience stores, full-service restaurants and fast food restaurants within both the residential ZIP code and school's ZIP code.	Change in BMI from kindergarten to grade one and from kindergarten to grade three.	Higher fruit and vegetable prices were linked to an increase in BMI between the times children went from kindergarten to first grade (β 0.054; $p<0.05$) and from kindergarten to third grade (β 0.114; $p<0.001$). Density of food outlets was not a significant predictor of weight gain.
Wang et al. (2008)	US	Multilevel cross-sectional analysis of 5,779 adults from 82 Californian neighbourhoods who participated in one of four surveys between 1981 and 1990.	Number and geographic density of multiple shop types (large chain supermarkets, medium independent grocers, small independent grocers, specialized stores, ethnic grocers, chain convenience stores, fast food restaurants, pizza, pastry and sweets, bakeries, and doughnuts) within 0.5-mile buffer of neighbourhood defined by administrative unit. The number and geographic density was calculated for each survey year.	Interviewer-administered diet questionnaire and clinical measures of health status.	The presence of stores selling unhealthy food options increased during the study period however no clear trend was found that could link the environmental change to a change in dietary behaviours or health status.
Wang et al. (2007)	US	Multilevel cross-sectional analysis of 7,595 adults from 82 Californian neighbourhoods who participated in one of five surveys between 1979 and 1990.	Geographic density of fast food restaurants, convenience stores, small grocery stores, ethnic markets, and supermarkets within 0.5 miles of neighbourhoods and Euclidean distance to the nearest stores.	Clinically measured BMI.	An increased BMI in women was associated with living in closer proximity to the nearest ethnic market (β -0.157; $p<0.05$) or supermarket (β -0.300; $p<0.05$) as was a higher density of small grocery stores (β 0.053; $p<0.05$).

Table 5.1 continued A brief overview of studies investigating objective measures of the food environment with individual dietary behaviours and health outcomes

Author (year)	Country	Sample/study design	Predictor	Outcome	Findings
Australia and New Zealand					
Ball et al. (2006a)	Australia	Multilevel cross-sectional analysis of 1,347 women from 45 suburbs in Melbourne.	Density (number per 10,000 people) of supermarkets and fruit and vegetable stores within suburbs.	Self-reported fruit and vegetable consumption.	The food environment was unrelated to fruit and vegetable intake.
Pearce et al. (2009)	New Zealand	Multilevel cross-sectional analysis of 12,529 individuals in 1,178 neighbourhood meshblocks in a nationwide sample.	Travel distance along a road network to nearest multinational and locally operated fast food outlet from the population weighted centroid of meshblocks.	Dietary quality and BMI calculated from self-reported height and weight.	Living further than 2.8 kilometres from a multinational fast food outlet was associated with eating the recommended intake of vegetables (OR 1.17; 95% CI 1.00-1.37) but also with a higher likelihood of being overweight (OR 1.17; 95% CI 1.03-1.32).
Pearce et al. (2008b)	New Zealand	Multilevel cross-sectional analysis of 12,529 individuals in 1,178 meshblocks in a nationwide sample.	Travel time along a road network to the nearest supermarket and convenience store from the population weighted centroid of meshblocks.	Self reported fruit and vegetable consumption.	Travel times to supermarkets and convenience stores were unrelated to fruit and vegetable consumption.

Table 5.1 continued A brief overview of studies investigating objective measures of the food environment with individual dietary behaviours and health outcomes

Author (year)	Country	Sample/study design	Predictor	Outcome	Findings
Timperio et al. (2008)	Australia	Cross-sectional analysis of 340 children within Melbourne and Geelong recruited from both State and catholic primary schools.	For each of the five store categories (greengrocers, supermarkets, convenience stores, fast food outlets and restaurants/cafes/takeaway) three measures of access were created from each child's household address: 1) the presence of at least one store within 800 metres of road network distance; 2) the total number of stores within 800 metres of road network distance; and 3) road network distance to the nearest store.	Child's fruit and vegetable consumption as reported by their parents.	Vegetable consumption was lower for those who live closer to the nearest supermarket (OR 1.27; 95% CI 1.07-1.51) and fast food outlet (OR 1.19; 95% CI 1.06-1.35) and for those who have at least one convenience stores (OR 0.75; 95% CI 0.57-0.99) or a higher number of convenience stores within 800 metres (OR 0.84; 95% CI 0.74-0.95). Fruit intake was lower for those with at least one fast food outlet within 800 metres (OR 0.62; 95% CI 0.40-0.95) and access to a greater number of either fast food outlets (OR 0.82; 95% CI 0.67-0.99) or convenience stores (OR 0.84; 95% CI 0.73-0.98).
Thornton et al. (2009)	Australia	Multilevel cross-sectional analysis of 2,547 individuals from 49 census collector districts in Melbourne.	Three measures of fast food access were calculated from each respondent's household location: 1) the total number of fast food restaurants within three kilometres of road network distance; 2) the number of different fast food chains within three kilometres of road network distance; and 3) the road network distance to the closest fast food restaurant. Five major fast food chains were examined (Red Rooster, McDonalds, Kentucky Fried Chicken, Hungry Jacks and Pizza Hut).	Self-reported frequency of fast food purchased for consumption at home within the previous month (never, monthly and weekly) from same five fast food chains used in the predictor variables.	Variety of fast food stores within three kilometres was a positive predictor of monthly fast food purchasing independent of individual characteristics (OR 1.13; 95% CI 1.02-1.25). Density and proximity were not found to be significant predictors of fast food purchasing after adjustment for individual and household socioeconomic predictors.

Table 5.1 continued   A brief overview of studies investigating objective measures of the food environment with individual dietary behaviours and health outcomes

Author (year)	Country	Sample/study design	Predictor	Outcome	Findings
Turrell and Giskes (2008)	Australia	Multilevel cross-sectional analysis of 1,001 individuals from 50 census collector districts in Brisbane.	For eight categories of takeaway food stores (major-chain franchise, general, Asian, other ethnic, café/coffee shop, 'healthier' takeaway, and 'sweet food' takeaway) three measures of access were created: 1) density of stores (number per 10,000 people) within buffers 2.5 kilometres (Euclidean distance) from the geographic centroid of CCDs); 2) the road network distance from the CCD centroid to nearest shop; and 3) the average road network from the CCD centroid to the every type of takeaway shop within the 2.5 kilometre buffer.	Self-reported frequency of purchasing takeaway foods for home use.	Access to takeaway food was unrelated to purchasing.
United Kingdom					
Pearson et al. (2005)	England	Multilevel cross-sectional study design of 426 individuals but analysed as a single-level regression.	Road network distance from each respondent's home to the nearest supermarket.	Self reported fruit and vegetable intake.	Distance to the nearest supermarkets was unrelated to fruit and vegetable intake.
Cummins et al. (2005a)	Glasgow	Prospective natural experiment involving 191 participants from an intervention community and 221 from a comparison community.	The provision a new food hypermarket into a community with previously poor food access.	The change in outcomes (fruit and vegetable consumption, self reported health, and psychological health) twelve months after the introduction of the new food retailer.	The introduction of the store had no overall effect on the dietary outcomes but amongst those who visited the new store some reduction in poor psychological health was detected.

This finding is important because it emphasizes that the provision of *both* healthy and unhealthy foods in supermarkets can potentially influence dietary behaviours of local residents.

Natural experiments investigating changes in food environments have found no evidence that improved supermarket access leads to improved dietary and health outcomes (Wang et al. 2008, Cummins et al. 2005a). For instance, Cummins et al. who evaluated the impact of the introduction of a large food 'hypermarket' in a poor community in Glasgow did not find evidence that fruit and vegetable consumption of residents changed after the store opened (Cummins et al. 2005a).

Independent grocery stores and convenience stores The presence of smaller grocery stores and convenience stores in the local food environment has been found to have a negative influence on health in some studies through links to a higher weight status (Powell et al. 2007a, Wang et al. 2007, Morland et al. 2006), while others were unable to establish an association with diet or health outcomes (Pearce et al. 2008b, Wang et al. 2008, Inagami et al. 2006, Sturm and Datar 2005, Morland et al. 2002a). Lower fruit and vegetable consumption in children has been linked to a higher concentration of convenience stores (Timperio et al. 2008) and although it would be parents/adults doing the bulk of the food shopping, this result may be indicative of the reduced variety and quality of fresh produce likely to be stocked by convenience stores.

Fast food restaurants Two studies linked better access to fast food with increased weight (Mehta and Chang 2008, Maddock 2004), although others found no association with weight status among adults (Wang et al. 2007, 2008, Morland et al. 2006) and children (Sturm and Datar 2005, Burdette and Whitaker 2004). It may well be that health differences amongst children are more difficult to detect while their bodies are still growing. In relation to dietary outcomes, studies in Australia and the US have mostly failed to find an association between fast food environments (Turrell and Giskes 2008, Wang et al. 2008, Jeffery et al. 2006, Morland et al. 2002a), although Thornton et al. (submitted) showed that the variety of fast food restaurants residents have access to is a positive predictor of fast food purchasing. Others have suggested that a higher exposure to fast food retailing is associated with a decrease in fruit and vegetable intake (Pearce et al. 2009, Timperio et al. 2008).

Methodological Considerations

In recent years there have been some methodological improvements in studies of food environments and health. For example, multilevel studies are more common and enable researchers to separate area-level effects from individual-level effects and to ascertain how much of the variation in dietary and health outcomes can be attributed to area-level and individual-level differences

(Pearce et al. 2009, Mehta and Chang 2008, Pearce et al. 2008b, Turrell and Giskes 2008, Giskes et al. 2007, Ball et al. 2006a, Inagami et al. 2006, Morland et al. 2002a, 2006). As a consequence of this research there has been a realisation that interventions aimed at the small area-level may benefit a larger population more than health promotion strategies targeted to individuals. However, there are also significant methodological problems that threaten the quality of the evidence from the studies reviewed. These issues relate to how food environments are conceptualized and measured; data sources; timing of individual and environmental data collection; the lack of specificity of exposure and outcome; self-selection into neighbourhoods; and insufficient variation in food environment exposures.

The advent of Geographic Information System (GIS) technologies has made the measuring of food environments easier. However, to date most measures have been driven by technological advances and not theory. This has resulted in measures that may not be a true reflection of accessibility perhaps biasing estimates towards the null. For example, a lack of theory has meant administrative units are used to define the geographic boundaries of access or a contextual exposure (Messer 2007, Lee et al. 2006). Often access is defined as the distance to the closest food store of interest (e.g. supermarket) from the geographic centroid, or a measure of density may be used in a spatial area defined from the centroid or boundary of an administrative unit. This introduces what is referred to as 'aggregation error' because the measures are not an accurate reflection of access for all households in that area unit (Hewko et al. 2002). Although measures from the centroid are fine for households near the centre of the spatial unit, inaccuracies will occur for households closer to the boundaries. This is particularly a problem when large spatial units are used. Issues related to the measurement of food environments using GIS will be discussed further in Chapter 13.

Many studies rely on data for business listings sourced from spatial data companies. Although these data sources are a time and cost effective means of creating a snapshot of community nutrition environments, location data is not always spatially accurate and have the potential to bias estimates, particularly given that these errors may not be randomly distributed (Boone et al. 2008). Further, there may be problems in multilevel studies if the period of data collection for the food stores and collection of data from individuals in areas do not match. In the instance of individual data later being supplemented with spatial data from commercial sources, it may be the case that recently opened food stores are shown to exist in areas when in fact they were not in existence during the period of individual data collection.

Further, the lack of specificity between the food environment (or exposure) and health behaviour or outcome may also obscure potential associations. For example, although Turrell and Giskes (2008) created specific categories of takeaway stores for the exposure variable, the outcome variable related to purchasing was asked as a general question and could not be specifically matched to any of the exposure categories.

Although industry classifications such as the Australian and New Zealand Standard Industrial Classification (ANZSIC) (Australian Bureau of Statistics 2006) and North American Industry Classification System (NAICS) (United States Census Bureau 2007) may be used to define a food store type (for example the 2006 ANZSIC includes the code *Class 4512 Takeaway Food Services*, with stores classified in this code needing to meet certain criteria such as the provision of food for immediate consumption), the provision of 'healthy' or 'unhealthy' foods will not always be consistent and comparable between stores of the same classification. For example, a franchised fast food chain (for example McDonald's) may be classified the same way as an independent takeaway store that sells fresh sushi. Additionally, although supermarkets stock a considerable range of healthy alternatives and fresh produce they also carry a large amount of unhealthy food options (such as chips/crisps, confectionary and soft drinks). It is also generally accepted that compared to smaller independent outlets, larger supermarkets often provide a larger variety of fresh produce and low fat products, are often cheaper, have longer opening hours and are more likely to be located in areas accessible by public transport (Latham and Moffat 2007, Morland and Filomena 2007, Jetter and Cassady 2006). Therefore easy access to them could be considered an important health benefit as they offer residents the choice to purchase healthier food alternatives at a cheaper price. Greengrocers or fruit and vegetable stores and markets are also considered as healthy food stores because they offer a wide variety of fresh fruit and vegetables. However, these stores often do not stock many of the core food items (for example bread and milk) meaning that their potential influence on healthy dietary behaviours may be more limited. Fast food and other takeaway food outlets are often used as proxies for unhealthy food stores. In contrast to supermarkets this classification may be more clear cut as products sold at fast food outlets are generally considered unhealthy due to their high energy density (Prentice and Jebb 2003). However, when exploring fast food as an unhealthy option in the local environment, there is a need to consider whether to only measure multinational chain brand fast food restaurants or to include all independent takeaway outlets. Either definition may be appropriate depending upon the research hypothesis and the outcome being explored. For example, when investigating the consumption of specific multinational fast foods then it may be appropriate to include only these same stores as the exposure, whereas in a study that aims to explore overall dietary behaviours it may be more appropriate to include all takeaway food outlets.

Environmental influences on diet are said to potentially involve two pathways: access to foods for home consumption from supermarkets and grocery stores, and access to readymade food for home and out-of-home consumption (for example takeaways, restaurants) (Cummins and Macintyre 2006). It may be for these reasons researchers so commonly investigate supermarkets and fast food restaurants. To date, investigations that include a limited number of store types may not provide an adequate reflection of total food environments. For example, studies may include only one store type such as supermarkets and fast food stores as a measure of healthy or unhealthy environments respectively. This approach assumes the availability of healthy foods to only be within supermarkets or that the availability within these stores reflects availability more generally in the local environment. This approach also assumes that the availability and quality of foods does not vary across supermarkets in different neighbourhoods (Moore et al. 2008a, 2008b). The collection of more detailed information to support or refute these assumptions is proposed under the framework heading of *Consumer Nutrition Environments* (see Figure 5.1).

Another problem with studies to date is that they are subject to selection effects: people, at least theoretically, can select themselves into a neighbourhood on the basis of its characteristics; that is some people choose to live within neighbourhoods that provide services that will facilitate health promoting behaviours. Information regarding why people chose to live in particular neighbourhoods is not collected in most studies (Riva et al. 2007, Dutton et al. 2005, Oakes 2004).

Additionally, one of the explanations for the concentration of fast food stores in disadvantaged areas is that fast food is more attractive to people living in these areas (that is they preferentially locate in areas where the socio–demographic and socioeconomic characteristics of residents closely matches the characteristics of the most frequent fast food consumers (Melaniphy 1992)). Therefore, in this instance, *demand* for fast food is likely to be driving supply rather than *supply* (or accessibility) influencing demand. This phenomenon is referred to as endogeneity. Subramanian et al. (2007) have argued that fast food chains potentially open in areas because of the taste preferences of local residents and previous research has supported an association between taste preferences and fast food consumption among adolescents (French et al. 2001b). These issues remain relatively unexplored and would be best tested through longitudinal data analysis and the collection of more detailed information on why people have chosen to live in particular neighbourhoods (see Chapter 12).

Finally, if the quality of food environments does not vary sufficiently between areas it may be difficult to detect associations between the environment and health outcomes even if the environment is an important predictor (Blakely and Woodward 2000).

Summary

Although the studies reviewed in Table 5.1 provide some evidence linking food environments to health behaviours and outcomes, the strength and direction of effects have not always been consistent. The inconsistent, or lack of, evidence should not necessarily be interpreted as an absence of effect because of the methodological inconsistencies and shortcomings of many of the studies. Future research needs to better conceptualize food environments before undertaking GIS analysis and to ensure that studies carefully consider the methodological issues described above.

Evidence Regarding Consumer Nutrition Environments

Within supermarkets, shoppers make purchasing decision which relate to planned purchases (that is those on a shopping list), and both unplanned and impulse purchases which are both made once within the store (Hawkes 2008). Consumer nutrition environments are therefore important as they can influence each of these purchasing behaviours, particularly unplanned and impulse purchasing.

Numerous within-store factors may influence food choice and these vary depending upon shop type. For instance, supermarkets may encourage the purchasing of items by placing them specifically near counters or at the end-of-aisles. Alternatively, fast food restaurants may promote products by offering incentives with their purchase such as toys. In this section we review the small body of evidence that has examined within-store factors and the potential links to dietary behaviours and health.

Variations in Within-store Characteristics

Within-store investigations have revealed that the price and availability of a shopping basket of healthy food items and fresh produce differed by shop type with larger supermarkets most likely to stock a wider variety of products and at cheaper prices than smaller independent grocery and convenience stores (Morland and Filomena 2007, Jetter and Cassady 2006, Cummins and Macintyre 2002b).

Some studies have revealed that variations may also exist between areas for the same shop type categories. In the US, differences that exist in the availability of healthy foods between large supermarkets and smaller food outlets were found to be attributable to the characteristics of areas. For example, food stores are likely to stock reduced amounts of healthy foods in areas with greater levels of disadvantage or a higher percentage of Black people (Franco et al. 2008, Glanz et al. 2007, Morland and Filomena 2007, Horowitz et al. 2004), while fresh produce was also found to be of a lower quality in low income neighbourhoods (Andreyeva et al. 2008, Glanz et al. 2007, Block and Kouba 2006).

In relation to both fast food and full-service restaurants, it was reported that in Los Angeles the poorer areas or areas with more African American people offered and promoted more unhealthy menu items than restaurants in more affluent areas (Lewis et al. 2005). Further, unhealthy foods in the form of fast foods were also found to be more expensive in areas with a greater proportion of African American people (Graddy 1997). Potential explanations for such findings included higher costs associated with running a business in these areas, although the possibility for racial discrimination cannot be discounted (Graddy 1997).

These US based findings have not been replicated internationally with Cummins and Macintyre (2002b) showing a greater availability of healthy food in more deprived areas of Glasgow and in some instances more deprived areas also had cheaper food prices. In Australia, it was found that although the availability of fruit and vegetables were similar across areas with different levels of disadvantage

(Kavanagh et al. 2007, Winkler et al. 2006) the prices were again cheaper in poorer areas (Kavanagh et al. 2007).

Other features of food stores and fast food outlets that may differ between stores and contribute to a less healthy dietary profile among customers include supersizing, point of sale advertising and the lack of readily available nutrition information. A comparison of sit down and fast food restaurants in the US on these and many other factors found that they differed in many respects but one was not consistently more healthy than the other (that is fast food restaurants were more likely to encourage the purchasing of larger portion sizes *and* more likely to offer nutrition information) (Saelens et al. 2007). A review of international literature by Hawkes (2008) suggested that the aisles near to the entrance of supermarkets are often used for displaying impulse items, while items popular with children are placed near to the checkout. In relation to product promotion, it is common for supermarkets to have twice as many in-store promotions for fatty and sugary foods relative to fruit and vegetables (Dibb 2005).

Consumer Nutrition Environments as a Determinant of Diet and Health

Within the US, Cheadle et al. (1991) reported that those who lived near grocery stores with a higher availability of low fat and high fibre products tended to have healthier diets. With regard to food costs, higher prices for fast food have been linked to an increased fibre intake, lower saturated fat intake and healthier overall diet (Beydoun et al. 2008), while a reduction in the price of low fat snacks in vending machines and fruit and vegetables within a cafeteria were linked to increased sales of these items (French 2003). A possible explanation for this is provided in a separate US study in which participants rated cost as the second most important factor (after taste) when choosing food (Glanz et al. 1998). These cost issues are important to understand when discussing obesity, as energy dense diets are reported to be cheaper (Drewnowski and Darmon 2005, Drewnowski et al. 2004). Therefore, the temptation to select unhealthy food alternatives because they are cheaper greatly increases the risk of becoming obese. This issue may be context dependent and linked to macro factors determined by governments and industry. To date, Australian research has not found an association between objectively measured availability and price of healthy food items within grocery stores and food purchasing choices (Giskes et al. 2007).

Supersizing (or the provision of excessively large portions) which is common in fast food restaurants is based on the premise that the more you buy the more you save. Researchers have speculated 'supersizing', resulting in larger portion sizes of calorie laden food (particularly at fast food restaurants), may be contributing to the obesity epidemic (Young and Nestle 2007, Rolls et al. 2002, 2004, 2006, Rolls 2003, Young and Nestle 2002, French et al. 2001a, Hill and Peters 1998). Although McDonald's have removed 'supersized' options from their menus in the US, other fast food chains have not followed suit (Young and Nestle 2007).

For those who do frequent fast food restaurants, healthier nutritional choices are made more difficult if the calorie or fat content of the product is not readily available. The *Nutrition Labelling and Education Act* which came into effect in the US in 1994 exempts restaurants from most food labelling requirements (Wootan and Osborn 2006) and fast food restaurants have taken advantage of these laws (Wootan and Osborn 2006, Wootan et al. 2006). Although Yamamoto et al. (2005) showed that the provision of nutrition information had a minimal effect on the fast food and restaurant ordering behaviours of adolescents, other have demonstrated that it encourages healthier dietary behaviours in adults (Bassett et al. 2008). Other health education material (such as shelf labelling and posters) within grocery stores has been linked to a healthier self-reported dietary profile in the US (Cheadle et al. 1991).

Winson (2004) discussed the common practice of placing 'commodity' food items (from which supermarkets do not make large profits) towards the back of stores so that in order to reach them customers have to pass items that are more profitable (and often less healthy). This is described as a deliberate tactic to maximize profits with little thought given to downstream health outcomes. Other factors that have been rarely investigated include: the relative proportion of healthy and unhealthy foods; presence of snacks at checkouts; buy-one get-one free promotion (usually on unhealthy snack foods); promotions that entice children and offer incentives such as toys; and a more prominent promotion of unhealthy food products over health promoting material such as those that promote fruit and vegetable consumption.

Methodological Considerations

Collection of data on the consumer nutrition environment requires internal audits which are resource intensive and the information is often difficult to collect and measure. Food types may vary in the size in which they are sold (for example 500 grams or 1 kilogram of rice) and measurement of quality is highly subjective and prone to poor intra- and inter-rater reliability. Furthermore, data collection is costly and time consuming and may result in voluminous data that is difficult to classify. It is perhaps for these reasons that studies of the consumer nutrition environment have been relatively uncommon.

The potential exists for the Nutrition Environment Measures Surveys (NEMS) (Glanz et al. 2007, Saelens et al. 2007) to guide future research on in-store factors. The NEMS are observational tools created to assess the availability of healthy options, price, and quality of foods in retails stores (NEMS-S) (Glanz et al. 2007) and the availability of more healthy foods, facilitators and barriers to healthful eating, pricing, and signage/promotion of healthy and unhealthy foods in restaurants (NEMS-R) (Saelens et al. 2007). At present, these measurement tools for the consumer nutrition environment have been developed for use in the US. However, with minor modifications these tools could be used to explore stores within other countries providing a consistent measurement tool to aid public health research.

The information obtained from these surveys could be analysed using multilevel analysis conducted at three levels: individuals, stores and neighbourhoods. To do this, studies would need to collect information on where individuals purchased food. Data of this nature have rarely been collected. Research may also benefit from the possibility of monitoring electronic sales data for individuals within supermarket to detect if within-store interventions have influenced food purchasing patterns (Ni Mhurchu et al. 2007).

Summary

Variations in the consumer nutrition environment do exist by store size, store type, and across neighbourhoods. These factors have the potential to strongly influence purchasing decisions and some factors may be more powerful for low income communities (for example the promotion of energy dense and cheap foods). To gain a better understanding of the role of the consumer nutrition environment on health behaviours and outcomes requires more thorough investigations that utilize new instruments such as NEMS.

Other Nutrition Environments

Although the environments around the home (neighbourhoods) remain the main source of foods for many individuals and families, people may be exposed to food environments in a variety of contexts throughout the day. In Figure 5.1, this is reflected under the heading 'organizational nutrition environments'. Among adults, a common contextual exposure is the workplace. Depending on the size of the organization, workers may be exposed to foods through vending machines, small cafeterias, or larger restaurant size food outlets. Each of these has the potential to increase an employee's exposure to unhealthy snacks and meals (Story et al. 2008, Devine et al. 2007, Pratt et al. 2007, French et al. 2001a). This may also be the case during recreational activities such as attending sporting events where the bulk of the food items available are of low nutritional quality. Exposure to supermarkets may also occur within people's homes with major chain-brand supermarkets offering online food shopping services, potentially influencing purchasing decisions through the promotion of certain products and discounts.

In the US the proximity of fast food restaurants around home and work addresses was explored (Jeffery et al. 2006) (Table 5.1). Oddly, this study found that for men, the density of restaurants around the work address was inversely associated with BMI. Further research is required to investigate the associations between food environments around work locations and associations with diet and health status.

Discussion

Guided by the framework for nutrition environments proposed in Figure 5.1 by Glanz and colleagues (2005), we investigated factors that related to the local food environment. Three environmental factors were proposed: the community nutrition environments; the consumer nutrition environment; and the organizational nutrition environment. Evidence shows that community and consumer nutrition environments tend to differ considerably between small areas characterized by different levels of socioeconomic disadvantage and racial composition. As a result, community and consumer nutrition environments have the potential to influence dietary behaviours and health outcomes above and beyond individual level determinants.

There is a continued need to identify neighbourhood level community nutrition environments that may make access to healthy foods difficult, particularly if such neighbourhoods contain a higher proportion of low income people who may have difficulty accessing vehicles and the resources necessary to purchase healthy alternatives. In such communities, it may also be important to ensure restrictions are placed on the amount of advertising (through billboards and so on) that promote less healthy food products. Studies examining the community nutrition environment may have important implications for planning agencies as it is a simple yet effective way to determine food access for residents in a particular area. In addition, data on the consumer nutrition environment could help inform and improve upon currently used definitions of 'healthy' and 'unhealthy' food stores. Finally, more sophisticated measures of access are required and the methodological tool GIS should be used to assist with the development of these.

With regard to the consumer nutrition environment, most variation occurs by store type with larger supermarkets providing customers with better access to a wider variety of healthy food items (fresh produce and low fat alternatives). However, supermarkets also have the ability to promote unhealthy dietary behaviours through price promotions, advertising and product placement. Such factors need to be monitored and perhaps guide policy at the macro-environment level to ensure that the number and promotion of unhealthy products does not become disproportionally higher than healthy food items.

It is also recognized that people spend much of their time outside of their local neighbourhood and therefore are exposed to food environments in multiple contexts such as at work or during recreational activities. There is much work to be done to describe these food environments and their impact on health which can then be used to guide policies that ensure access to healthy alternatives within these environments.

A more consistent approach to studying each of the environmental factors might be achieved with the use of conceptual frameworks such as the one proposed by Glanz et al. (2005). Future research may wish to build on this to ensure consistency across studies. For example, with regard to the community nutrition environment, the current framework may require a more detailed approach to the classification of food stores and measures of access. Within-store factors could be better measured using tools such as NEMS (Glanz et al. 2007, Saelens et al. 2007) which may need to be altered based on the country and local context.

Further, improvements to the measurement of outcomes such as obesity are required. At present much of the evidence is based on self-reported data and on one half of the body weight equation: energy in. Further data on the links between food environments and health outcomes such as cardiovascular risk factors (for example blood pressure, cholesterol) would greatly improve the evidence base.

Conclusion

To date there has been a considerable rhetoric about the importance of place based interventions for disease outcomes. With regard to local food environments, the evidence is yet to show consistent links to features that promote or prevent disease. For example, we do not know whether introducing legislation to limit fast food chains setting up in areas will reduce the risk of adverse health outcomes. Research is required that will enable policy makers to move beyond rhetoric by providing evidence about the impact of environmental characteristics on health outcomes such as obesity, cardiovascular disease and diabetes. This evidence can then inform place based interventions for the primary prevention of these diseases.

References

Andreyeva, T., Blumenthal, D.M., Schwartz, M.B., Long, M.W. and Brownell, K.D. 2008. Availability and prices of foods across stores and neighborhoods: the case of New Haven, Connecticut. *Health Affairs* 27(5), 1381-8.

Apparicio, P., Cloutier, M.S. and Shearmur, R. 2007. The case of Montreal's missing food deserts: evaluation of accessibility to food supermarkets. *International Journal of Health Geographics* 6, 4.

Australian Bureau of Statistics. 2006. *Australian and New Zealand Standard Industrial Classification (ANZSIC)*. Canberra: Australian Bureau of Statistics.

Ball, K., Crawford, D. and Mishra, G. 2006a. Socio-economic inequalities in women's fruit and vegetable intakes: a multilevel study of individual, social and environmental mediators. *Public Health Nutrition* 9(5), 623-30.

Ball, K., Timperio, A.F. and Crawford, D.A. 2006b. Understanding environmental influences on nutrition and physical activity behaviors: where should we look and what should we count? *International Journal of Behavioral Nutrition and Physical Activity* 3, 33.

Bassett, M.T., Dumanovsky, T., Huang, C., Silver, L.D., Young, C., Nonas, C., Matte, T.D., Chideya, S. and Frieden, T.R. 2008. Purchasing behavior and calorie information at fast-food chains in New York City, 2007. *American Journal of Public Health* 98(8), 1457-9.

Beydoun, M.A., Powell, L.M. and Wang, Y. 2008. The association of fast food, fruit and vegetable prices with dietary intakes among US adults: is there modification by family income? *Social Science and Medicine* 66(11), 2218-29.

Blakely, T.A. and Woodward, A.J. 2000. Ecological effects in multi-level studies. *Journal of Epidemiology and Community Health* 54(5), 367-74.

Block, D. and Kouba, J. 2006. A comparison of the availability and affordability of a market basket in two communities in the Chicago area. *Public Health Nutrition* 9(7), 837-45.

Block, J.P., Scribner, R.A. and DeSalvo, K.B. 2004. Fast food, race/ethnicity, and income – A geographic analysis. *American Journal of Preventive Medicine* 27(3), 211-7.

Boone, J.E., Gordon-Larsen, P., Stewart, J.D. and Popkin, B.M. 2008. Validation of a GIS facilities database: quantification and implications of error. *Annals of Epidemiology* 18(5), 371-7.

Burdette, H.L. and Whitaker, R.C. 2004. Neighborhood playgrounds, fast food restaurants, and crime: relationships to overweight in low-income preschool children. *Preventive Medicine* 38(1), 57-63.

Burns, C.M. and Inglis, A.D. 2007. Measuring food access in Melbourne: access to healthy and fast foods by car, bus and foot in an urban municipality in Melbourne. *Health and Place* 13(4), 877-85.

Cheadle, A., Psaty, B.M., Curry, S., Wagner, E., Diehr, P., Koepsell, T. and Kristal, A. 1991. Community-level comparisons between the grocery store environment and individual dietary practices. *Preventive Medicine* 20(2), 250-61.

Cummins, S., Curtis, S., Diez-Roux, A.V. and Macintyre, S. 2007. Understanding and representing 'place' in health research: a relational approach. *Social Science and Medicine* 65(9), 1825-38.

Cummins, S. and Macintyre, S. 2002a. 'Food deserts' – evidence and assumption in health policy making. *British Medical Journal* 325(7361), 436-8.

Cummins, S. and Macintyre, S. 2002b. A systematic study of an urban foodscape: the price and availability of food in Greater Glasgow. *Urban Studies* 39(11), 2115-30.

Cummins, S. and Macintyre, S. 2006. Food environments and obesity – neighbourhood or nation? *International Journal of Epidemiology* 35(1), 100-4.

Cummins, S., Petticrew, M., Higgins, C., Findlay, A. and Sparks, L. 2005a. Large scale food retailing as an intervention for diet and health: quasi-experimental evaluation of a natural experiment. *Journal of Epidemiology and Community Health* 59(12), 1035-40.

Cummins, S.C., McKay, L. and Macintyre, S. 2005b. McDonald's restaurants and neighborhood deprivation in Scotland and England. *American Journal of Preventive Medicine* 29(4), 308-10.

Devine, C.M., Nelson, J.A., Chin, N., Dozier, A. and Fernandez, I.D. 2007. 'Pizza is cheaper than salad': assessing workers' views for an environmental food intervention. *Obesity* 15 Suppl 1, 57S-68S.

Dibb, S. 2004. *Rating retailers for health: how supermarkets can affect your chances of a healthy diet*. London: National Consumer Council.

Dibb, S. 2005. *Healthy competition: how supermarkets can affect your chances of a healthy diet*. London: National Consumer Council.

Diez Roux, A.V., Link, B.G. and Northridge, M.E. 2000. A multilevel analysis of income inequality and cardiovascular disease risk factors. *Social Science and Medicine* 50(5), 673-87.

Diez-Roux, A.V., Nieto, F.J., Caulfield, L., Tyroler, H.A., Watson, R.L. and Szklo, M. 1999. Neighbourhood differences in diet: the Atherosclerosis Risk in Communities (ARIC) Study. *Journal of Epidemiology and Community Health* 53(1), 55-63.

Diez Roux, A.V., Nieto, F.J., Muntaner, C., Tyroler, H.A., Comstock, G.W., Shahar, E., Cooper, L.S., Watson, R.L. and Szklo, M. 1997. Neighborhood environments and coronary heart disease: a multilevel analysis. *American Journal of Epidemiology* 146(1), 48-63.

Do, D.P., Dubowitz, T., Bird, C.E., Lurie, N., Escarce, J.J. and Finch, B.K. 2007. Neighborhood context and ethnicity differences in body mass index: a multilevel analysis using the NHANES III survey (1988-1994). *Economics and Human Biology* 5(2), 179-203.

Drewnowski, A. and Darmon, N. 2005. Food choices and diet costs: an economic analysis. *Journal of Nutrition* 135(4), 900-4.

Drewnowski, A., Darmon, N. and Briend, A. 2004. Replacing fats and sweets with vegetables and fruits – a question of cost. *American Journal of Public Health* 94(9), 1555-9.

Dutton, T., Turrell, G. and Oldenburg, B. 2005. *Measuring socioeconomic position in population health monitoring and health research. Health Inequalities Monitoring Series No. 3*. Brisbane: Queensland University of Technology.

Ellaway, A., Anderson, A. and Macintyre, S. 1997. Does area of residence affect body size and shape? *International Journal of Obesity Related Metabolic Disorders* 21(4), 304-8.

Ford, E.S., Mokdad, A.H., Giles, W.H., Galuska, D.A. and Serdula, M.K. 2005. Geographic variation in the prevalence of obesity, diabetes, and obesity-related behaviors. *Obesity Research* 13(1), 118-22.

Franco, M., Diez Roux, A.V., Glass, T.A., Caballero, B. and Brancati, F.L. 2008. Neighborhood characteristics and availability of healthy foods in Baltimore. *American Journal of Preventive Medicine* 35(6), 561-7.

French, S.A. 2003. Pricing effects on food choices. *Journal of Nutrition* 133(3), 841S-43S.

French, S.A., Story, M. and Jeffery, R.W. 2001a. Environmental influences on eating and physical activity. *Annual Review of Public Health* 22, 309-35.

French, S.A., Story, M., Neumark-Sztainer, D., Fulkerson, J.A. and Hannan, P. 2001b. Fast food restaurant use among adolescents: associations with nutrient intake, food choices and behavioral and psychosocial variables. *International Journal of Obesity Related Metabolic Disorders* 25(12), 1823-33.

Giles-Corti, B. and Donovan, R.J. 2002. Socioeconomic status differences in recreational physical activity levels and real and perceived access to a supportive physical environment. *Preventive Medicine* 35(6), 601-11.

Giskes, K., Van Lenthe, F.J., Brug, J., Mackenbach, J.P. and Turrell, G. 2007. Socioeconomic inequalities in food purchasing: the contribution of respondent-perceived and actual (objectively measured) price and availability of foods. *Preventive Medicine* 45(1), 41-8.

Glanz, K., Basil, M., Maibach, E., Goldberg, J. and Snyder, D. 1998. Why Americans eat what they do: taste, nutrition, cost, convenience, and weight control concerns as influences on food consumption. *Journal of the American Dietetic Association* 98(10), 1118-26.

Glanz, K., Sallis, J.F., Saelens, B.E. and Frank, L.D. 2005. Healthy nutrition environments: concepts and measures. *American Journal of Health Promotion* 19(5), 330-3.

Glanz, K., Sallis, J.F., Saelens, B.E. and Frank, L.D. 2007. Nutrition Environment Measures Survey in stores (NEMS-S): development and evaluation. *American Journal of Preventive Medicine* 32(4), 282-9.

Graddy, K. 1997. Do fast-food chains price discriminate on the race and income characteristics of an area? *Journal of Business and Economic Statistics* 15(4), 391-401.

Hawkes, C. 2008. Dietary implications of supermarket development: A global perspective. *Development Policy Review* 26(6), 657-92.

Hewko, J., Smoyer-Tomic, K.E. and Hodgson, M.J. 2002. Measuring neighbourhood spatial accessibility to urban amenities: does aggregation error matter? *Environment and Planning A* 34, 1185-206.

Hill, J.O. and Peters, J.C. 1998. Environmental contributions to the obesity epidemic. *Science* 280(5368), 1371-4.

Horowitz, C.R., Colson, K.A., Hebert, P.L. and Lancaster, K. 2004. Barriers to buying healthy foods for people with diabetes: Evidence of environmental disparities. *American Journal of Public Health* 94(9), 1549-54.

Inagami, S., Cohen, D.A., Finch, B.K. and Asch, S.M. 2006. You are where you shop: grocery store locations, weight, and neighborhoods. *American Journal of Preventive Medicine* 31(1), 10-7.

Jeffery, R.W., Baxter, J., McGuire, M. and Linde, J. 2006. Are fast food restaurants an environmental risk factor for obesity? *International Journal of Behavioral Nutrition and Physical Activity* 3, 2.

Jetter, K.M. and Cassady, D.L. 2006. The availability and cost of healthier food alternatives. *American Journal of Preventive Medicine* 30(1), 38-44.

Kavanagh, A.M., Goller, J.L., King, T., Jolley, D., Crawford, D. and Turrell, G. 2005. Urban area disadvantage and physical activity: a multilevel study in Melbourne, Australia. *Journal of Epidemiology and Community Health* 59(11), 934-40.

Kavanagh, A., Thornton, L., Tattam, A., Thomas, L., Jolley, D. and Turrell, G. 2007. *Place does matter for your health: a report of the Victorian Lifestyle and Neighbourhood Environment Study*. Parkville: University of Melbourne.

King, T., Kavanagh, A.M., Jolley, D., Turrell, G. and Crawford, D. 2006. Weight and place: a multilevel cross-sectional survey of area-level social disadvantage and overweight/obesity in Australia. *International Journal of Obesity* 30(2), 281-7.

Larsen, K. and Gilliland, J. 2008. Mapping the evolution of 'food deserts' in a Canadian city: supermarket accessibility in London, Ontario, 1961-2005. *International Journal of Health Geographics* 7, 16pp.

Latham, J. and Moffat, T. 2007. Determinants of variation in food cost and availability in two socioeconomically contrasting neighbourhoods of Hamilton, Ontario, Canada. *Health and Place* 13(1), 273-87.

Lee, C., Moudon, A.V. and Courbois, J.Y. 2006. Built environment and behavior: spatial sampling using parcel data. *Annals of Epidemiology* 16(5), 387-94.

Lewis, L.B., Sloane, D.C., Nascimento, L.M., Diamant, A.L., Guinyard, J.J., Yancey, A.K. and Flynn, G. 2005. African Americans' access to healthy food options in South Los Angeles restaurants. *American Journal of Public Health* 95(4), 668-73.

Macdonald, L., Cummins, S. and Macintyre, S. 2007. Neighbourhood fast food environment and area deprivation-substitution or concentration? *Appetite* 49(1), 251-4.

Maddock, J. 2004. The relationship between obesity and the prevalence of fast food restaurants: State-level analysis. *American Journal of Health Promotion* 19(2), 137-43.

Mehta, N.K. and Chang, V.W. 2008. Weight status and restaurant availability a multilevel analysis. *American Journal of Preventive Medicine* 34(2), 127-33.

Melaniphy, J.C. 1992. *Restaurant and Fast Food Site Selection*. New York: John Wiley and Sons, Inc.

Messer, L.C. 2007. Invited commentary: Beyond the metrics for measuring neighborhood effects. *American Journal of Epidemiology* 165(8), 868-71; discussion 72-3.

Moore, L.V., Diez Roux, A.V. and Brines, S. 2008a. Comparing Perception-Based and Geographic Information System (GIS)-based characterizations of the local food environment. *Journal of Urban Health* 85(2), 206-16.

Moore, L.V., Diez Roux, A.V., Nettleton, J.A. and Jacobs, D.R., Jr. 2008b. Associations of the local food environment with diet quality – a comparison of assessments based on surveys and geographic information systems: the multi-ethnic study of atherosclerosis. *American Journal of Epidemiology* 167(8), 917-24.

Morland, K., Diez Roux, A.V. and Wing, S. 2006. Supermarkets, other food stores, and obesity: the atherosclerosis risk in communities study. *American Journal of Preventive Medicine* 30(4), 333-9.

Morland, K. and Filomena, S. 2007. Disparities in the availability of fruits and vegetables between racially segregated urban neighbourhoods. *Public Health Nutrition* 10(12), 1481-9.

Morland, K., Wing, S. and Roux, A.D. 2002a. The contextual effect of the local food environment on residents' diets: The atherosclerosis risk in communities study. *American Journal of Public Health* 92(11), 1761-7.

Morland, K., Wing, S., Roux, A.D. and Poole, C. 2002b. Neighborhood characteristics associated with the location of food stores and food service places. *American Journal of Preventive Medicine* 22(1), 23-9.

Nestle, M. 2002. *Food Politics.* Berkeley and Los Angeles, CA: University of California Press.

Ni Mhurchu, C., Blakely, T., Wall, J., Rodgers, A., Jiang, Y. and Wilton, J. 2007. Strategies to promote healthier food purchases: a pilot supermarket intervention study. *Public Health Nutrition* 10(6), 608-15.

Oakes, J.M. 2004. The (mis)estimation of neighborhood effects: causal inference for a practicable social epidemiology. *Social Science and Medicine* 58(10), 1929-52.

Pearce, J., Blakely, T., Witten, K. and Bartie, P. 2007a. Neighborhood deprivation and access to fast-food retailing: a national study. *American Journal of Preventive Medicine* 32(5), 375-82.

Pearce, J., Day, P. and Witten, K. 2008a. Neighbourhood provision of food and alcohol retailing and social deprivation in urban New Zealand. *Urban Policy and Research* 26(2), 213-27.

Pearce, J., Hiscock, R., Blakely, T. and Witten, K. 2008b. The contextual effects of neighbourhood access to supermarkets and convenience stores on individual fruit and vegetable consumption. *Journal of Epidemiology and Community Health* 62(3), 198-201.

Pearce, J., Hiscock, R., Blakely, T. and Witten, K. 2009. A national study of the association between neighbourhood access to fast-food outlets and the diet and weight of local residents. *Health and Place* 15(1), 193-7.

Pearce, J., Witten, K., Hiscock, R. and Blakely, T. 2007b. Are socially disadvantaged neighbourhoods deprived of health-related community resources? *International Journal of Epidemiology* 36(2), 348-55.

Pearson, T., Russell, J., Campbell, M.J. and Barker, M.E. 2005. Do 'food deserts' influence fruit and vegetable consumption? – A cross-sectional study. *Appetite* 45(2), 195-7.

Powell, L.M., Auld, M.C., Chaloupka, F.J., O'Malley, P.M. and Johnston, L.D. 2007a. Associations between access to food stores and adolescent body mass index. *American Journal of Preventive Medicine* 33(4 Suppl), S301-7.

Powell, L.M., Chaloupka, F.J. and Bao, Y. 2007b. The availability of fast-food and full-service restaurants in the United States: associations with neighborhood characteristics. *American Journal of Preventive Medicine* 33(4 Suppl), S240-5.

Powell, L.M., Slater, S., Mirtcheva, D., Bao, Y. and Chaloupka, F.J. 2007c. Food store availability and neighborhood characteristics in the United States. *Preventive Medicine* 44, 189-95.

Pratt, C.A., Lemon, S.C., Fernandez, I.D., Goetzel, R., Beresford, S.A., French, S.A., Stevens, V.J., Vogt, T.M. and Webber, L.S. 2007. Design characteristics of worksite environmental interventions for obesity prevention. *Obesity* 15(9), 2171-80.

Prentice, A.M. and Jebb, S.A. 2003. Fast foods, energy density and obesity: a possible mechanistic link. *Obesity Reviews* 4(4), 187-94.

Reidpath, D.D., Burns, C., Garrard, J., Mahoney, M. and Townsend, M. 2002. An ecological study of the relationship between social and environmental determinants of obesity. *Health and Place* 8(2), 141-5.

Riva, M., Gauvin, L. and Barnett, T.A. 2007. Toward the next generation of research into small area effects on health: a synthesis of multilevel investigations published since July 1998. *Journal of Epidemiology and Community Health* 61(10), 853-61.

Robert, S.A. and Reither, E.N. 2004. A multilevel analysis of race, community disadvantage, and body mass index among adults in the US. *Social Science and Medicine* 59(12), 2421-34.

Rolls, B.J. 2003. The Supersizing of America: Portion Size and the Obesity Epidemic. *Nutrition Today* 38(2), 42-53.

Rolls, B.J., Morris, E.L. and Roe, L.S. 2002. Portion size of food affects energy intake in normal-weight and overweight men and women. *American Journal of Clinical Nutrition* 76(6), 1207-13.

Rolls, B.J., Roe, L.S., Kral, T.V., Meengs, J.S. and Wall, D.E. 2004. Increasing the portion size of a packaged snack increases energy intake in men and women. *Appetite* 42(1), 63-9.

Rolls, B.J., Roe, L.S. and Meengs, J.S. 2006. Larger portion sizes lead to a sustained increase in energy intake over 2 days. *Jornal of the American Dietetic Association* 106(4), 543-9.

Saelens, B.E., Glanz, K., Sallis, J.F. and Frank, L.D. 2007. Nutrition Environment Measures Study in restaurants (NEMS-R): development and evaluation. *American Journal of Preventive Medicine* 32(4), 273-81.

Smoyer-Tomic, K.E., Spence, J.C. and Amrhein, C. 2006. Food deserts in the Prairies? Supermarket accessibility and neighborhood need in Edmonton, Canada. *Professional Geographer* 58(3), 307-26.

Smoyer-Tomic, K.E., Spence, J.C., Raine, K.D., Amrhein, C., Cameron, N., Yasenovskiy, V., Cutumisu, N., Hemphill, E. and Healy, J. 2008. The association between neighborhood socioeconomic status and exposure to supermarkets and fast food outlets. *Health and Place* 14(4), 740-54.

Story, M., Kaphingst, K.M., Robinson-O'Brien, R. and Glanz, K. 2008. Creating healthy food and eating environments: policy and environmental approaches. *Annual Review of Public Health* 29, 253-72.

Sturm, R. and Datar, A. 2005. Body mass index in elementary school children, metropolitan area food prices and food outlet density. *Public Health* 119(12), 1059-68.

Subramanian, S.V., Glymour, M. and Kawachi, I. 2007. Identifying causal ecological effects on health: A methodological assessment, in *Macrosocial Determinants of Population Health*, edited by S. Galea. New York: Springer, 301-31.

Sundquist, J., Malmstrom, M. and Johansson, S.E. 1999. Cardiovascular risk factors and the neighbourhood environment: a multilevel analysis. *International Journal of Epidemiology* 28(5), 841-5.

Swinburn, B., Egger, G. and Raza, F. 1999. Dissecting obesogenic environments: the development and application of a framework for identifying and prioritizing environmental interventions for obesity. *Preventive Medicine* 29(6 Pt 1), 563-70.

Thornton, L.E., Bentley, R.J. and Kavanagh, A.M. 2009. Fast food purchasing and access to fast food restaurants: a multilevel analysis of VicLANES. *International Journal of Behavioral Nutrition and Physical Activity* 6, 28.

Timperio, A., Ball, K., Roberts, R., Campbell, K., Andrianopoulos, N. and Crawford, D. 2008. Children's fruit and vegetable intake: associations with the neighbourhood food environment. *Preventive Medicine* 46(4), 331-5.

Turrell, G., Bentley, R., Thomas, L.R., Jolley, D., Subramanian, S.V. and Kavanagh, A.M. 2009. A multilevel study of area socio-economic status and food purchasing behaviour. *Public Health Nutrition* 12(11), 2074-83.

Turrell, G. and Giskes, K. 2008. Socioeconomic disadvantage and the purchase of takeaway food: a multilevel analysis. *Appetite* 51, 69-81.

Turrell, G., Kavanagh, A. and Subramanian, S.V. 2006. Area variation in mortality in Tasmania (Australia): the contributions of socioeconomic disadvantage, social capital and geographic remoteness. *Health and Place* 12(3), 291-305.

United States Census Bureau. 2007. *North American Industry Classification System (NAICS)*. Washington, DC: United States Census Bureau.

van Lenthe, F.J. and Mackenbach, J.P. 2002. Neighbourhood deprivation and overweight: the GLOBE study. *International Journal of Obesity* 26(2), 234-40.

Wang, M.C., Cubbin, C., Ahn, D. and Winkleby, M.A. 2008. Changes in neighbourhood food store environment, food behaviour and body mass index, 1981-1990. *Public Health Nutrition* 11(9), 963-70.

Wang, M.C., Kim, S., Gonzalez, A.A., MacLeod, K.E. and Winkleby, M.A. 2007. Socioeconomic and food-related physical characteristics of the neighbourhood environment are associated with body mass index. *Journal of Epidemiology and Community Health* 61(6), 491-8.

Winkler, E., Turrell, G. and Patterson, C. 2006. Does living in a disadvantaged area entail limited opportunities to purchase fresh fruit and vegetables in terms of price, availability, and variety? Findings from the Brisbane Food Study. *Health and Place* 12(4), 741-8.

Winson, A. 2004. Bringing political economy into the debate on the obesity epidemic. *Agriculture and Human Values* 21, 299-312.

Wootan, M.G. and Osborn, M. 2006. Availability of nutrition information from chain restaurants in the United States. *American Journal of Preventive Medicine* 30(3), 266-8.

Wootan, M.G., Osborn, M. and Malloy, C.J. 2006. Availability of point-of-purchase nutrition information at a fast-food restaurant. *Preventive Medicine* 43(6), 458-9.

Wrigley, N., Warm, D. and Margetts, B. 2003. Deprivation, diet, and food-retail access: findings from the Leeds 'food deserts' study. *Environment and Planning A* 35(1), 151-88.

Yamamoto, J.A., Yamamoto, J.B., Yamamoto, B.E. and Yamamoto, L.G. 2005. Adolescent fast food and restaurant ordering behavior with and without calorie and fat content menu information. *Journal of Adolescent Health* 37(5), 397-402.

Yancey, A.K., Cole, B.L., Brown, R., Williams, J.D., Hillier, A., Kline, R.S., Ashe, M., Grier, S.A., Backman, D. and McCarthy, W.J. 2009. A cross-sectional prevalence study of ethnically targeted and general audience outdoor obesity-related advertising. *Milbank Quarterly* 87(1), 155-84.

Young, L.R. and Nestle, M. 2002. The contribution of expanding portion sizes to the US obesity epidemic. *American Journal of Public Health* 92(2), 246-9.

Young, L.R. and Nestle, M. 2007. Portion sizes and obesity: responses of fast-food companies. *Journal of Public Health Policy* 28(2), 238-48.

Zenk, S.N., Schulz, A.J., Israel, B.A., James, S.A., Bao, S.M. and Wilson, M.L. 2005. Neighborhood racial composition, neighborhood poverty, and the spatial accessibility of supermarkets in metropolitan Detroit. *American Journal of Public Health* 95(4), 660-7.

Chapter 6
Childhood Obesity and the Food Environment

Mat Walton and Louise Signal

Introduction

Childhood overweight and obesity is a concern to health professionals and policy makers in many countries (Lang and Rayner 2005, World Health Organization 2004), not only because of the impacts on children's morbidity, social and education outcomes, but also because of the downstream impacts of childhood obesity on rates of chronic diseases in adulthood (Wang and Beydoun 2007, Datar and Sturm 2006, Reilly 2005). It has been argued that children have less control than adults over the environments they live in, and therefore are more susceptible to environments that promote excess food consumption and limited physical activity (Swinburn 2008, Caraher and Coveney 2004, Department of Health 2004). In this context, it becomes important to examine the places children inhabit and the multiple influences on their diets and food preferences, such as parents, peers, marketing and food availability. This chapter examines influences on children's, but not infant's, diets in home, community and school environments, focussing on developed countries, using a social ecological theory of obesity causation.

Defining Childhood Overweight and Obesity

Obesity can be defined as an excess of body fat (Ogden et al. 2007). Within most epidemiological studies of childhood obesity, both overweight and obesity are measured, as disease risk begins increasing with overweight. Body fat is difficult to measure directly (Ogden et al. 2007), so often it is estimated using the Body Mass Index (BMI) (Anderson and Butcher 2006, Reilly 2005), calculated by dividing weight (in kilograms) by height (in metres) squared (Sattar and Lean 2007). A normal BMI for children changes with age, and differs between sexes, therefore, overweight is usually defined as a BMI equal or greater than the 85th percentile of a reference population, while obesity is defined as a BMI equal to or greater than the 95th percentile (Sattar and Lean 2007, Reilly 2005).

The Burden of Childhood Overweight and Obesity

Childhood overweight and obesity is a widespread issue. In a study comparing overweight and obesity prevalence data from over 20 countries, Wang and Lobstein (2006: 13) conclude that 'the prevalence of childhood obesity is increasing in almost all industrialized countries for which data are available, and in several lower-income countries'. According to statistics compiled by the International Obesity Taskforce, the United States (US) tops all countries for prevalence of overweight and obesity, with 35 per cent and 36 per cent of 6-17 year old boys and girls respectively classified as such in 2003/04 (International Obesity Taskforce 2007). The World Health Organization regions of the Americas, Mediterranean, Europe, and the Pacific all show over 20 per cent prevalence of combined overweight and obesity amongst boys and girls (International Obesity Taskforce 2007). Popkin (Chapter 2) notes that child obesity is emerging in all countries of the world, both developed and developing, and that between 0.3 and 0.6 per cent of children in each country are becoming overweight each year.

The trends in childhood overweight and obesity vary within countries by age, gender, ethnicity, socioeconomic status, and geographic location. These variations are not always straight forward, and demographic factors may interact in some cases. In a recent detailed analysis of US statistics, Wang and Beydoun (2007) show that while the prevalence of overweight and obesity increased in all age groups since the 1960s, the rate was slower for younger children than adolescents. The same study reported fairly consistent differences in overweight and obesity prevalence by ethnic group, but the trends differed across genders. Socioeconomic status was inversely related to prevalence of obesity amongst Whites, but not among African Americans or Hispanics (Wang and Beydoun 2007).

The impacts of obesity in children include risk factors for cardiovascular disease (Flynn et al. 2006), such as increased hypertension, raised total cholesterol, and hyperinsulinaemia (Sattar and Lean 2007, Lobstein and Jackson-Leach 2006, Reilly 2005). In the shorter term, obesity has been associated with low self esteem and behaviour problems (Reilly 2005). Worldwide there are also emerging trends of a small (but significant in terms of impact on child health), growth in type 2 diabetes amongst obese children and adolescents (Ogden et al. 2007, Flynn et al. 2006, Lobstein and Jackson-Leach 2006). The morbidity experienced by obese children is also likely to negatively impact on academic achievement (Datar and Sturm 2006, Story et al. 2006, Ball et al. 2005). Wang and Beydoun (2007) estimate that in the US, about one half of all obese school aged children become obese adults. Continuation of obesity from childhood into adulthood is more likely as the severity of childhood obesity increases and where there has been a continuation of obesity from childhood to adolescence (Sattar and Lean 2007, Reilly 2005).

Causes of Childhood Obesity

At its simplest, the cause of overweight and obesity can be considered an imbalance between energy intake (through food eaten), and energy out (through physical activity) (World Health Organization 2003). As detailed by Popkin (Chapter 2), the foods being eaten by much of the world are increasingly energy-dense, sweet, and processed, with more food from animal sources being consumed and a reduction in fruit and vegetables. Children, for the most part, consume the same foods as adults, and are affected by societal changes in the availability and consumption patterns of food. Physical activity patterns have also changed over recent decades, with reductions in the amount of walking and cycling by children in developed countries, and an increase in screen-based activities such as watching TV (Dollman et al. 2005) (see Oliver and Schofield, Chapter 9).

To understand imbalances between energy intake and energy expenditure, the complex factors that contribute to changes in food and physical activity environments need to be understood. Swinburn et al. (1999) suggest that people find healthy lifestyles difficult within 'obesogenic' environments that promote high energy intake and sedentary behaviours, and that 'systems-based, environmental interventions are therefore needed to increase the rather modest impact of individual and public education programs' (Swinburn et al. 1999: 563). An 'obesogenic' environment, according to Swinburn et al. (1999) is made up of physical, sociocultural, economic and political components. As Hoek and McLean discuss (Chapter 4), political and economic forces have strongly influenced what foods are produced, how available these are, and how they are marketed. Over time these changes are likely to impact on sociocultural aspects of food, such as acceptability and food preferences.

When considering childhood obesity, the ability for children to positively engage with an obesogenic food environment should be considered. In a European Union based review, Lang and Rayner (2005) note that policy responses should not rely on food and activity choices made by children, as 'their choices are for the most part determined by features of the adult-framed environment, such as transport, culture, education, and eating habits' (pp. 307-308).

Social ecological theory suggests that the food an individual adult or child eats cannot be understood without reference to the multiple and diverse physical, economic, political, and sociocultural influences operating (Swinburn et al. 1999). By explicitly considering physical and social environments, the ecological model incorporates both 'context' and 'collective' properties of space (Macintyre et al. 2002). The model assumes that individuals move between multiple physical and social environments, and that there is 'reciprocity' between elements within an environment, and between different environments (Green et al. 2000). Through the process of reciprocity, the actions of individuals can be considered both shaped by their physical and social environments, and shaping of these environments. Actions can reproduce existing structures, or create change through interaction across physical and social environments.

Children's Food Environment

In this chapter we use a social ecological theory to explore an understanding of the food environment of children, focusing on influences on children's food consumption across three environments within which children participate: the home, school and community. Figure 6.1 represents possible interactions between home, school and community environments with each other and with children. Each descriptor within the environments represents an association with children's diets or weight reported in the literature. While each environment is discussed in detail below, the purpose of Figure 6.1 is to emphasize that children move between environments, and environments influence each other. Therefore to fully understand the causes of children's diets and weight gain, an appreciation of the different environments children inhabit and the reciprocity between them is required.

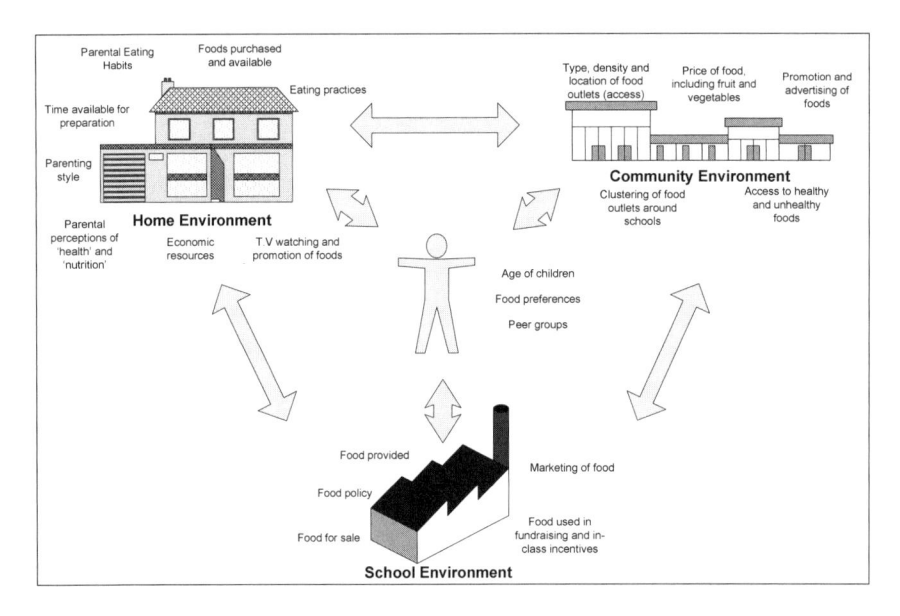

Figure 6.1 Children's food environment

Home Environment

The foods purchased and available in the home are a key determinant of children's food environment (Rosenkranz and Dzewaltowski 2008) and impact on children's consumption patterns, and preference development. Food practices within a household are largely a result of history. Foods that are available, affordable, acceptable and fit with household routines will be regularly consumed. While children can influence the home food environment over time, they are also heavily influenced by the existing food environment in a household, which can be affected in numerous ways.

Commonly reported influences on foods in the home include the time available for food preparation (Inglis et al. 2005), parenting style (Scaglioni et al. 2008, Jain et al. 2001), parental eating habits and perceptions of how diet influences health and weight (Coveney 2005, Campbell and Crawford 2001), and access to foods and food outlets (Reidpath et al. 2002, Jain et al. 2001). An association with television watching and eating foods commonly advertised on television has also been shown (Utter et al. 2006). Eating practices within the household, such as eating as a family at a table rather than in front of the television, have been associated with foods eaten and food preferences of children (Brown et al. 2008).

In general, the evidence for the influences mentioned above having an impact on children's diets within the home environment is good. However no study can demonstrate, or in fact attempts to demonstrate, that any one influence is sufficient or necessary to cause overweight or obesity in children. Many of the factors mentioned above, such as parental perceptions, parental eating habits, and time pressures show some patterning by household socioeconomic status (Signal et al. 2008b, Coveney 2005, Inglis et al. 2005), and may therefore be partly a response to the economic resources of a household.

There is good evidence that household economic resources play a large role in determining the foods purchased and consumed in households. Unlike rent or mortgage payments, the amount of money a household spends on food is to some degree discretionary (Ricciuto et al. 2006, Turrell and Kavanagh 2006, Turrell 1996). Drewnowski and Darmon (2005) suggest that low-income families, in the face of diminishing income, will attempt to maintain food costs as a fixed percentage of income. In these circumstances, it seems likely that parents will be more concerned that children are not hungry, rather than ensuring a nutritionally adequate diet (Drewnowski and Specter 2004, Jain et al. 2001). This may foster a reliance on cheaper, energy-dense and often nutritionally poor foods, which are increasingly available (see Popkin, Chapter 2).

Aspects of the home environment, such as parents' feeding practices, television watching, and economic resources, have been extensively researched (Rosenkranz and Dzewaltowski 2008). The type and size of study varies considerably, with qualitative and quantitative, cross-sectional and longitudinal designs employed. From this mix of studies it is difficult to identify one or two aspects of the home food environment crucial to children's diets and weight status, and there have been relatively few attempts to consider the home food environment as a whole. One fairly consistent finding however is the influence of socioeconomic status, including economic resources, on all identified aspects of the home food environment (Rosenkranz and Dzewaltowski 2008).

Much of the food children consume in other physical environments, such as at school, is brought from home (Dowler 2008, Sanigorski et al. 2005). In New Zealand, for example, the 2002 Children's Nutrition Survey indicated that 33 per cent of children aged 5-14 years brought all the food they consume at school from home (Utter et al. 2007). The amount of food provided within school for children will likely impact on the amount and type of food brought from home, and is one example of reciprocity between home and school environments.

School Environment

The geographic and socioeconomic patterning of children's weight and diets seen in community epidemiological studies (discussed above), is also evident when comparing schools by location or the socioeconomic status of the area they serve (Minaker et al. 2006). A US study has shown systematic variations in food practices within schools can vary by location and socioeconomic position, with rural and more deprived schools having less healthy options, and increased access to high-energy foods compared to urban and less deprived schools. (Nanney et al. 2008).

The school food environment may include: food for sale; fruit, breakfast, or lunches provided to students; food provided within classrooms as part of parties or lessons; school food policies; and marketing of food (French and Wechsler 2004). A small US study by Kubik et al. (2005b) found an association between higher BMI of 7th and 8th graders and the number of food practices that included energy-dense foods within schools, such as use in fundraising and in-classroom incentives. This suggests that increasing availability of energy-dense foods in a school setting may increase BMI of students. Further research would be valuable to confirm this finding.

While few qualitative studies of school food environments have been undertaken, results suggest that parents and teachers consider the school food environment to be an important, but not the only, influence in children's diets (Kubik et al. 2005a). In general, international literature suggests that when food is available for sale within a school, and that sale of food is a revenue stream, the food for sale is likely to be less healthy (Finkelstein et al. 2008, Fleischhacker 2007, Nollen et al. 2007, Kubik et al. 2003). There is also good evidence that limiting the availability of less healthy foods in schools, and providing good quality and appealing healthy foods, leads to modest improvements in consumption of healthy foods (Moore and Tapper 2008, Fleischhacker 2007, Fogarty et al. 2007, Anderson et al. 2005). For students from socioeconomically deprived households, the food provided for free within schools is likely to account for a higher proportion of daily energy intake, than for children from less deprived households (Fleischhacker 2007, Nelson et al. 2007) and therefore has greater importance for their health.

Nollen et al. (2007) identify a number of barriers to schools improving their food environments, such as lack of financial and staffing resources available to make changes. Kubik at al. (2005a) also note that nutrition must compete with curricula and other social activities for these resources. It has been suggested that where schools generate income from the sale of food, for example through à la carte options and vending machines, any reduction in sales resulting from an increase in healthy food availability could cause a loss of income to support school operations (Nollen et al. 2007, Wojcicki and Heyman 2006, Bauer et al. 2004, Sallis et al. 2003). There has been little research to test this assertion, although a recent review of seven US reports showed no reduction in revenue to date when healthier foods were introduced (Wharton et al. 2008).

It is common for schools to sell food to raise revenue. A recent US study of a nationally representative sample of schools found 17 per cent of elementary schools, 82 per cent of middle schools, and almost all secondary schools had vending machines, which usually offered low-nutrient, energy-dense foods (Finkelstein et al. 2008). It not surprising that elementary schools would have fewer opportunities for students to purchase foods, as younger children are likely to have less money to spend. In a representative sample of primary school (equivalent of elementary schools in the US) food environments in New Zealand, Carter and Swinburn (2004) reported that 37 per cent of schools surveyed ran a food service for profit, while almost all schools used food for fundraising purposes. The impact of reducing 'unhealthy' foods for sale on school finances requires further study.

While the evidence base for interventions that change the food environment of schools is dominated by small to medium sized studies based in the US, the emerging picture suggests that reducing the availability of high energy foods, and increasing the availability of fruit, vegetables, and other nutritious foods, will lead to modest improvements in students' diets. Such interventions are more likely to be successful with school wide support and food policies in place (Moore and Tapper 2008, Lytle et al. 2006, Kubik et al. 2003). The link between modest changes in student diets and weight gain have not been consistently shown, so changes to the school food environment by themselves may not reduce prevalence of overweight and obesity.

Community Environment

Communities provide a setting for both the provision and promotion of food. There is a somewhat mixed picture emerging from studies that have examined the geographical distribution of food outlets across neighbourhoods and associations with BMI or diet quality. Cummins and Macintyre (2006) summarize this literature as presenting a fairly consistent picture in the US that poorer people and ethnic minorities have worse access to 'healthier' foods, while in the United Kingdom (UK), studies have not found consistent associations. Studies in Australia and New Zealand tend to support the UK findings rather than North American (Crawford et al. 2008, Pearce et al. 2008b).

There have been fewer studies conducted outside the US on associations between access to food in the community and children's diets or weight status. Within the US, Powell et al. (2007) found an association between lower BMI amongst adolescents and the presence of large supermarkets in their neighbourhoods, and higher BMI with more convenience stores in their neighbourhood. Sturm and

Datar (2005) found that higher prices of fruit and vegetables within an urban area were associated with higher BMI amongst third grade school children. However, density of food outlets was not associated with BMI, and poorer neighbourhoods tended to have slightly more food outlets (Sturm and Datar 2005). This last finding has also been shown in New Zealand, where more socioeconomically deprived neighbourhoods tend to have a higher prevalence of food and alcohol outlets (Pearce et al. 2008a).

Crawford et al. (2008) examined the association between location of fast food outlets with children's obesity in Melbourne, Australia. They found the likelihood of being obese reduced with each additional fast food outlet within 2 km of children's homes. This study is interesting given that for children, associations have been found between consumption of fast food with increased energy intake and reduced quality of diet, implying that regular consumption of fast food over time will increase the likelihood of overweight and obesity (Taveras et al. 2005, Bowman et al. 2004). It may be that proximity of fast food outlets to children's home does not relate well to consumption. Other studies suggest that fast food outlets offer a socially acceptable and affordable place for older children and adolescents to socialize with peers (Story et al. 2002). The influence of peers on diets is likely to increase as children age (Borra et al. 2003). In further work looking at the geographical distribution of fast food outlets, there is some evidence of clustering around schools (Zenk and Powell 2008, Pearce et al. 2007, Austin et al. 2005). A possible explanation linking these research findings could be that proximity of fast food outlets to children's home is relatively unimportant, with parents largely controlling what foods are consumed in the home. For adolescents, the location of fast food outlets in areas where they socialize with peers may have more of an influence on regularity of consumption and energy intake. For example, clustering of fast food outlets around schools may facilitate easy access for groups of children to meet.

Advertising of foods in the community, through outdoor food advertisements, may also influence children's food preferences and intentions, as has been shown with outdoor advertising of alcohol and cigarettes (Henriksen et al. 2008; Pasch et al. 2007). In a pilot study, Maher et al. (2005) found an average of 87 outdoor food advertisements within a 1 km radius of ten secondary schools, with 70.2 per cent for products categorized as unhealthy, based on New Zealand government nutrition guidelines for adolescents (Ministry of Health 1998). Within the UK, expenditure on outdoor food advertising increased from 3.6 per cent of all food advertising expenditure in 1994, to 8.8 per cent in 2002 (Hastings et al. 2003).

Availability and advertising of unhealthy food products may also cluster around other environments children and adolescents frequent. A small study on the provision and marketing of food and beverages available within 16 council owned swimming pools and libraries, in the same geographical region as Maher et al.'s study (2005), found 73 per cent of products provided were 'unhealthy', while almost half of the advertisements displayed were 'unhealthy'. No health promoting advertisements were evident (Al-Shehri et al. 2007). In an internet survey of sponsorship displayed on websites for the five sports most commonly played by children and adolescents in New Zealand, 640 sponsors were found over 107 websites (Maher et al. 2006). Of these sponsors, 32.7 per cent were for products classified as 'unhealthy' (alcohol, gambling or unhealthy food), compared to 15.5 per cent classified as 'healthy' (health promoting messages, healthy foods).

The research conducted into the community food environment and children's diets and weight consists of mostly small studies that focus on a limited aspect of the environment. Caution must be exercised when drawing generalizations from this literature. However, a picture is emerging from research looking at availability and promotion of food in community environments, which suggests that 'unhealthy' options will outweigh 'healthy' food options. Children, and particularly adolescents, are likely to be exposed to both regular opportunities to consume less healthy foods, and to marketing messages for less healthy foods in multiple community settings.

Conclusion

In this chapter we have highlighted that childhood overweight and obesity is increasing internationally (International Obesity Taskforce 2007) with significant negative short and long term health consequences. In order to understand the causes of this phenomena we have taken a social ecological approach (Story et al. 2008). We have explored the literature on children's food environments, identifying the factors that constrain or promote children's food choices in the home, school and the community. Factors in the home environment that appear significant are: foods purchased and available in the home; time available for food preparation; parental eating habits; parental perceptions of how diet influences health and weight; television watching and the promotion of foods; eating practices within the household; and, perhaps most importantly, household economic resources.

Factors in the school food environment may include: food for sale; fruit, breakfast, or lunches provided to students; school food policies; marketing of food; use of food in fundraising and in-classroom incentives. Key factors in the community appear to be the availability and marketing of food. There is debate in the literature about the association of geographic distribution of food outlets with BMI or diet quality. However, there is some evidence that the type, density and location of food outlets, and the price of fruit and vegetables are associated with children's BMI. Clustering of fast food outlets around schools, for example, may facilitate opportunities for young people to use fast food outlets as a place to socialize. There is some evidence that the availability and promotion of 'unhealthy' food options outweighs 'healthy' food options in community environments that children frequent.

Children move between home, community and school environments, and as they do factors such as their age, food preferences, and the preferences of their peers influence the foods eaten (as illustrated in Figure 6.1). They not only react to the influences in these environments, they interact to reinforce or change them; the notion of reciprocity. No one factor in the children's food environment can be held responsible for increasing rates of childhood overweight and obesity. This means that no one intervention can be championed as the key to improving children's diets and preventing excess weight gain. Given the vulnerability of children to their environment, we believe there can be no dispute about the critical need to change children's food environments to ensure they promote children's health.

A social ecological model of obesity causation clearly demands intervention within multiple environments and across sociocultural, physical, economic, and political components of these environments. Developing interventions to impact across environments is not a new idea, and is central to health promotion as outlined in the Ottawa Charter (World Health Organization 1986). The process of reciprocity, the interaction of environments on individuals and individuals on environments, also needs to be considered (Green et al. 2000). For example, interventions could focus on mobilising children and adults to improve obesogenic environments rather than focusing on them as passive recipients of information. Theories of behaviour change that emphasize reciprocity, such as social cognitive theory, may assist (Bandura 1986).

Political arguments for interventions around childhood obesity tend to support either a simple individually-focused energy equation, or a more complex ecological argument of causation (Lang and Rayner 2005). The individual focus tends to support social marketing campaigns, education based interventions, and shows a reluctance to intervene in markets through regulation (Lang and Rayner 2005, 2007). This individual position tends to be supported by the food and advertising industries and more conservative political groups (Schwartz and Brownell 2007, White 2007).

A focus on ecological determinants of obesity, while potentially including educational and social marketing interventions, attempts to intervene across sociocultural, physical, economic, and political environments. This may mean restricting advertising of food products, differential taxes on food products, regulating standards of food products, or regulating food environments such as schools (Sacks et al. 2008), as discussed by Giskes (Chapter 10). These types of policies tend to be supported by public health advocates and more liberal political groups (White 2007, Lang and Rayner 2005).

Given the geographic, ethnic and socioeconomic patterning of obesity, and factors that influence it, interventions must be carefully designed to reduce inequalities. For example, attempts to reduce the availability of less healthy foods in the community may impact on accessibility of all foods for households in neighbourhoods where transport options are limited and more shopping is conducted at convenience stores. On the other hand, this same intervention may reduce the amount of less healthy food purchased by children. How these impacts balance out is likely to depend on the individual situation of households and neighbourhoods prior to the intervention. Tools such as health impact assessment (Scott-Samuel 1996) and the Health Equity Assessment Tool (Signal et al. 2008a) may help the design of equitable interventions.

While the research evidence to support the design of environmental interventions to address childhood overweight and obesity may not yet be conclusive, the health burden of these conditions on children necessitates urgent action, albeit with careful evaluation and intervention modification as evidence becomes more robust.

References

Al-Shehri, K., Buchanan, S., Douglas, N., Knobloch, T., Kumar, K., Morgan, T., et al. 2007. *30 Minutes a Day and Giving it Away: healthy activities with poor food choices.* Unpublished paper. Wellington: University of Otago.

Anderson, A.S., Porteous, L.E.G., Foster, E., Higgins, C., Stead, M., Hetherington, M., et al. 2005. The impact of a school-based nutrition education intervention on dietary intake and cognitive and attitudinal variables relating to fruits and vegetables. *Public Health Nutrition,* 8, 650-6.

Anderson, P.M. and Butcher, K.E. 2006. Childhood obesity: trends and potential causes. *Future of Children,* 16, 19-45.

Austin, S.B., Melly, S.J., Sanchez, B.N., Patel, A., Buka, S. and Gortmaker, S.L. 2005. Clustering of Fast-Food Restaurants Around Schools: A Novel Application of Spatial Statistics to the Study of Food Environments. *American Journal of Public Health,* 95, 1575-81.

Ball, J., Watts, C. and Quigley, R. 2005. *A Rapid Review of the Literature on the Association Between Nutrition and School Pupil Performance.* Wellington: Obesity Action Coalition.

Bandura, A. 1986. *Social Foundations of Thought and Action: A Social Cognitive Theory*. Englewood Cliffs, NJ: Prentice-Hall.

Bauer, K.W., Yang, Y.W. and Austin, S.B. 2004. 'How Can We Stay Healthy when you're Throwing All of this in Front of Us?' Findings from Focus Groups and Interviews in Middle Schools on Environmental Influences on Nutrition and Physical Activity. *Health Education and Behaviour,* 31, 34-46.

Borra, S.T., Kelly, L., Shirreffs, M.B., Neville, K. and Greiger, C.J. 2003. Developing health messages: Qualitative studies with children, parents, and teachers help identify communications opportunities for healthful lifestyles and the prevention of obesity. *Journal of the American Dietetic Association,* 103, 721-8.

Bowman, S.A., Gortmaker, S.L., Ebbeling, C.B., Pereira, M.A. and Ludwig, D.S. 2004. Effects of Fast-Food Consumption on Energy Intake and Diet Quality Among Children in a National Household Survey. *Pediatrics,* 113, 112-18.

Brown, R., Scragg, R. and Quigley, R. 2008. *Does the family environment contribute to food habits or behaviours and physical activity in children?* Wellington: Agencies for Nutrition Action.

Campbell, K. and Crawford, D. 2001. Family food environments as determinants of preschool-aged children's eating behaviours: implications for obesity prevention policy. A review. *Australian Journal of Nutrition and Dietetics,* 58, 19.

Caraher, M. and Coveney, J. 2004. Public health nutrition and food policy. *Public Health Nutrition,* 7, 591-598.

Carter, M.A. and Swinburn, B. 2004. Measuring the 'obesogenic' food environment in New Zealand primary schools. *Health Promotion International,* 19, 15-20.

Coveney, J. 2005. A qualitative study exploring socio-economic differences in parental lay knowledge of food and health: implications for public health nutrition. *Public Health Nutrition,* 8, 290-7.

Crawford, D.A., Timperio, A.F., Salmon, J.A., Baur, L., Giles-Corti, B., Roberts, R.J., et al. 2008. Neighbourhood fast food outlets and obesity in children and adults: the CLAN Study. *International Journal of Pediatric Obesity,* iFirst DOI: 10.1080/17477160802113225.

Cummins, S. and Macintyre, S. 2006. Food environments and obesity – neighbourhood or nation? *International Journal of Epidemiology,* 35, 100-4.

Datar, A. and Sturm, R. 2006. Childhood overweight and elementary school outcomes. *International Journal of Obesity,* 30, 1449-60.

Department of Health. 2004. *Choosing Health: Making healthy choices easier.* London: HM Government.

Dollman, J., Norton, K. and Norton, L. 2005. Evidence for secular trends in children's physical activity behaviour. *British Journal of Sports Medicine,* 39, 892-7.

Dowler, E. 2008. Symposium on 'Intervention policies for deprived households' Policy initiatives to address low-income households' nutritional needs in the UK. *Proceedings of the Nutrition Society,* 67, 289.

Drewnowski, A. and Darmon, N. 2005. Food choices and diet costs: an economic analysis. *Journal of Nutrition,* 135, 900-4.

Drewnowski, A. and Specter, S. 2004. Poverty and Obesity: the role of energy density and energy costs. *American Journal of Clinical Nutrition,* 79, 6-16.

Finkelstein, D.M., Hill, E.L. and Whitaker, R.C. 2008. School Food Environments and Policies in US Public Schools. *Pediatrics,* 122, e251-259.

Fleischhacker, S. 2007. Food Fight: The Battle Over Redefining Competitive Foods. *Journal of School Health,* 77, 147.

Flynn, M.A.T., McNeil, D.A., Maloff, B., Mutasingwa, D., Wu, M., Ford, C., et al. 2006. Reducing obesity and related chronic disease risk in children and youth: a synthesis of evidence with 'best practice' recommendations. *Obesity Reviews,* 7(Suppl 1), 7-66.

Fogarty, A.W., Antoniak, M., Venn, A.J., Davies, L., Goodwin, A., Salfield, N., et al. 2007. Does participation in a population-based dietary intervention scheme have a lasting impact on fruit intake in young children? *International Journal of Epidemiology,* 36, 1080-5.

French, S.A. and Wechsler, H. 2004. School-based research and initiatives: fruit and vegetable environment, policy, and pricing workshop. *Preventive Medicine,* 39, 101-7.

Green, L.W., Poland, B.D. and Rootman, I. 2000. The settings approach to health promotion, in *Settings for Health Promotion: Linking theory and practice*, edited by B.D. Poland et al. London: Sage.

Hastings, G., Stead, M., McDermott, L., Forsyth, A., MacKintosh, A.M., Rayner, M., et al. 2003. Review of research on the effects of food promotion to children: Final Report prepared for the Food Standards Agency. Glasgow: Food Standards Agency.

Henriksen, L., Feighery, E.C., Schleicher, N.C., Cowling, D.W., Kline, R.S. and Fortmann, S.P. 2008. Is adolescent smoking related to the density and proximity of tobacco outlets and retail cigarette advertising near schools? *Preventive Medicine,* 47, 210-14.

Inglis, V., Ball, K. and Crawford, D. 2005. Why do women of low socioeconomic status have poorer dietary behaviours than women of higher socioeconomic status? A qualitative exploration. *Appetite*, 45, 334-43.

International Obesity Taskforce. 2007. Childhood Overweight (inlcuding obesity) statistics by WHO region. Available at: www.iotf.org/database/childhoodoverweightglobal.htm [accessed: 21 August 2008].

Jain, A., Chamberlin, L.A., Carter, Y., Powers, S.W. and Whitaker, R.C. 2001. Why Don't Low-Income Mothers Worry About Their Preschoolers Being Overweight? *Pediatrics,* 107, 1138.

Kubik, M.Y., Lytle, L.A., Hannan, P.J., Perry, C.L. and Story, M. 2003. The Association of the School Food Environment With Dietary Behaviors of Young Adolescents. *American Journal of Public Health,* 93, 1168-73.

Kubik, M.Y., Lytle, L.A. and Story, M. 2005a. Soft drinks, candy, and fast food: what parents and teachers think about the middle school food environment. *Journal of the American Dietetic Association*, 105, 233-9.

Kubik, M.Y., Lytle, L.A. and Story, M. 2005b. Schoolwide food practices are associated with body mass index in middle school students. *Archives of Pediatrics and Adolescent Medicine,* 159, 1111-4.

Lang, T. and Rayner, G. 2005. Obesity: a growing issue for European policy? *Journal of European Social Policy,* 15, 301-27.

Lang, T. and Rayner, G. 2007. Overcoming policy cacophony on obesity: an ecological public health framework for policymakers. *Obesity Reviews,* 8(Suppl 1), 165-181.

Lobstein, T. and Jackson-Leach, R. 2006. Estimated burden of paediatric obesity and co-morbidities in Europe. Part 2. Numbers of children with indicators of obesity-related disease. *International Journal of Pediatric Obesity*, 1, 33-41.

Lytle, L.A., Kubik, M.Y., Perry, C., Story, M., Birnbaum, A.S. and Murray, D.M. 2006. Influencing healthful food choices in school and home environments: Results from the TEENS study. *Preventive Medicine,* 43, 8-13.

Macintyre, S., Ellaway, A. and Cummins, S. 2002. Place effects on health: how can we conceptualise, operationalise and measure them? *Social Science and Medicine* 55, 125-39.

Maher, A., Wilson, N. and Signal, L. 2005. Advertising and availability of 'obesogenic' foods around New Zealand secondary schools: a pilot study. *New Zealand Medical Journal,* 118, U1556.

Maher, A., Wilson, N., Signal, L. and Thomson, G. 2006. Patterns of sports sponsorship by gambling, alcohol and food companies: an Internet survey. *BMC Public Health,* 6, 95.

Minaker, L.M., McCargar, L., Lambraki, I., Jessup, L., Driezen, P., Calengor, K., et al. 2006. School region socio-economic status and geographic locale is associated with food behaviour of Ontario and Alberta adolescents. *Canadian Journal of Public Health,* 97, 357-61.

Ministry of Health. 1998. *Food and Nutrition Guidelines for Healthy Adolescents: A background paper*. Wellington: Ministry of Health.

Moore, L. and Tapper, K. 2008. The impact of school fruit tuck shops and school food policies on children's fruit consumption: A cluster randomised trial of schools in deprived areas. *Journal of Epidemiology and Community Health,* Doi: 10.1136/jech.2007.070953.

Nanney, M.S., Bohner, C. and Friedrichs, M. 2008. Poverty-Related Factors Associated with Obesity Prevention Policies in Utah Secondary Schools. *Journal of the American Dietetic Association,* 108, 1210-15.

Nelson, M., Lowes, K. and Hwang, V. 2007. The contribution of school meals to food consumption and nutrient intakes of young people aged 4-18 years in England. *Public Health Nutrition,* 10, 652-62.

Nollen, N.L., Befort, C.A., Snow, P., Daley, C.M., Ellerbeck, E.F. and Ahluwalia, J.S. 2007. The school food environment and adolescent obesity: qualitative insights from high school principals and food service personnel. *International Journal of Behavioral Nutrition and Physical Activity,* 4, doi:10.1186/1479-5868-1184-1118.

Ogden, C.L., Yanovski, S.Z., Carroll, M.D. and Flegal, K.M. 2007. The epidemiology of obesity. *Gastroenterology,* 132, 2087-102.

Pasch, K.E., Komro, K.A., Perry, C.L., Hearst, M.O. and Farbakhsh, K. 2007. Outdoor alchohol advertising near schools: what does it advertise and how is it related to intentions and use of alcohol among young adolescents? *Journal of Studies on Alchohol and Drugs,* 68, 587-96.

Pearce, J., Blakely, T., Witten, K. and Bartie, P. 2007. Neighbourhood Deprivation and Access to Fast-Food Retailing. *American Journal of Preventive Medicine,* 32, 375-82.

Pearce, J., Day, P. and Witten, K. 2008a. Neighbourhood Provision of Food and Alcohol Retailing and Social Deprivation in Urban New Zealand. *Urban Policy and Research,* 26, 213-227.

Pearce, J., Hiscock, R., Blakely, T. and Witten, K. 2008b. The contextual effects of neighbourhood access to supermarkets and convenience stores on individual fruit and vegetable consumption. *Journal of Epidemiology and Community Health,* 62, 198-201.

Powell, L.M., Auld, M.C., Chaloupka, F.J., O'Malley, P.M. and Johnston, L.D. 2007. Associations Between Access to Food Stores and Adolescent Body Mass Index. *American Journal of Preventive Medicine,* 33, S301-S307.

Reidpath, D.D., Burns, C., Garrard, J., Mahoney, M. and Townsend, M. 2002. An ecological study of the relationship between social and environmental determinants of obesity. *Health and Place,* 8, 141-5.

Reilly, J.J. 2005. Descriptive epidemiology and health consequences of childhood obesity. *Best Practice and Research Clinical Endocrinology and Metabolism,* 19, 327-41.

Ricciuto, L., Tarasuk, V. and Yatchew, A. 2006. Socio-demographic influences on food purchasing among Canadian households. *European Journal of Clinical Nutrition,* 60, 778-90.

Rosenkranz, R.R. and Dzewaltowski, D.A. 2008. Model of the home food environment pertaining to childhood obesity. *Nutrition Reviews,* 66, 123-40.

Sacks, G., Swinburn, B.A. and Lawrence, M.A. 2008. A systematic policy approach to changing the food system and physical activity environments to prevent obesity. *Australia and New Zealand Health Policy,* 5.

Sallis, J.F., McKenzie, T.L., Conway, T.L., Elder, J.P., Prochaska, J.J., Brown, M., et al. 2003. Environmental interventions for eating and physical activity: A randomized controlled trial in middle schools. *American Journal of Preventive Medicine,* 24, 209-17.

Sanigorski, A.M., Bell, A.C., Kremer, P.J. and Swinburn, B.A. 2005. Lunchbox contents of Australian school children: room for improvement. *European Journal of Clinical Nutrition,* 59, 1310-16.

Sattar, N. and Lean, M. 2007. *ABC of Obesity.* Oxford: BMJ Books.

Scaglioni, S., Salvioni, M. and Galimberti, C. 2008. Influence of parental attitudes in the development of children eating behaviour. *British Journal of Nutrition,* 99, S22-S25.

Schwartz, M.B. and Brownell, K.D. 2007. Actions necessary to prevent childhood obesity: creating the climate for change. *Journal of Law, Medicine and Ethics,* 35, 78-89.

Scott-Samuel, A. 1996. Health impact assessment. *British Medical Journal,* 313, 183-4.

Signal, L., Martin, J., Cram, F. and Robson, B. 2008a. *The Health Equity Assessment Tool: a user's guide.* Wellington: Ministry of Health.

Signal, L.N., Lanumata, T., Robinson, J., Tavila, A., Wilton, J. and Mhurchu, C.N. 2008b. Perceptions of New Zealand Nutrition Labels in New Zealand by Maori, Pacific and Low-income Shoppers. *Public Health Nutrition,* 11, 706-13.

Story, M., Kaphingst, K.M. and French, S. 2006. The role of schools in obesity prevention. *Future of Children,* 16, 109-42.

Story, M., Kaphingst, K.M., Robinson-O'Brien, R. and Glanz, K. 2008. Creating Healthy Food and Eating Environments: Policy and Environmental Approaches. *Annual Review of Public Health,* 29, 253-72.

Story, M., Neumark-Sztainer, D. and French, S. 2002. Individual and Environmental Influences on Adolescent Eating Behaviors. *Journal of the American Dietetic Association*, 102, S40-S51.

Sturm, R. and Datar, A. 2005. Body mass index in elementary school children, metropolitan area food prices and food outlet density. *Public Health*, 119, 1059-68.

Swinburn, B. 2008. Obesity prevention: the role of policies, laws and regulations. *Australia and New Zealand Health Policy,* 5, 12.

Swinburn, B., Egger, G. and Raza, F. 1999. Dissecting Obesogenic Environments: The Development and Application of a Framework for Identifying and Prioritizing Environmental Interventions for Obesity. *Preventive Medicine*, 29, 563-70.

Taveras, E.M., Berkey, C.S., Rifas-Shiman, S.L., Ludwig, D.S., Rockett, H.R.H., Field, A.E., et al. 2005. Association of Consumption of Fried Food Away From Home With Body Mass Index and Diet Quality in Older Children and Adolescents. *Pediatrics*, 116, e518-24.

Turrell, G. 1996. Structural, material and economic influences on the food-purchasing choices of socioeconomic groups. *Australian and New Zealand Journal of Public Health,* 20, 611-17.

Turrell, G. and Kavanagh, A.M. 2006. Socio-economic pathways to diet: modelling the association between socio-economic position and food purchasing behaviour. *Public Health Nutrition,* 9, 375-83.

Utter, J., Schaaf, D., Mhurchu, C.N. and Scragg, R. 2007. Food choices among students using the school food service in New Zealand. *New Zealand Medical Journal*, 120.

Utter, J., Scragg, R. and Schaaf, D. 2006. Associations between television viewing and consumption of commonly advertised foods among New Zealand children and young adolescents. *Public Health Nutrition*, 9, 606-12.

Wang, Y. and Beydoun, M.A. 2007. The Obesity Epidemic in the United States – Gender, Age, Socioeconomic, Racial/Ethnic, and Geographic Characteristics: A Systematic Review and Meta-Regression Analysis. *Epidemiologic Reviews,* 29, 6-28.

Wang, Y. And Lobstein, T. 2006. Worldwide trends in childhood overweight and obesity. *International Journal of Pediatric Obesity*, 1(1), 11-25.

Wharton, C.M., Long, M. and Schwartz, M.B. 2008. Changing nutrition standards in schools: The emerging impact on school revenue. *Journal of School Health*, 78, 245-51.

White, J. 2007. The Health Select Committee Inquiry into Obesity and Type Two Diabetes in New Zealand: An initial analysis of submissions. Wellington: Fight the Obesity Epidemic New Zealand Incorporated.

Wojcicki, J.M. and Heyman, M.B. 2006. Healthier choices and increased participation in a middle school lunch program: Effects of nutrition policy changes in San Francisco. *American Journal of Public Health,* 96, 1542-7.

World Health Organization. 1986. *Ottawa Charter for Health Promotion 1986*. Geneva: World Health Organization.

World Health Organization. 2003. *Diet, Nutrition and the Prevention of Chronic Diseases*. Geneva: World Health Organization.

World Health Organization. 2004. *Global Strategy on Diet, Physcial Activity and Health*. Geneva: World Health Organization.

Zenk, S.N. and Powell, L.M. 2008. US secondary schools and food outlets. *Health and Place,* 14, 336-46.

PART III
Physical Activity, Environment and Obesity (Energy Out)

Chapter 7
The Role of the Changing Built Environment in Shaping Our Shape

Billie Giles-Corti, Jennifer Robertson-Wilson,
Lisa Wood and Ryan Falconer

Introduction

> When Jem and I raced each other up the sidewalk to meet Atticus coming home from work, I didn't give him much of a race. It was our habit to run and meet Atticus the moment we saw him round the post office corner in the distance…
>
> *To Kill a Mocking Bird* (1960 p. 34) Harper Lee

This quote from Harper Lee's Pulitzer Prize-winning novel, set in the US state of Alabama in 1936, highlights a number of changes observed in the developed world post-World War II (WWII). In this brief passage, we read of children running unaccompanied in the street, a parent (Atticus) walking home from work, and the presence of local destinations (the post office) proximate to homes. In a few words, the quote underscores changes observed in the way we work, live, shop and commute, and speaks to the restrictions placed on children's independent mobility in the twenty-first century.

Since WWII – and particularly the last three decades – technological advances have dramatically influenced the design and development of buildings and communities, the way populations are mobilized and fed, the nature of work, and methods of communication. These advances have shaped both the amount of human energy expended and the amount of fossil fuel consumed in the course of undertaking activities of daily living. While often subtle, it is likely that together these changes have had a profound impact on human populations.

Globally, around two billion adults and 20 million children under the age of five years are either overweight or obese (World Health Organization 2006). The estimated direct and indirect economic costs of this epidemic are immense, and are likely to continue to grow unless obesity levels can be curbed (Access Economics 2008, Allender and Rayner 2007). Moreover, societal patterns of overweight and obesity and their health consequences often reflect social inequalities and cluster with, and are compounded by, other health disparities in some disadvantaged population groups (Friel et al. 2007, Pickett et al. 2005, Eckersley 2001).

Increasingly, an imbalance in three key domains – physical inactivity, sedentary behaviour (Healy et al. 2008) and poor eating behaviours (Joint WHO/FAO Expert Consultation on Diet Nutrition and the Prevention of Chronic Diseases 2003) – are recognized as responsible for increases in weight status. While the focus of this chapter is on sedentary behaviours and physical activity, the impact of the food environment on food availability and food choices is an emerging field of research (Story et al. 2009, Papas et al. 2007). Rapid changes in the food industry in the last two decades have been paralleled by changes in eating behaviours (Story et al. 2008) and patterns. Over the next decade, this area of research will undoubtedly grow and is already becoming an important field of endeavour (Story et al. 2009).

Globally, 60 per cent of adults are insufficiently active to benefit their health (World Health Organization 2002) increasing the risk of major chronic diseases and their risk factors, as well as disability and burden of disease (World Health Organization 2004). Given the prevalence of inactivity, even a modest increase in the number in those physically active is likely to produce large population-wide impacts (Stephenson and Bauman 2000). Sedentary behaviour includes low energy expenditure (1-1.5 METS) behaviours such as TV/video watching, computer use and driving (Owen et al. 2009), which are independently associated with weight status (Healy et al. 2008, Brown et al. 2003) and metabolic profiles (Healy et al. 2008, Dunstan et al. 2007). Hence, there are now calls for a public health response to sedentariness, including the need for guidelines for sedentary behaviour (Brown et al. 2009, Owen et al. 2009).

To help understand broader factors that may be contributing to the obesity epidemic and the changing landscape of health compromising behaviours, this chapter reflects on technological, environmental and societal changes that have occurred since WWII. It considers how these changes may be contributing to the obesity epidemic through their influence on the built environment which in turn, may be contributing to levels of sedentariness and physical inactivity. While at first glance individual changes may appear subtle, the collective impact over the last five to six decades appears to be massive. This chapter considers whether these changes are important from a weight status perspective, and what can be done to ameliorate their impacts.

Background

Never before in human history have so many people in the developed world been able to be so sedentary in the course of meeting their needs for daily living (also see Sallis and Owen (1999) for discussion). Industrial and home labour saving devices – including washing machines, leaf blowers (replacing the humble garden rake), snow blowers (replacing shovels), motorized lawn mowers, and television remote controls – maximize convenience while minimizing effort (Kelty et al. 2008). Similarly, rapid advances in computer technology (Statistics Canada 2006) and telephony even in the last decade – including the advent of computer games, computerized sporting matches (for example hockey and golf), 3D virtual worlds (for example *Second Life*) and web-based social networking tools (for example Facebook, My Space) – maximize our ability to communicate with, 'meet,' maintain contact and recreate with friends, colleagues and family, with no more effort than the tap of a few fingers. Compared with our grandparents, feeding and clothing ourselves have become effortless, often involving a one-stop shopping centre trip in a motor vehicle, facilitated by well-designed, readily available shopping trolleys and convenient parking. However, these days people don't even need to leave their house to order clothing, fast food, and groceries – these can all be ordered by telephone or the internet and delivered door-to-door. Equally, bills once paid in-person or by post – perhaps involving a short walk to the post-office or post-box – can now be paid via the internet, with its 24/7 'anywhere, anytime' convenience. In a society where we increasingly regard ourselves as 'consumers' rather than 'citizens' (Kunstler 1998), we now have the option of paying someone to wash our car, walk our dog, clean our house, tidy our garden, or paint our houses. In the twenty-first century, we have traded opportunities to expend energy in day-to-day life for sedentary alternatives, many of which involve consuming fossil fuel (including queuing in motor vehicles while waiting for drive-in car washes, banks and coffee shops). Our changing lifestyles have been facilitated by a technological revolution, preceded by the twentieth century transport revolution.

Impacts of Motorized Travel on the Obesity Epidemic

Perhaps the biggest lifestyle impact of the twentieth century was produced by the advent of the motor vehicle (Falconer and Giles-Corti 2008). Henry Ford's dream to make motor vehicles affordable for the masses (Kjellstrom and Hinde 2007) transformed society in both intended and unintended ways. The private transportation revolution influenced where and how we live; where we work, recreate and shop; social norms about desirable methods of mobilizing populations; how we build communities; and ultimately, the amount of human and fossil fuel energy expenditure required for activities of daily living. In short, the realization of Henry Ford's dream nurtured people's evolutionary drive to conserve energy expenditure and, as a consequence, appears to have contributed to our ever-burgeoning waist lines.

The advent and rapid uptake of the motor vehicle resulted in a redesign of cities. In the early 1800s, cities were densely settled and land uses were mixed, with a clear distinction between city and country. Cities were spatially cohesive, orderly and efficient to facilitate ease of movement (Falconer and Giles-Corti 2008). With transportation limited to foot or animal power, residences, services and places of employment were proximate to reduce time (Newman 2009) with connected street networks that maximized circulation (Falconer and Giles-Corti 2008). However, over time, the compact city became associated with the ills of industrialization. The lack of separation between residences and noxious land uses, and associated noise, produced city landscapes characterized by congestion, pollution (Transportation Research Board 2005), overcrowding and disease (Frumkin et al. 2004). Planners and public health officials recognized the potential benefits of segregation compared with agglomeration of land uses (Falconer and Giles-Corti 2008).

'Zoning' of land uses was introduced as a mechanism to separate housing from other noxious or noisy land uses. The practice of zoning and separation of land uses had a profound impact on urban development. It distanced residences from services, facilities and places of work, thereby increasing travel times and rendering some travel impractical except by motor vehicle. By the 1900s, zoning had become an important element of planning strategy and enthusiastically embraced as a mechanism to preserve local character, social stability and property values (Hall and Porterfield 2000).

In many North American and Australasian cities, zoning continues to be an integral part of planning practice used to 'ensure orderly urban development and to protect basic residential amenity' (Gleeson 2006: 21). Single use urban zones are built into Town Planning Schemes, meaning mixed use development proposals are often in contravention of statutory policy. However, while arguably creating order and protecting the public from environmental pollution, zoning has contributed to an increased need and demand for privatized transportation (Frumkin et al. 2004), as the separation of land uses increases distances between activities of daily living. The model of human habitat created by zoning also impacts on the social and cultural vitality and 'soul' of cities and townships (Kunstler 1998).

In many cities, zoning laws have also constrained housing density within a suburb or area (Kunstler 1998), often with good intention but with inadvertent consequences for walkability. Suburban sprawl was spawned by zoning laws that enshrine the sanctity of the ¼ acre block for a house with a large backyard. There are numerous zoning and regulatory hurdles now to jump over to replicate the closely nestled terrace housing iconic of the UK and that characterized small pockets of Australian architecture from around the turn of last century. Density, which is beneficial for walkability and for environmental sustainability, is sometimes tarnished with fears about congestion or the potential for urban slums (Kunstler 1998).

Private motorized mobility therefore contributed to profound change in many cities. Initially the introduction of public transport systems revolutionized cities, allowing workers to live in suburbs and commute to work and other essential services (Frumkin et al. 2004). However, it was the mass production and distribution of motor vehicles that had the biggest impact. Automobiles increased residents' mobility and independence even further, allowing people to live great distances away from employment centres and further encouraging suburbanization (Falconer and Giles-Corti 2008). Affordable automobiles increased mobility, and fuelled the great Australasian and North American dream of the ¼ acre suburban lot which has resulted 'urban sprawl'. 'Urban sprawl' is characterized by low population densities, curvilinear street networks, limited mixed use and high dependence on individual motor vehicles to travel from place to place (Frumkin et al. 2004).

Until the end of World War II, there was little car ownership as urban areas were relatively concentrated and public transport was the principal mode of transportation (Laird and Newman 2001). In Melbourne, for example, only one in every four households owned a car at the war's end (Davison 2004). Moreover, streets tended to be multi-functional: they were play spaces for children and places for public transport, walking and bicycling, as well as for cars (Davison 2004).

However, as land uses were separated, increasing the need for a private motor vehicle, pressures grew to design communities to improve motor vehicle mobility, at the expense of pedestrians and cyclists and related infrastructure. The unintended impacts on local residents went unnoticed for decades (Giles-Corti and King 2009).

However, it is now recognized that neighbourhood environments have multiple direct and indirect impacts on health, including the physical activity and obesity levels of residents (Kopelman et al. 2007, Transportation Research Board 2005). Studies consistently show that adults are more likely to walk for transport in compact, pedestrian-friendly neighbourhoods characterized by connected street networks, access to mixed use planning, the presence of destinations and higher densities (Duncan and Mummery 2005, Transportation Research Board 2005, Owen et al. 2004). Conversely, indices of weight status (BMI, overweight and obesity) are positively associated with living in urban sprawl (Robertson-Wilson and Giles-Corti in press, Black and Macinko 2008, Papas et al. 2007, Booth et al. 2005). Residents living in urban sprawl spend more time driving (Pendola and Gen 2007, Lopez-Zutina et al. 2006, Wen et al. 2006, Frank et al. 2004), and engage in sedentary pursuits such as television viewing (Sugiyama et al. 2007), and less time walking (Frank et al. 2004). However, as discussed later, broader built environment factors that constrain physical activity (for example, graffiti) (Foster and Giles-Corti 2008, Papas et al. 2007) also appear to be important correlates of weight status.

Designing communities for motor vehicles has also impacted children's use of their neighbourhood and their active transport behaviour (Giles-Corti et al. 2009). Active transport (AT) includes travel by foot, bicycle and other non-motorized vehicles (National Public Health Partnership 2001). Increasing daily activities such as walking to school and doing errands with or for parents have been identified as means of increasing physical activity in children and adolescents (Jago and Baranowski 2004, Dietz and Gortmaker 2001). Yet evidence consistently shows that in developed countries, children's AT has declined in the last two decades (Harten and Olds 2004, Bradshaw 2001, French et al. 2001, Roberts 1996).

Time and distance are key factors that influence walking and cycling for transport (Anonymous 2002, French et al. 2001). Thus, neighbourhood design has a powerful influence on young people's AT options. While adults are more likely to walk in more compact traditional neighbourhoods characterized by connected street networks with more proximate destinations than in low density suburbs with separated land uses (Owen et al. 2007, Leslie et al. 2005, Owen et al. 2004, Saelens et al. 2003), children's AT is influenced by the presence of traffic and parental concerns about safety (Jago and Baranowski 2004, Anonymous 2002, Lam 2001a, 2001b).

In primary school children, parental concerns about traffic danger (Harten and Olds 2004, Anonymous 2002), lack of safe crossings infrastructure (Timperio et al. 2004) and concerns for personal safety (Anonymous 2002, DiGuiseppi et al. 1998) influence the willingness of parents to allow children to use AT modes. Parental concerns about traffic safety are perhaps understandable (Giles-Corti et al. 2009). Pedestrian motor vehicle collisions are the third leading cause of pediatric mortality in Australia (Moon et al. 1998), after motor vehicle accidents and accidental drowning, and a child's risk of pedestrian injury is associated with overall traffic exposure (Macpherson et al. 1998, Carlin et al. 1997). Thus, the design of communities and exposure to traffic influences the feasibility and safety of young people using active modes, and the preparedness of their parents to permit children to walk or cycle.

The sticker 'mum's taxi' that adorns the back window of many a family car exemplifies the way in which the location and unwalkability of children's schools and activities, coupled with parental fears about traffic safety have combined with the normalization of getting in the car to go somewhere, thus contributing to the decline of AT among children and young people.

Societal Changes Fuelling the Obesity Epidemic

Numerous social trends are implicated in the obesity epidemic. The increasing number of households with both parents working, time-poor lifestyles and pressures of work, are disruptions to family life that erode shared mealtimes and leisure time, and foster greater consumption of fast and convenience food (von Kardorff and Ohibrecht 2008). Moreover, multiple role juggling (McBride 1990) and work-family spillover are faced by many women and families in managing home, work and community life, which impacts on leisure time. Thus, time and energy for physical activity or food preparation are often the casualties of modern lifestyles.

However, changes in the built environment and their impact on methods of transport have also impacted the social realm of society in numerous ways. For example, visiting friends and relatives (and sometimes even neighbours) now often entails a trip in the car. Children are now less likely to play on the street with other neighbourhood children, and are often either chauffeured to structured play and activity regimes after school, or stay at home. The nature of the modern built environment, along with the 'harried isolating nature of modern life' (Walljasper 2007: 10) has not only increased inactivity, but has eroded opportunities for human contact, with potentially detrimental consequences for sense of community, social capital and mental wellbeing (Wood and Giles-Corti 2008).

Other social trends have also fuelled the obesity epidemic by increasing sedentary behaviour. As articulated by Desmond Morris, author of *The Human Animal*, 'our community is now our pocket book' or in contemporary parlance, our contacts list in our cell phone, or even our internet. People's notion of community is increasingly less geographically defined, and as noted earlier, often includes internet-based affiliations and social networks, or 'interest-based' groups with widely scattered membership. Social isolation, loneliness and social support are increasingly on the radar in relation to the growing burden of depression and anxiety in the developed world (World Health Organization 2003), but also impact on obesity and physical activity. Inequality and low social status for example can increase anxiety and stress and reduce people's ability to exercise control over their lives (Pickett et al. 2005). This is evident in the bi-directional relationship between eating and psychological wellbeing (for example, comfort eating in times of distress) (Polivy and Herman 2005) and can impede motivation and capacity to be physically active. For people who are already overweight or obese, self-consciousness about their body has been identified as a barrier to physical activity among overweight children as well as adults (Canadian Institute for Health Information 2006).

Psycho-social mechanisms can however be harnessed positively for physical activity, as reflected in studies documenting the role of social support in encouraging and sustaining physical activity (McNeill et al. 2006, Felton et al. 2002, Kahn et al. 2002). In a built environment context, walking groups and recreational centres often incorporate social support strategies. In addition, the presence of areas for recreation, physical activity and play are not only important for physical activity and obesity prevention, but also promote social capital (Baum and Palmer 2002), which in turn has positive repercussions for social connectedness and mental wellbeing (Araya 2006, Whitley and McKenzie 2005).

Social trends have a symbiotic relationship with the built environment in their impact on physical activity and obesity. For example, as noted earlier, both actual and perceived safety and its psychological correlate fear, may constrain physical activity and actual and perceived safety appears to be impacted by the design of the built environment. Studies to date indicate an inverse relationship between fear and both physical activity and obesity (Stafford et al. 2007, Boslaugh et al. 2004). Fear and safety concerns can hinder outdoor activity and walking (Parkes and Kearns 2006), and for children, parental perceptions of safety and concerns about 'stranger danger' restrict children's independent mobility within neighbourhoods (Veitch et al. 2006, Timperio et al. 2004). A negative spiral can result, as fewer people out and about walking and using parks can further erode perceptions of safety and fuel fear, and at the community level, impede the development of trust and sense of community (Sooman and Macintyre 1995, Perkins et al. 1992). On the other hand, the more people 'out walking', the safer the neighbourhood is for those who walk (Wood and Giles-Corti 2008) and built environments that foster pedestrian activity through destinations to walk to also enhance safety on the streets (Kunstler 1998). It has also been hypothesized that fear and threatening environments can contribute to obesity through physiological responses associated with 'allostatic load' (Stafford et al. 2007). Allostatic load is a measure of wear and tear on the body resulting from efforts to maintain stability in response to stressors. This is an avenue of research that merits further investigation.

So How Important are these Changes as Contributors to the Obesity Epidemic?

A decade ago, Sallis and Owen (1999: 124) challenged us to '[C]onsider how human-constructed environments now make it possible, and even encourage people, to lead extremely sedentary lifestyles'. Reflecting on the evidence presented above, the authors highlighted how environmental changes have reduced physical activity levels by influencing how we move from one place to another (for example, increased reliance on motor vehicles cars rather than bicycles or walking); where we live (for example, the constraints to physical activity in suburban neighbourhoods); the nature of our work (for example, increasingly sedentary and screen-based); and how we play (for example, increasingly sedentary pursuits such as TV, video games and computer-based social networking). Individually and collectively, these changes are negatively impacting levels of physical activity and sedentariness, and hence our levels of obesity. Essentially, 'we have "engineered" physical activity out of our lives' (for example, Hill in Barclay (2004), Kohl in Roser (2008), Lincoln in National Heart Forum (2007)).

Thus over the last decade, an ever-growing body of evidence under the umbrella of 'active living research' has emerged (Schilling and Linton 2005). Although a multitude of 'associations' between environmental factors such as urban sprawl and obesity (Robertson-Wilson and Giles-Corti (in press), Black and Macinko 2008, Papas et al. 2007, Wells et al. 2007, Booth et al. 2005) have emerged, a major criticism of the evidence to date is that it is cross-sectional, which prevents causal inferences being drawn. For example, it is possible that obese people choose to live in urban sprawl, rather than urban sprawl causing obesity (Black and Macinko 2008, Frank et al. 2007) (see Chapter 12).

As randomized controlled trials, the gold standard for health intervention evidence, are not feasible for studies of the impact of urban design on obesity, there are calls for the evaluation of 'natural experiments' of environment-related policy (for example, urban growth policy) or design initiatives (for example, park zoning in new neighbourhoods or urban renewal projects) implemented in a community by a given sector (for example, local government) (Black and Macinko 2008, Brownson et al. 2008, Ramanathan et al. 2008, Ogilvie et al. 2007, Wells et al. 2007, Ball et al. 2006, Ogilvie et al. 2006, Petticrew et al. 2005). These studies appear to be feasible (see for example Giles-Corti et al. 2008, Merom et al. 2003) to design and undertake, and will assist in answering the cause-effect questions of causality.

Despite the lack of longitudinal and experimental studies, leading US, UK, Canadian and Australian health and transport organizations have published position papers on the potential impact of the built environment on physical activity (National Institute for Health and Clinical Excellence 2008, Heart and Stroke Foundation of Canada 2007, Transportation Research Board 2005) and obesity (National Preventative Health Taskforce 2008, Kopelman et al. 2007). These reports agree that the built environment – land use patterns, transportation systems, building design and social infrastructure – can create conditions that optimize or deter health (including obesity outcomes) and positive health behaviours (including physical activity), and conclude that there is sufficient evidence to warrant public health action.

The sum total of built environment and society changes on levels of physical activity and sedentariness are difficult to estimate, however studies of populations that have remained relatively unexposed to these forces, help to quantify their impact. For example, Bell and colleagues (2002), studied the impact of the uptake of motorized transport on Chinese men and women and found that after adjustment for known confounders, the odds of becoming obese doubled in men (but not women) living in households that attained a motor vehicle. Moreover, in a series of studies, physical activity and obesity were assessed among Canadian Old Order Amish and Old Order Mennonite youth and Old Order Amish adults (Tremblay et al. 2008, Bassett 2008, Bassett et al. 2007, Esliger et al. 2006, Tremblay et al. 2005, Bassett et al. 2004). These two population groups 'continue to live a traditional agrarian lifestyle' (Tremblay et al. 2008: 838) and they share some traditional lifestyle practices such as children walking to school and having chores (Bassett 2008, Tremblay et al. 2008, Bassett et al. 2007). Of the two groups, the practices of the Old Order Amish are stricter (Bassett 2008, Tremblay et al. 2008, Bassett et al. 2007), with many 'purposefully refrain[ing] from adopting modern technologies such as highline electricity and automobile ownership' (Bassett et al. 2007: 410-411).

A number of studies have examined levels of physical activity and obesity in these groups, and at times compared them with the general population. Among a sample of Old Order Amish Canadian adults, one quarter of the men and 27 per cent of women were classified as overweight with only 9 per cent of women and no men classified as obese (Bassett et al. 2004). Interestingly Bassett and colleagues (2008) report national obesity rates of between 8 per cent to over 30 per cent depending on country and measurement. Further, average daily step counts of over 14,000 were observed in women and over 18,000 among men. This is a striking difference with non-Amish adults who, according to a recent meta-analysis, accumulate less than 10,000 steps per day (Bohannon, 2007) and more than half are overweight or obese.

Tremblay and colleagues (2005) also compared differences in physical activity and fitness among Old Order Mennonite Canadian children and non-Mennonite Canadian children. Using accelerometry, they found Old Order Mennonite youth spent more time in moderate-to-vigorous daily physical activity compared with either rural or urban non-Mennonite youth, although no differences in weight status were observed (Tremblay et al. 2005). Bassett et al.'s (2007) subsequent study of Old Order Amish Canadian youth reported step counts comparable to those found for the Amish adults (Bassett et al. 2004) with these youth accumulating a weekly average of 15,563 daily steps, with specific averages of over 17,000 for boys and over 13,000 for girls. Step counts were also higher during the week versus the weekend (Bassett et al. 2007). Further, the obesity prevalence was 1.4 per cent with just over 7 per cent of the youth being classified as overweight. Compared with non-Amish North American youth, the prevalence of overweight and obesity in Amish youth is much lower and daily steps much higher (Bassett 2008).

Recent comparisons across the three groups of Old Order Amish, Old Order Mennonite, and non-Mennonite/Amish youth reveal 'a progressive temporal gradient suggesting that contemporary living is associated with lower levels of moderate to vigorous physical activity (that is OOA> OOM> CL)' (Tremblay et al. 2008: 838, Esliger et al. 2006). This gradient is also apparent for some fitness indicators (Tremblay et al. 2008, Esliger et al. 2006).

So What Can Be Done About These Trends?

Researchers studying Old Order Amish or Mennonite communities are quick to quell the idea that physical activity interventions should focus on everyone adopting a 'traditional agrarian lifestyle' (Bassett 2008). However, the health benefits of these traditional lifestyles appear self-evident (particularly for youth), and Tremblay and colleagues (2008: 839) have suggested that at the very least, interventions could be considered that explore ways of 'reintroduc[ing] additional lifestyle-embedded physical activity'. This notion is consistent with the active living research movement which aims to increase incidental activity throughout the course of the day (Schilling et al. 2009).

This is not without challenges as contemporary lifestyles rely on labour-saving devices, technology and motorized travel. For example, apart from concerns about traffic and personal safety, children's mode choice for trips to school is often influenced by scheduling considerations associated with parental work commitments. Motor vehicle trips to school often reflect the time at which children need to be at school and parents need to be at work (DiGuiseppi et al. 1998). Similarly, the employment status of the mother (DiGuiseppi et al. 1998) and motor vehicle ownership (that is, only one motor vehicle) (Black et al. 2001, DiGuiseppi et al. 1998); or no motor vehicle (Roberts 1996) are associated with mode choice. Likewise, the hours worked by mothers also appear to be important. Children are more likely to walk or cycle from school if there is no adult at home after school (Begg et al. 2007, Black et al. 2001). Broader trends in the educational sector also impact on the capacity of children to walk or ride to school, including the growing number of parents opting for private schooling not within their local area and the growing number of schools offering specialist programmes (for example, music, drama or art scholarships) that draw students from a much wider geographic catchment (Kwong 2000, Schneider and Buckley 2002).

Moreover, to facilitate change there clearly needs to be a commitment to rethink what we consider to be 'default' city planning; that is the way we develop and use land and transportation systems, including the coordination of homes, workplaces and shopping facilities. While researchers should continue to seek causal evidence, they should also be advocating for change. The imperatives to do so now go well beyond human individual health. Not only is excessive motor vehicle travel detrimental to human health (Woodcock et al. 2007), but the transport sector as a whole produces about 17 per cent of Australia's green house gases (Commissioner for Environmental Sustainability 2008).

Concluding Comments

The unintended consequences of efforts to protect the public from communicable disease and environmental pollution in the nineteenth and twentieth centuries appear to have contributed to diseases of over-consumption in the twenty-first century. However, in this century, there are synergies between healthy, planning, transportation and environmental sectors that can be exploited with the aim of reducing fossil fuel consumption while increasing human energy consumption and improving the public's health. As the developing world rushes towards the convenience of privatized motor vehicle travel (for example, the introduction of India's low-cost $2,000 Nano Car in March 2009, targeting towards its massive middle class), there is an urgent need for both the developed and developing worlds to rethink the way we build cities and communities and the way we mobilize populations. Without compromising the need to protect the public from air and noise pollution and noxious substances, there is a need for the health sector to advocate for healthy, sustainable, well designed, higher density, transit-oriented, leafy communities that encourage non-motorized forms of transport at the expense of the motor vehicle. As argued in this chapter, this is likely to be good not only for human health, but also for environmental, social and community health.

Acknowledgements

Wood is supported by an NHMRC Capacity Building Grant (#458668). Giles-Corti is supported by a NHMRC Senior Research Fellowship (#513702). Editorial assistance provided by Ms Gina Wood and Ms Lisa Bayly (both supported by the NHMRC Capacity Building Grant (#458668)) is gratefully acknowledged.

References

Access Economics. 2008. *The Growing Cost of Obesity in 2008: Three Years On.* Melbourne: Diabetes Australia.

Allender, S. and Rayner, M. 2007. The burden of overweight and obesity-related ill health in the UK. *Obesity Reviews,* 8(5), 467-73.

Anonymous. 2002. Barriers to children walking and biking to school-United States. 1999. *Morbidity and Mortality Weekly Reports,* 51(32), 701-4.

Araya, R. 2006. Perceptions of social capital and the built environment and mental health. *Social Science and Medicine,* 62(12), 3072-83.

Ball, K., Salmon, J., Giles-Corti, B. and Crawford, D. 2006. How can socio-economic differences in physical activity among women be explained? A qualitative study. *Women and Health,* 43(1), 93-113.

Barclay, L. 2004. Obesity and the built environment: a newsmaker interview with James O. Hill, PhD [Online: Medscape Medical News]. Available at: http://www.medscape.com/viewarticle/478315_print [accessed 10 August 2008].

Bassett, D. 2008. Physical activity of Canadian and American children: a focus on youth in Amish, Mennonite and modern cultures. *Applied Physiology, Nutrition, and Metabolism,* 33(4), 831-5.

Bassett, D., Pucher, J., Buehler, R., Thompson, D.L. and Crouter, S.E. 2008. Walking, cycling and obesity rates in Europe, North America and Australia. *Journal of Physical Activity and Health,* 5(6), 795-814.

Bassett, D., Schneider, P. and Huntington, G. 2004. Physical activity in an old order Amish community. *Medicine and Science in Sports and Exercise,* 36(1), 79-85.

Bassett, D., Tremblay, M., Esliger, D., Copeland, J., Barnes, J. and Huntington, G. 2007. Physical activity and body mass index of children in an old order Amish community. *Medicine and Science in Sports and Exercise,* 39(3), 410-15.

Baum, F. and Palmer, C. 2002. 'Opportunity structures'; urban landscape, social capital and health promotion in Australia. *Health Promotion International,* 17(4), 351-61.

Begg, S., Vos, T., Barker, B., Stevenson, C., Stanley, L. and Lopez, A. 2007. *The Burden of Disease and Injury in Australia, 2003.* Canberra: AIHW.

Bell, A., Ge, K. and Popkin, B. 2002. The road to obesity or the path to prevention: motorized transportation and obesity in China. *Obesity Research,* 10(4), 277-83.

Black, C., Collins, A. and Snell, M. 2001. Encouraging walking: the case of journey-to-school trips in compact urban areas. *Urban Studies,* 38(7), 1121-41.

Black, J.L. and Macinko, J. 2008. Neighborhoods and obesity. *Nutrition Reviews,* 66(1), 2-20.

Bohannon, R. 2007. Number of Pedometer-Assessed Steps Taken Per Day by Adults: A Descriptive Meta-Analysis. *Physical Therapy,* 87(12), 1642-50.

Booth, K.M., Pinkston, M.M. and Poston, W.S. 2005. Obesity and the built environment. *Journal of the American Dietetic Association,* 105(5, Suppl 1), S110-117.

Boslaugh, S.E., Luke, D.A., Brownson, R.C., Naleid, K.S. and Kreuter, M.W. 2004. Perceptions of neighborhood environment for physical activity: is it 'who you are' or 'where you live'? *Journal of Urban Health,* 81(4), 671-81.

Bradshaw, R. 2001. School children's travel – the journey to school. *Geography,* 86, 77-8.

Brown, W.J., Bauman, A.E. and Owen, N. 2009. Stand up, sit down, keep moving: turning circles in physical activity research? *British Journal of Sports Medicine,* 43(2), 86-8.

Brown, W.J., Miller, Y.D. and Miller, R. 2003. Sitting time and work patterns as indicators of overweight and obesity in Australian adults. *International Journal of Obesity and Related Metabolic Disorders,* 27(11), 1340-6.

Brownson, R.C., Kelly, C.M. and Eyler, A.A. 2008. Environmental and policy approaches for promoting physical activity in the United States: a research agenda. *Journal of Physical Activity and Health,* 5(4), 488-503.

Canadian Institute for Health Information. 2006. *Improving the Health of Canadians: Promoting Healthy Weights.* Ottawa, Canada: CIHI.

Carlin, J.B., Stevenson, M.R., Roberts, I., Bennett, C.M., Gelman, A. and Nolan, T. 1997. Walking to school and traffic exposure in Australian children. *Australian and New Zealand Journal of Public Health,* 21(3), 286-92.

Commissioner for Environmental Sustainability. 2008. *Public Transport's Role in Reducing Greenhouse Emissions: A Position Paper.* Melbourne: CES.

Davison, G. 2004. *Car Wars: How The Car Won Our Hearts And Conquered Our Cities.* Crows Nest, NSW: Allen and Unwin.

Dietz, A. and Gortmaker, S. 2001. Preventing obesity in children and adolescents. *Annual Review of Public Health,* 22, 337-53.

DiGuiseppi, C., Roberts, I., Li, L. and Allen, D. 1998. Determinants of car travel on daily journeys to school: cross sectional survey of primary school children. *British Medical Journal,* 316(7142), 1426-8.

Duncan, M. and Mummery, K. 2005. Psychosocial and environmental factors associated with physical activity among city dwellers in regional Queensland. *Preventive Medicine,* 40(4), 363-72.

Dunstan, D.W., Salmon, J., Healy, G.N., Shaw, J.E., Jolley, D., Zimmet, P.Z. and Owen, N., 2007. Association of television viewing with fasting and 2-h postchallenge plasma glucose levels in adults without diagnosed diabetes. *Diabetes Care,* 30(3), 516-22.

Eckersley, R.M. 2001. Losing the battle of the bulge: causes and consequences of increasing obesity. *Medical Journal of Australia,* 174(11), 590-2.

Esliger, D., Tremblay, M., Bassett, D., Barnes, J., Copeland, J. and Huntington, G. 2006. Physical activity profile of old order Amish, old order Mennonite and contemporary-living children, in *Physical Activity and Obesity International Congress Satellite Conference Handbook,* edited by A.P. Hills, N.A. King and N.M. Byrne. Brisbane, Australia: Queensland University of Technology.

Falconer, R. and Giles-Corti, B. 2008. Smart development: designing the built environment for improved access and health outcomes, in *Transitions: Pathways Towards Sustainable Urban Development in Australia,* edited by P.W. Newman. Collingwood, Victoria: CSIRO.

Felton, G.M., Dowda, M., Ward, D.S., Dishman, R.K., Trost, S.G., Saunders, R. and Pate, R.R. 2002. Differences in physical activity between black and white girls living in rural and urban areas. *Journal of School Health,* 72(6), 250-5.

Foster, S. and Giles-Corti, B. 2008. The built environment, neighborhood crime and constrained physical activity: an exploration of inconsistent findings. *Preventive Medicine,* 47(3), 241-51.

Frank, L.D., Andresen, M.A. and Schmid, T.L. 2004. Obesity relationships with community design, physical activity, and time spent in cars. *American Journal of Preventive Medicine,* 27(2), 87-96.

Frank, L.D., Saelens, B.E., Powell, K.E. and Chapman, J.E. 2007. Stepping towards causation: do built environments or neighborhood and travel preferences explain physical activity, driving, and obesity? *Social Science and Medicine,* 65(9), 1898-1914.

French, S., Story, M. and Jeffery, R. 2001. Environmental influences on eating and physical activity. *Annual Review of Public Health,* 22, 309-35.

Friel, S., Chopra, M. and Satcher, D. 2007. Unequal weight: equity oriented policy responses to the global obesity epidemic. *BMJ,* 335(7632), 1241-3.

Frumkin, H., Frank, L. and Jackson, R. 2004. *Urban Sprawl and Public Health: Designing, Planning and Building for Healthy Communities.* Washington DC: Island Press.

Giles-Corti, B., Kelty, S.F., Zubrick, S.R. and Villanueva, K.P. 2009. Encouraging walking for transport and physical activity in children and adolescents: how important is the built environment? *Sports Med,* 39, 995-1009.

Giles-Corti, B. and King, A.C. 2009. Creating active environments across the life course: 'thinking outside the square'. *British Journal of Sports Medicine,* 43(2), 109-113.

Giles-Corti, B., Knuiman, M., Timperio, A., Van Niel, K., Pikora, T.J., Bull, F.C., Shilton, T. and Bulsara, M. 2008. Evaluation of the implementation of a state government community design policy aimed at increasing local walking: design issues and baseline results from RESIDE, Perth Western Australia. *Preventive Medicine,* 46(1), 46-54.

Gleeson, B. 2006. *Australian Heartlands: Making Space for Hope in the Suburbs,* Crows Nest, NSW: Allen and Unwin.

Hall, K. and Porterfield, G. 2000. *Community by Design: New Urbanism for Suburbs and Small Communities.* New York: McGraw Hill.

Harten, N. and Olds, T. 2004. Patterns of active transport in 11-12 year old Australian children. *Australian and New Zealand Journal of Public Health,* 28, 167-72.

Healy, G.N., Wijndaele, K., Dunstan, D.W., Shaw, J.E., Salmon, J., Zimmet, P.Z. and Owen, N. 2008. Objectively measured sedentary time, physical activity, and metabolic risk: the Australian diabetes, obesity and lifestyle study AusDiab. *Diabetes Care,* 31(2), 369-71.

Heart and Stroke Foundation of Canada. 2007. *Position Statement: The Built Environment, Physical Activity, Heart Disease and Stroke.* Ottawa, Canada: HSF Canada.

Jago, R. and Baranowski, T. 2004. Non-curricular approaches for increasing physical activity in youth: a review. *Preventive Medicine,* 39(1), 157-63.

Joint WHO/FAO Expert Consultation on Diet Nutrition and the Prevention of Chronic Diseases. 2003. *Diet, Nutrition and the Prevention of Chronic Diseases: Report of a Joint WHO/FAO Expert Consultation.* WHO Technical Report Series; 916. Geneva, Switzerland: WHO.

Kahn, E.B., Ramsey, L.T., Brownson, R.C., Heath, G.W., Howze, E.H., Powell, K.E., Stone, E.J., Rajab, M.W. and Corso, P. 2002. The effectiveness of interventions to increase physical activity. A systematic review. *American Journal of Preventive Medicine,* 22(4 Suppl), 73-107.

Kelty, S., Giles-Corti, B. and Zubrick, S.R. 2008. Healthy body, healthy mind: why physically active children are healthier physically, psychologically and socially, in *Physical Activity and Children: New Research,* edited by N. Beaulieu. Hauppauge, NY: Nova Science Publishers Inc.

Kjellstrom, T. and Hinde, S. 2007. Car culture, transport policy, and public health, in *Globalization and Health,* edited by I. Kawachi and S. Wamala. New York: Oxford University Press.

Kopelman, P., Jebb, S.A. and Butland, B. 2007. Executive summary: foresight 'tackling obesities: future choices' project. *Obesity Reviews,* 8 (Suppl 1), VI-IX.

Kunstler, J. 1998. *Home from Nowhere; Remaking Our Everyday World for the 21st Century.* New York: Simon and Schuster.

Kwong, J. 2000. Introduction: marketization and privatization in education. *International Journal of Educational Development,* 20(2), 87-92.

Laird, P. and Newman, P. 2001. How we got here: the role of transport in the development of Australia and New Zealand, in *Back On Track: Rethinking Transport Policy in Australia and New Zealand,* edited by P. Laird, P. Newman, M. Bachels and J. Kenworthy. Sydney: UNSW Press.

Lam, L.T. 2001a. Factors associated with parental safe road behavior as a pedestrian with young children in metropolitan New South Wales, Australia. *Accident Analysis and Prevention,* 33(2), 203-10.

Lam, L.T. 2001b. Parental risk perceptions of childhood pedestrian road safety. *Journal of Safety Research,* 32(4), 465-78.

Leslie, E., Saelens, B., Frank, L., Owen, N., Bauman, A., Coffee, N. and Hugo, G. 2005. Residents' perceptions of walkability attributes in objectively different neighbourhoods: a pilot study. *Health and Place,* 11(3), 227-36.

Lopez-Zutina, J., Lee, H. and Friisa, R. 2006. The link between obesity and the built environment: evidence from an ecological analysis of obesity and vehicle miles of travel in California. *Health and Place,* 12(4), 656-64.

Macpherson, A., Roberts, I. and Pless, I. 1998. Children's exposure to traffic and pedestrian injuries. *American Journal of Public Health,* 88(12), 1840-3.

McBride, A. 1990. Mental health effects of women's multiple roles. *American Psychologist,* 45(3), 381-4.

McNeill, L.H., Kreuter, M.W. and Subramanian, S.V. 2006. Social environment and physical activity: a review of concepts and evidence. *Social Science and Medicine,* 63(4), 1011-22.

Merom, D., Bauman, A., Vita, P. and Close, G. 2003. An environmental intervention to promote walking and cycling – the impact of a newly constructed rail trail in western Sydney. *Preventive Medicine,* 36(2), 235-42.

Moon, L., Rahman, N. and Bhatia, K. 1998. *Australia's Children: Their Health and Wellbeing.* Canberra: AIHW.

National Heart Forum. 2007. *Sprawling Suburbs, Spreading Waistlines: Transport and Planning Discouraging Physical Activity* [Online: National Heart Forum]. Available at: http://www.heartforum.org.uk/News_Media_pressreleases_1856.aspx [accessed: 20 January 2009].

National Institute for Health and Clinical Excellence 2008. *Promoting or Creating Built or Natural Environments That Encourage and Support Physical Activity.* London, UK: NICE.

National Preventative Health Taskforce. 2008. *Australia: The Healthiest Country by 2020: A Discussion Paper.* Canberra: Commonwealth of Australia.

National Public Health Partnership. 2001. *Promoting Active Transport: An Intervention Portfolio to Increase Physical Activity as a Means of Transport.* Melbourne: NPHP.

Newman, P., Beatley, T. and Boyer, H. 2009. *Resilient Cities: Responding to Peak Oil and Climate Change.* Washington DC: Island Press.

Ogilvie, D., Foster, C.E., Rothnie, H., Cavill, N., Hamilton, V., Fitzsimons, C.F. and Mutrie, N. 2007. Interventions to promote walking: systematic review. *BMJ,* 334(7605), 1204.

Ogilvie, D., Mitchell, R., Mutrie, N., Petticrew, M. and Platt, S. 2006. Evaluating health effects of transport interventions: methodologic case study. *American Journal of Preventive Medicine,* 31(2), 118-26.

Owen, N., Bauman, A. and Brown, W. 2009. Too much sitting: a novel and important predictor of chronic disease risk? *British Journal of Sports Medicine,* 43(2), 81-3.

Owen, N., Cerin, E., Leslie, E., Dutoit, L., Coffee, N., Frank, L.D., Bauman, A.E., Hugo, G., Saelens, B.E. and Sallis, J.F. 2007. Neighborhood walkability and the walking behavior of Australian adults. *American Journal of Preventive Medicine,* 33(5), 387-95.

Owen, N., Humpel, N., Leslie, E., Bauman, A. and Sallis, J. 2004. Understanding environmental influences on walking: review and research agenda. *American Journal of Preventive Medicine*, 27(1), 67-76.

Papas, M.A., Alberg, A.J., Ewing, R., Helzlsouer, K J., Gary, T.L. and Klassen, A.C. 2007. The built environment and obesity. *Epidemiologic Reviews*, 29(1), 129-43.

Parkes, A. and Kearns, A. 2006. The multi-dimensional neighbourhood and health: a cross-sectional analysis of the Scottish household survey, 2001. *Health and Place*, 12(1), 1-18.

Pendola, R. and Gen, S. 2007. BMI, auto use, and the urban environment in San Francisco. *Health and Place*, 13(2), 551-6.

Perkins, D.D., Mekks, J.W. and Taylor, R.B. 1992. The physical environment of street blocks and resident perception of crime and disorder: implications for theory and measurement. *Journal of Environmental Psychology*, 12(1), 21-34.

Petticrew, M., Cummins, S., Ferrell, C., Findlay, A., Higgins, C., Hoyd, C., Kearns, A. and Spark, L. 2005. Natural experiments: an underused tool for public health? *Public Health*, 119(9), 751-7.

Pickett, K.E., Kelly, S., Brunner, E., Lobstein, T. and Wilkinson, R.G. 2005. Wider income gaps, wider waistbands? An ecological study of obesity and income inequality. *Journal of Epidemiology and Community Health*, 59(8), 670-74.

Polivy, J. and Herman, C.P. 2005. Mental health and eating behaviours: a bi-directional relation. *Canadian Journal of Public Health*, 96(Supplement 3), S43-S46.

Ramanathan, S., Allison, K.R., Faulkner, G. and Dwyer, J.J.M. 2008. Challenges in assessing the implementation and effectiveness of physical activity and nutrition policy interventions as natural experiments. *Health Promotion International*, 23(3), 290-7.

Roberts, I. 1996. Safely to school? *Lancet*, 347(9016), 1642.

Robertson-Wilson, J. and Giles-Corti, B. In Press. Walkabililty, neighbourhood design, and obesity, in *Obesogenic Environments: Complexities, Perceptions and Objective Measures,* edited by T. Townsend, S. Alvanides and A. Lake. UK: Wiley-Blackwell.

Roser, M. 2008. Mueller design may encourage fitness, expert says. *American Statesman* [Online: Hailey Group News, 29 November 2008]. Available at: http://haileygroup.com/news/?p=37 [accessed: 24 March 2009].

Saelens, B., Sallis, J. and Frank, L. 2003. Environmental correlates of walking and cycling: findings from the transportation, urban design, and planning literatures. *Annals of Behavioral Medicine*, 25(2), 80-91.

Sallis, J. and Owen, N. 1999. *Physical Activity and Behavioral Medicine.* Thousand Oaks, CA: Sage Publications.

Schilling, J.M., Giles-Corti, B. and Sallis, J.F. 2009. Connecting active living research and public policy: transdisciplinary research and policy interventions to increase physical activity. *Journal of Public Health Policy*, 30(Suppl 1), S1-S15.

Schilling, J. and Linton, L.S. 2005. The public health roots of zoning: in search of active living's legal genealogy. *American Journal of Preventive Medicine*, 28 (2, Suppl 2), 96-104.

Schneider, M. and Buckley, J. 2002. What do parents want from schools? Evidence from the internet. *Educational Evaluation and Policy Analysis*, 24(2), 133-44.

Sooman, A. and Macintyre, S. 1995. Health and perceptions of the local environment in socially contrasting neighbourhoods in Glasgow. *Health and Place*, 1(1), 15-26.

Stafford, M., Cummins, S., Ellaway, A., Sackcra, A., Wiggins, R. and Macintyre, S. 2007. Pathways to obesity: identifying local, modifiable determinants of physical activity and diet. *Social Science and Medicine*, 65(9), 1882-97.

Statistics Canada. 2006. *General Social Survey: The Internet and the Way We Spend Our Time.* Ottawa, Canada: Statistics Canada.

Stephenson, J. and Bauman, A. 2000. *The Cost of Illness Attributable to Physical Inactivity in Australia.* Canberra: CDHAC and Australian Sports Commission.

Story, M., Giles-Corti, B., Yaroch, A.M., Cummins, S., Frank, L.D,, Huang, T.T.-K. and Lewis, L.B. 2009. Work Group IV: Future directions for measures of the food and physical activity environments. *American Journal of Preventive Medicine,* 36(S4):S182-S188.

Story, M., Kaphingst, K.M., Robinson-O'Brien, R. and Glanz, K. 2008. Creating healthy food and eating environments: policy and environmental approaches. *Annual Review of Public Health*, 29(1), 253-72.

Sugiyama, T., Salmon, J., Dunstan, D.W., Bauman, A.E. and Owen, N. 2007. Neighborhood walkability and TV viewing time among Australian adults. *American Journal of Preventive Medicine*, 33(6), 444-9.

Timperio, A., Crawford, D., Telford, A. and Salmon, J. 2004. Perceptions about the local neighborhood and walking and cycling among children. *Preventive Medicine*, 38(1), 39-47.

Transportation Research Board. 2005. *Does the Built Environment Influence Physical Activity? Examining the Evidence.* Washington, DC: TRB.

Tremblay, M., Barnes, J., Copeland, J. and Esliger, D. 2005. Conquering childhood inactivity: is the answer in the past? *Medicine and Science in Sports and Exercise*, 37(7), 1187-94.

Tremblay, M., Esliger, D., Copeland, J., Barnes, J. and Bassett, D. 2008. Moving forward by looking back: lessons learned from long-lost lifestyles. *Applied Physiology, Nutrition, and Metabolism*, 33(4), 836-42.

Veitch, J., Bagley, S., Ball, K. and Salmon, J. 2006. Where do children usually play? A qualitative study of parents' perceptions of influences on children's active free-play. *Health and Place* 12(4), 383-93.

Von Kardorff, E. and Ohibrecht, H. 2008. Overweight, obesity and eating disorders in adolescents – a socio-somatic reaction to social change? *Journal of Public Health,* 16(6), 429-38.

Walljasper, J. 2007. *The Great Neighborhood Book.* Gabriola Island, BC, Canada: New Society.

Wells, N.M., Ashdown, S.P., Davies, E.H.S., Cowett, F.D. and Yang, Y. 2007. Environment, design and obesity: opportunities for interdisciplinary collaborative research. *Environment and Behavior,* 39(1), 6-33.

Wen, L.M., Orr, N., Millett, C. and Rissel, C. 2006. Driving to work and overweight and obesity: findings from the 2003 New South Wales health survey, Australia. *International Journal of Obesity*, 30(5), 782-6.

Whitley, R. and McKenzie, K. 2005. Social capital and psychiatry: review of the literature. *Harvard Review of Psychiatry*, 13(2), 71-84.

Wood, L. and Giles-Corti, B. 2008. Is there a place for social capital in the psychology of health and place? *Journal of Environmental Psychology*, 28(2), 154-63.

Woodcock, J., Banister, D., Edwards, P., Prentice, A.M. and Roberts, I. 2007. Energy and transport. *Lancet*, 370(9592), 1078-88.

World Health Organization. 2002. *The World Health Report 2002: Reducing Risks, Promoting Healthy Life.* Geneva: WHO.

World Health Organization. 2003. *Investing In Mental Health.* Geneva: WHO.

World Health Organization. 2004. *Global Strategy on Diet, Physical Activity and Health.* Geneva: WHO.

World Health Organization. 2006. *Obesity and Overweight.* Geneva: WHO.

Wood, L. and Giles-Corti, B. 2008. Is there a place for social capital in the psychology of health and place? *Journal of Environmental Psychology*, 28(2), 154-63.

Woodcock, J., Banister, D., Edwards, P., Prentice, A.M. and Roberts, I. 2007. Energy and transport. *Lancet*, 370(9592), 1078-88.

Chapter 8
Understanding the Local Physical Activity Environment and Obesity

Gavin Turrell

Introduction

Since 1997, a number of position papers have promulgated the notion that the obesity epidemic confronting developed societies is in part the result of neighbourhood environments that limit opportunities for physical activity (Hill et al. 2003, Jeffery and Utter 2003, Nestle and Jacobson 2000, Poston and Foreyt 1999; Swinburn et al. 1999, Hill and Peters 1998, Egger and Swinburn 1997). Over this same period, a large number of studies have investigated the association between the neighbourhood environment and physical activity (Saelens and Handy 2008, Wendel-Vos et al. 2007, McCormack et al. 2004, Owen et al. 2004, Humpel et al. 2002). Much of this work is premised on the assumption that neighbourhood environments which are not conducive to physical activity are more likely to exhibit higher levels of overweight and obesity (and by extension, poorer health). In 2008, Mujahid and colleagues published an article which made the following observation: 'Although neighbourhood environments are often identified as potentially important factors in understanding the obesity epidemic, little research provides evidence of this importance' (Mujahid et al. 2008: 1356). Similar conclusions have been reached by writers of other recent articles (Burdette and Hill 2008, Kipke et al. 2007, Papas et al. 2007).

Clearly, based on the foregoing, there are differing views about the contribution of the neighbourhood environment to obesity: on the one hand, there are claims and assumptions being made about the importance of the neighbourhood, and on the other hand, there are those who are in effect challenging these claims by arguing that our knowledge and understanding of the issue is still at a nascent stage and hence indeterminate. The aim of this chapter is to provide a resource that can help inform all perspectives on the topic. Specifically, the chapter provides a review of the evidence on the relationship between bodyweight and the neighbourhood physical activity environment (hereafter referred to as simply the 'neighbourhood environment'). The chapter is divided into three sections. The first provides details about the review methodology – how the articles were identified, the criteria used in their selection, how the review was delimited, and the process used to assess and report on the articles. Section two presents the results of the review. The third section provides a discussion of the evidence, a critique of the extant studies, and makes some suggestions for future research.

The Review Methodology

Identification and Selection of Articles

The review covers all years up to and including 2007, as well as articles published or 'in Press' (available as pre-prints) during January to November 2008. Articles for inclusion in the review were identified using three databases: PubMed, Medline, and Web of Science. A broad range of search terms was initially used including (but not limited to) 'neighbourhood environment', 'built environment', 'neighbourhood', 'local area', 'environment', 'physical activity environment', 'bodyweight', 'overweight', 'obesity', 'body mass index (BMI)', 'waist-hip-ratio', and 'obesogenic'. These terms were used in various combinations. The reference sections of articles were also searched.

After this initial search, a more refined search was performed using different aspects of bodyweight in combination with specific features of the neighbourhood environment previously shown to be related to physical activity, including 'urban design', 'street connectivity', 'walkability', 'crime', 'incivilities', 'density', 'land-use mix', 'social capital', 'opportunity structures', 'transport' and 'aesthetics' (among others).

Once the candidate articles had been identified, they were screened on the basis of one primary selection criterion: the study must have included a measure of the neighbourhood environment and examined this in relation to a measure of bodyweight.

Despite this fairly thorough search and selection process, it is likely that relevant articles were missed hence the review is comprehensive in its coverage but not exhaustive.

Delimitation of the Review

The review is delimited in four main ways. First, it is restricted to research that examines the neighbourhood environment and bodyweight among adults, thus it excludes children and adolescents (which is the topic of Chapter 9).

Second, studies were excluded if they focused on area-units at a higher level than the neighbourhood, such as the state, county, province, or metropolitan area. Clearly, 'neighbourhood' is a slippery concept, and one that is not often defined or discussed in detail in area-based studies. Typically, 'neighbourhood' is defined in terms of smaller-scale administrative units (e.g. census tracts, block groups) or in relation to catchments or buffers around respondents' homes. Sometimes, no actual area-unit akin to a neighbourhood is included in a study: rather, respondents are required to use their own subjective definition of the term, for example, when answering survey items such as 'which of the following recreation facilities are in your neighbourhood'.

Third, neighbourhood environments are likely to influence bodyweight indirectly via physical activity (and diet). The focus of this chapter is on the link between the neighbourhood environment and bodyweight and the now large volume of work examining neighbourhoods and physical activity is only used as supporting material and is not reviewed in depth.

Fourth, the relationship between neighbourhood socioeconomic disadvantage and bodyweight has been investigated in a number of studies, and the findings of this work typically show that residents of disadvantaged areas are more likely to be overweight or obese than their counterparts in advantaged areas (Regidor et al. 2008, McLaren 2007, King et al. 2006, van Lenthe and Mackenbach 2002). Aspects of the neighbourhood environment are likely to contribute to these inequalities: however, this relationship falls outside the scope of the review, which focuses exclusively on the neighbourhood environment and its association with bodyweight more generally.

The Review Process

When assessing and reporting the evidence the following 'rules' were adhered to. First, if the study presented its data stratified by gender, the results are reported for men and women separately. Similarly, if the study presents its data using different measures of bodyweight (e.g. overweight, obesity, or BMI) the results for each measure are reported.

Second, all results are reported – significant and non-significant. To report only the significant associations is likely to contribute to publication bias and give a distorted picture of the relationship between the neighbourhood environment and bodyweight.

Third, study results are often presented without and with adjustment for a range of individual- and area-level covariates (e.g. age, sex, socioeconomic position, race, or ethnicity). When this occurs, only the results of the fully adjusted models are reported. A central issue for the review is whether the neighbourhood environment is associated with bodyweight independent of the characteristics of the individuals who live in the neighbourhoods. Reporting the results of the fully adjusted models is one way of assessing this, although this approach is not without potential inherent problems of over-adjustment (Diez Roux 2004).

Results of the Review

The search and selection processes outlined above identified 29 empirical studies that examined the relationship between the neighbourhood environment and bodyweight. Twenty-three (79 per cent) of these studies were conducted in the US and one each in Australia, Canada, the UK, China, New Zealand, and Europe. Seven studies were published between 2002 and 2005, and 22 between 2006 and 2008. Twenty-eight studies were cross-sectional and one was based on a two-wave cohort study. In addition, three previous reviews were identified

(Black and Macinko 2008, Papas et al. 2007, Booth et al. 2005). These reviews assessed articles published between 1966 and 2006, they focused on both the neighbourhood physical activity environment and the food environment, children and adults, and they included studies that operationalized the environment on the basis of large-scale area-units such as states, metropolitan areas, and counties (as well as neighbourhoods). Given the foregoing, the degree of overlap between these earlier reviews and this present review is limited.

For review purposes, the environmental measures used in the empirical studies were categorized into two broad groupings: one reflecting the influence of the physical environment on bodyweight, and the other the influence of the social environment (see Table 8.1). Each of these two groupings was further subdivided into narrower categories that reflected similar environmental influences on bodyweight, and the evidence is organized and reviewed using these categories.

Table 8.1   Physical and social characteristics of the neighbourhood environment that have been examined in relation to obesity

Physical environment	Social environment
Street connectivity	Crime and safety
Street characteristics	Incivilities
Land-use mix	Social capital
Residential density	
Walkability	
Recreation facilities and opportunity structures	
Transport	
Neighbourhood aesthetics	

Street Connectivity and Bodyweight

Neighbourhoods characterized by a well-connected street network with numerous intersections, few cul-de-sacs, and small block-sizes are defined as having high levels of street connectivity. Town and transport planners have long-argued the benefits of street connectivity such as shorter travel trips, a greater number of destinations within easy reach, traffic calming (a large number of intersecting streets tends to slow traffic), less reliance on private transport and increased travel by public transport, and potentially more interactions among neighbourhood residents and therefore increased social capital. Importantly, street connectivity has been shown in a number of studies to facilitate walking in the neighbourhood (Frank et al. 2005, Saelens et al. 2003a, 2003b, Handy et al. 2002) and hence may be associated with lower levels of overweight and obesity.

The relationship between street connectivity and bodyweight was examined in six studies (Grafova et al. 2008; Smith et al. 2008, Lopez 2007, Rundle et al. 2007, Rutt and Coleman 2005, Frank et al. 2004). Each study used an 'objective' measure of connectivity: these were most commonly derived using Geographic Information Systems (GIS) analysis, and they tended to quantify the number (or density) of intersections contained within a proxy neighbourhood unit such as a census tract, or within a network buffer of varying distances from each respondent's home.

Of the six studies, four found no association between connectivity and bodyweight (Lopez 2007, Rundle et al. 2007, Rutt and Coleman 2005, Frank et al. 2004). Smith et al. (2008) found that for men, greater connectivity was associated with a lower risk of overweight and obesity, but not BMI. For women, greater street connectivity was associated with a lower risk of overweight, but not obesity or BMI. Grafova et al. (2008) found no association between street connectivity and obesity or overweight/obesity among men. For women, connectivity was not associated with obesity: however, women living in areas with high connectivity were less likely to be overweight/obese.

Street Characteristics and Bodyweight

Numerous aspects of the street environment within a neighbourhood are conducive to higher levels of physical activity, and walking and cycling in particular. These include (but are not limited to) the type of street one lives on (e.g. highway or suburban road), the amount and speed of traffic on the streets, whether or not the streets have sidewalks, the quality of those sidewalks (e.g. paved, evenness), whether the streets have lights, and street safety (Cleland et al. 2008, Hong and Farley 2008, Nagel et al. 2008, Ogilvie et al. 2008, Owen et al. 2004). Given these links, we might also expect to observe a relationship between the street environment and overweight and obesity.

The relationship between street characteristics and bodyweight was examined in six studies (Joshu et al. 2008, Boehmer et al. 2007, Wilson et al. 2007, Rutt and Coleman 2005, Catlin et al. 2003, Giles-Corti et al. 2003). Four of these used measures that were based on the respondents' perceptions of the neighbourhood environment (Joshu et al. 2008, Boehmer et al. 2007, Wilson et al. 2007, Catlin et al. 2003) and three assessed characteristics of the local streets using objective (often GIS derived) measures (Boehmer et al. 2007, Rutt and Coleman 2005, Giles-Corti et al. 2003).

Four studies examined the relationship between perceptions of sidewalk availability and bodyweight: three of these found no association (Joshu et al. 2008, Wilson et al. 2007, Catlin et al. 2003). The fourth study found that respondents who disagreed that sidewalks were on most streets in the neighbourhood were more likely to be obese (Boehmer et al. 2007).

Joshu et al. (2008) examined the association between obesity and perceptions of the availability of streetlights and heavy traffic. The likelihood of obesity was significantly higher among those who viewed heavy traffic as a problem in their neighbourhood; however, there was no association with perceptions of streetlight availability.

Rutt and Coleman (2005) used GIS to measure the total length of sidewalks in each respondent's neighbourhood and found no association between sidewalk length and BMI. Boehmer et al. (2007) conducted environmental audits within 400 metres of each respondent's home and found that the likelihood of obesity was *higher* in areas where the greatest proportion of sidewalk segments were flat and even (a counterintuitive result). In addition this study utilized an audit measure of street safety (an index based on the number of traffic lanes, slowing devices, aggressive drivers, crossing aids, and street lighting) and found no association with obesity. Giles-Corti et al. (2003) used objective measures of the neighbourhood streets and found that overweight respondents were more likely to live on a highway and to live in a street with no sidewalks or sidewalks on only one side of the street: however, obesity was not associated with the type of street or the presence of sidewalks.

Land-use Mix and Bodyweight

Land-use mix reflects the extent to which commercial, institutional, office, and residential land uses are situated within close proximity of each other (Frank et al. 2006, Frank 2000). In neighbourhoods comprised of a diversity of land uses, where a variety of facilities and services are centralized, destinations tend to be closer and travel distances shorter, hence reducing reliance on motorized transport and increasing the likelihood of travel by walking or cycling. Mixed use neighbourhoods often attract a critical mass of varied people to an area (e.g. residents, workers, visitors) thus helping to generate social capital, a sense of community, and increased public safety. Importantly, a number of studies show that greater land-use mix is positively associated with both walking for recreation and transport (Frank et al. 2008, Saelens and Handy 2008, Leslie et al. 2005). For these and other reasons, mixed use neighbourhoods are potentially more likely to be associated with lower levels of overweight and obesity.

The relationship between land-use mix and bodyweight was examined in five studies (Li et al. 2008, Rundle et al. 2007, Mobley et al. 2006, Rutt and Coleman 2005, Frank et al. 2004). All of the studies used an objective measure of land-use mix: typically, this consisted of an index that captured the evenness of distribution of different types of land use (e.g. residential, commercial, or institutional) within a network buffer, a proxy neighbourhood unit, or a prescribed distance from a respondent's home. Frank et al. (2004), Mobley et al. (2006), Rundle et al. (2007) and Li et al. (2008) each reported a negative association between land-use mix and bodyweight which is in accord with the theory: the likelihood of being overweight or obese, or of having a higher BMI, was significantly lower in neighbourhoods with a more even or increasing land-use mix. A study by Rutt and Coleman (2005) found the opposite: increasing land-use mix was positively associated with BMI in that residents who lived in neighbourhoods with more commercial and residential buildings had higher BMI.

Residential Density and Bodyweight

Residential density refers to the spatial concentration of people, dwellings, or establishments (e.g. retail) within a given land area. From a town planning and urban design perspective, increased residential density is seen as desirable as it often results in destinations such as shops and workplaces being closer to people's homes, it maximizes the use of neighbourhood infrastructure, saves energy, and provides support for local businesses and services. From a public health perspective, increased residential density sustains alternative modes of travel by facilitating pedestrian activity such as walking and cycling, and the use of public transport (Saelens and Handy 2008). Given this, neighbourhoods with higher levels of residential density should be associated with lower levels of overweight and obesity.

Seven studies examined the relationship between residential density and bodyweight (Smith et al. 2008, Lopez 2007, Pendola and Gen 2007, Ross et al. 2007, Rundle et al. 2007, Rutt and Coleman 2005, Frank et al. 2004). Six of these measured population density (usually population per square mile) and three found no association with obesity (Frank et al. 2004) or BMI (Pendola and Gen 2007, Rutt and Coleman 2005). Rundle et al. (2007) found that higher rates of population density were associated with significantly lower levels of BMI, and Lopez (2007) observed a similar relationship with obesity. Smith et al. (2008) found that higher population density was related to a lower risk of overweight among men, but not for obesity or BMI. Among women, higher population density was associated with a *greater* risk of obesity (a counterintuitive result) but no relationship was found between density and BMI or overweight. Ross et al. (2007) found no association between dwelling density and BMI for either men or women, and Lopez (2007) observed that the likelihood of obesity was significantly lower in neighbourhoods with higher total establishment density, although there was no association between retail density and obesity.

Walkable Neighbourhoods and Bodyweight

As the term implies, a 'walkable' neighbourhood is one that facilitates pedestrian activity such as walking and cycling for recreation or leisure, or to get to and from places (i.e. for transport). Walkable neighbourhoods are usually characterized by the following features: they have an identifiable 'core' such as a main street, public space, or shopping centre; they have higher levels of resident density and are comprised of a mix of different types of land-use; they have open spaces and parks for people to congregate and recreate; they have high levels of street connectivity and smaller block sizes and as a result destinations such as shops, schools, and workplaces are located close to home; and it is an environment that is likely to have high levels of social capital (Wood et al. 2008, Leyden 2003) and be perceived as safe by the resident population. A number of studies have observed that 'walkable' neighbourhoods promote higher levels of recreational and transport-related physical activity (Saelens et al. 2003a, 2003b, Frank et al. 2005, Owen et al. 2007, Ross et al. 2007) hence these types of neighbourhoods are also likely to be conducive to lower levels of overweight and obesity.

The relationship between 'walkable' neighbourhoods and bodyweight was examined in six studies (Berke et al. 2007, Boehmer et al. 2007, Frank et al. 2007, Doyle et al. 2006, Frank et al. 2006, Saelens et al. 2003a, 2003b). Most of the studies used a GIS-derived 'walkability' index that (for example) combined measures of density, land-use mix, street connectivity, block-sizes, and retail floor area. Often, index scores were calculated for a network buffer around each respondent's place of residence. Saelens et al. (2003a) and Doyle et al. (2006) each found that BMI was not significantly lower in more walkable neighbourhoods, and Berke et al. (2007) reported a similar null finding for overweight/obesity. However, Frank and colleagues found that residents of more walkable neighbourhoods had significantly lower levels of BMI (Frank et al. 2006) and obesity (Frank et al. 2007).

Boehmer et al. (2007) used both an audit measure of the number of non-residential destinations within a 400m buffer of each respondent's home, and survey items that captured peoples' perceptions of walking distance to destinations. No association was found between obesity and the observed number of destinations, and perceptions of destinations within a 5-minute walk from home. However, those who perceived no destinations within a 10-minute walk from home were 2.2 times more likely to be obese.

Recreational Facilities, Opportunity Structures and Bodyweight

There is now a large and growing literature that has examined the relationship between recreational infrastructure and opportunity structures and physical activity. This research has investigated a diverse range of factors, but the main focus has been on the availability of recreational facilities and opportunity structures within the neighbourhood (e.g. bike paths, gyms, recreation centres, parks, playgrounds, and open spaces) and their accessibility and convenience. The findings of this body of work have been the subject of a number of reviews (Wendel-Vos et al. 2007, Davison and Lawson 2006, Duncan et al. 2005, McCormack et al. 2004, Owen et al. 2004, Humpel et al. 2002). Whilst the quality and completeness of these reviews varies considerably (Gebel et al. 2007) some evidence suggests that the neighbourhood recreational infrastructure plays a role in facilitating participation in physical activity (although interestingly, the review articles reach somewhat different conclusions about the nature and extent of that role). Moreover, positive associations between the recreational environment and physical activity have been reported for both adults (Diez Roux et al. 2007, Ewing 2005, Owen et al. 2004, Humpel et al. 2002) and children (Davison and Lawson 2006) and in terms of respondents' perceptions and the objective attributes of the neighbourhood (Davison and Lawson 2006, Duncan et al. 2005, McCormack et al. 2004). Accepting that recreational facilities (variously defined) make at least some contribution to physical activity, we would by extension expect to observe lower rates of overweight and obesity in neighbourhoods with a 'better' recreational infrastructure and greater opportunity structures.

Relationships between recreational facilities, opportunity structures and bodyweight were examined in nine studies (Joshu et al. 2008, Witten et al. 2008, Boehmer et al. 2007, Wilson et al. 2007, Mobley et al. 2006, Poortinga 2006, Rutt and Coleman 2005, Catlin et al. 2003, Giles-Corti et al. 2003). Five of these used an objective indicator of facilities in the neighbourhood. Mobley et al. (2006) reported lower BMI in areas with more fitness facilities. Giles-Corti et al. (2003) found that obese respondents were more likely to have poor access to four or more recreation facilities. Boehmer et al. (2007) reported that having fewer recreational facilities within close proximity was associated with a lower likelihood of obesity among women but not men, and Rutt and Coleman (2005) found that BMI was not related to either the number of facilities or their distance from the respondent's home. Witten et al. (2008) examined the association between BMI and access to parks and beaches measured in minutes of car travel. No association was found between BMI and access to a park; however, respondents with the best beach access had significantly lower BMI.

Six studies included a measure of perceived availability or access to recreation facilities. Joshu et al. (2008) and Poortinga (2006) found that perceived availability was associated with a significantly lower likelihood of obesity. Catlin et al. (2003) found that the perceived absence of public outdoor exercise facilities was associated with an increased likelihood of overweight, but no relationship was observed for the perceived presence of walking or biking trails, parks, or indoor exercise facilities. Giles-Corti et al. (2003) reported that overweight respondents were more likely to perceive there were no walking or cycling paths within walking distance, although no association was found with obesity. Wilson et al. (2007) found that the perceived availability of recreation facilities was not associated with obesity (vs. underweight/normal weight); however, those who perceived facilities to be available were more likely to be overweight than obese. Boehmer et al. (2007) observed no relationship between perceptions of availability and distance to recreation facilities in the neighbourhood and obesity.

Transport and Bodyweight

Studies from the transportation and planning literatures have shown that many aspects of neighbourhood design (e.g. street connectivity, block-size, density, and land-use mix) play an important role in influencing transport patterns and travel choices (Frank et al. 2007, Saelens et al. 2003a, 2003b, Frank 2000). In particular, it is argued that land-use policies have often produced neighbourhood environments that fail to facilitate the use of public transport or pedestrian-related physical activity and, instead, inadvertently promote the use of private vehicles. By extension, neighbourhoods with high levels of private vehicle ownership and/ or more restricted access to public transport nodes (e.g. bus stops, train stations) should theoretically be characterized by higher levels of overweight and obesity.

Six studies examined the association between transportation and bodyweight (Boehmer et al. 2007, Pendola and Gen 2007, Rundle et al. 2007, Frank et al. 2004, Giles-Corti et al. 2003, Bell et al. 2002). Rundle et al. (2007) found that increasing accessibility of both bus stops and subways (i.e. density per square kilometre) were associated with lower levels of BMI. However, Boehmer et al. (2007) found no association between obesity and the percentage of sidewalk segments with a bus or transit stop within a 400m buffer of each respondent's home. This same study also found no association between obesity and perceptions about the availability of transit stops in the neighbourhood. Pendola and Gen (2007) found that BMI was significantly higher among respondents who reported using an automobile 'most or all of the time' to travel to the grocery store, and to work or school; however, there was no association between BMI and automobile use to visit friends or for other shopping purposes. Giles-Corti et al. (2003) found that respondents who had access to a motor vehicle 'all of the time' were *less* likely to be obese (a counterintuitive result). Bell et al. (2002) examined the influence of changes in motorized vehicle ownership between 1989-1997 on obesity and weight gain in China as a consequence of recent trends towards greater urbanization and economic development. Thirteen per cent of households acquired a motor vehicle between 1989 and 1997. Compared with those whose vehicle ownership did not change, men who acquired a vehicle experienced an average 1.8kg weight gain and were twice as likely to become obese. No significant effects of vehicle acquisition were observed for women. In 1997, household ownership of a motorized vehicle was associated with a significantly greater likelihood of being obese in men and women compared with those who did not own a vehicle. Based on data collected using a 2-day travel diary, Frank et al. (2004) found that time spent in a car as a passenger or driver was positively related with obesity. An extra hour in the car was associated with a 6 per cent increased likelihood of being obese.

Neighbourhood Aesthetics and Bodyweight

A number of studies (Santos et al. 2008, Strach et al. 2007, Cerin et al. 2006) and reviews (Brennan Ramirez et al. 2006, McCormack et al. 2004) have demonstrated that aesthetically pleasing environments, and green spaces in particular are conducive to higher levels of participation in both recreational and transport-related physical activity (Li et al. 2008, Tilt et al. 2007, Takano et al. 2002). Based on this evidence, we might reasonably expect to find that neighbourhood aesthetics and greenery are related to lower levels of overweight and obesity.

Five studies examined the association between neighbourhood aesthetics and bodyweight (Joshu et al. 2008, Boehmer et al. 2007, Tilt et al. 2007, Wilson et al. 2007, Ellaway et al. 2005). Three of these were based on respondents' perceptions about neighbourhood aesthetics or pleasantness. Boehmer et al. (2007) found that the likelihood of obesity was higher among respondents who reported that their neighbourhood was 'not pleasant' for physical activity, and who disagreed that there were interesting things to look at while walking in the neighbourhood. By contrast, Joshu et al. (2008) found no association between obesity and perceptions of the availability of enjoyable scenery in the neighbourhood, and Wilson et al. (2007) found no relationship between perceptions of neighbourhood pleasantness and either overweight or obesity.

Two studies used an objective measure of the level of vegetation and greenery in the neighbourhood (based on residential audits or satellite imagery). Ellaway et al. (2005) found that respondents living in greener neighbourhoods were about 40 per cent less likely to be overweight/obese, and Tilt et al. (2007) found that higher levels of neighbourhood greenness were associated with significantly lower levels of BMI.

Neighbourhood Crime, Safety and Bodyweight

Fear of crime and concerns about personal safety could reasonably be expected to predict whether or not a person uses their neighbourhood environment for recreational and/or transport-related physical activity. During the last few years, a number of quantitative studies have investigated the association between crime, safety, and physical activity, and a recent review of this evidence concluded that findings to date are often mixed and inconsistent (Foster and Giles-Corti 2008). However, this same review also noted that for certain subgroups of the population such as women and the elderly, crime-related safety may inhibit participation in physical activity. Possibly then, crime and personal safety may show an association with bodyweight.

Six studies examined the association between bodyweight, crime, and safety (Burdette and Hill 2008, Mujahid et al. 2008, Wilson et al. 2007, Doyle et al. 2006, Glass et al. 2006, Mobley et al. 2006). Three studies used an objective measure of crime and/or safety. Doyle et al. (2006) found that lower crime rates (all types) were associated with significantly lower levels of BMI among adults aged 18 years and older, and Mobley et al. (2006) found that higher rates of robbery arrests per 100,000 population were associated with significantly higher BMI among women aged 30-78 years. Glass et al. (2006: 457) created a composite 12-item scale using census variables and non-census measures of 'neighbourhood psychosocial hazards', defined as 'stable and visible features of neighbourhood environments that give rise to a heightened state of vigilance, alarm, or fear in residents'. In a sample of adults aged 50-70 years, the likelihood of obesity increased significantly in a linear manner across each quartile of the scale relative to the lowest quartile of psychosocial hazard.

Three studies were based on respondents' perceptions of crime and safety (and social disorder more generally). Wilson et al. (2007) found no association between general perceptions of neighbourhood safety and the likelihood of being either overweight or obese among adults aged 18-75 years. Burdette and Hill (2008) used a 3-item 'neighbourhood disorder' scale that captured perceptions of crime, noise, and cleanliness and found that heightened perceptions of disorder were associated with an increased risk of obesity. Mujahid et al. (2008) created a neighbourhood social environment scale that combined measures of safety, violent crime, aesthetic quality, and social cohesion. For women, living in a better social environment was not associated with BMI. For men, the findings were significant but in the opposite direction to that expected: males living in a better social environment had a higher BMI than their counterparts residing in worse environments.

Neighbourhood Incivilities and Bodyweight

Neighbourhoods characterized by extant incivilities such as graffiti, abandoned buildings, pollution, unattended dogs, or the presence of litter or broken glass in the local streets are likely to communicate a sense of social and physical disorder to the resident population, and hence invoke concerns about personal safety. As a consequence, these types of neighbourhoods may be less conducive to physical activity, as a number of studies have suggested (Heinrich et al. 2007, Lee et al. 2005, Brownson et al. 2001). By extension, neighbourhoods with visible incivilities may be associated with higher levels of overweight and obesity.

Four studies examined the association between incivilities and bodyweight (Joshu et al. 2008, Boehmer et al. 2007, Poortinga 2006, Ellaway et al. 2005). Two of these measured incivilities using an objective indicator. Boehmer et al. (2007) used a physical disorder scale that measured the presence of beer/liquor bottles or cans, cigarette/cigar butts or packages, condoms, drug-related paraphernalia, garbage/litter/broken glass, abandoned cars, graffiti, and broken windows, within a 5-minute walk (400m) from each respondent's home. Respondents in the top quartile of this measure (i.e. most disorder) were four times more likely to be obese than their counterparts in the bottom quartile (i.e. least disorder). In addition, an environmental audit was used to estimate the percentage of sidewalk segments with some or a lot of garbage, litter, or broken glass; respondents with the highest proportion of segments (i.e. 50 or more) were 3.7 times more likely to be obese than those with the least segments (i.e. 0-5). Ellaway et al. (2005) used a residential audit to measure the amount of graffiti, litter, and dog mess in the neighbourhood and found that respondents who lived in areas with high levels of incivilities were about 40 per cent more likely to be overweight/obese.

Three studies were based on respondents' perceptions of incivilities. Boehmer et al. (2007) collected information on perceptions of the presence of garbage, litter, or broken glass in the neighbourhood (1 item), and whether the neighbourhood was maintained (1 item); neither perception was associated with obesity. Joshu et al. (2008) found no association between obesity and respondent's perceptions of whether the neighbourhood had unattended dogs and foul air from car/factories. Poortinga (2006) however found that the likelihood of obesity was higher if teenagers and vandalism were perceived as being a problem in the local area.

Social Capital and Bodyweight

Social capital has been defined as 'features of social organization such as trust, norms, and networks that can improve the efficiency of society by facilitating coordinated actions' (Putnam 1993: 167). The concept has been measured in myriad ways, including (but not limited to) perceptions of trust and reciprocity, altruism, social integration, civic participation, and organizational memberships (Lochner et al. 1999). Studies have examined the relationship between social capital and mortality (Lochner et al. 2003, Skrabski et al. 2003, Kawachi et al. 1997), self-rated health (Subramanian et al. 2001, 2002, Kawachi et al. 1999) and health service use (Hendryx et al. 2002) and each of these show that health and well-being are better in areas with higher levels of social capital. Not all studies, however, have demonstrated a relationship between social capital and health (e.g. Turrell et al. 2006; Kelleher et al. 2004), and its contribution to population health has been contested (Lynch et al. 2000) and extensively debated (Muntaner et al. 2001, Pearce and Davey Smith 2003). In recent years, researchers have raised the possibility that social capital might be important for physical activity (McNeill et al. 2006). Walkable neighbourhoods are characterized by higher levels of social capital (Wood et al. 2008, Leyden 2003) and social capital and physical activity have been found to be positively associated (King 2008, Mummery et al. 2008). By extension, higher levels of social capital might therefore be related to lower levels of overweight and obesity.

Three studies examined the association between social capital and bodyweight (Wilson et al. 2007, Poortinga 2006, Cohen et al. 2006). Cohen et al. (2006) used a collective-efficacy scale that measured social cohesion and informal social control and found that both BMI and overweight were significantly lower in neighbourhoods with higher collective efficacy. Poortinga (2006) measured social capital on the basis of perceived neighbourhood friendliness (1 item) and social trust (1 item). The former was not associated with obesity; however, higher levels of trust were associated with a lower likelihood of obesity. Wilson et al. (2007) used a single-item measure of perceived trust in neighbours and found no association with the likelihood of being obese (vs. underweight/normal weight); however, those who reported trust in neighbours were more likely to be overweight than obese.

Discussion

Based on the available evidence, can we conclude that the neighbourhood physical activity environment influences the likelihood of being overweight or obese? In one respect, answering this question with any certainty or confidence is difficult. The evidence was mixed and inconsistent, and for each statistically significant association that was reported there was often an accompanying null finding. Moreover, this pattern tended to be observed irrespective of whether physical or social environmental factors were examined, or whether these were operationalized using objective indicators or respondents' perceptions. However, there is another way of interpreting the pattern of evidence which lends itself to the conclusion that the neighbourhood environment influences bodyweight. When statistically significant differences were found, they were usually (although not always) in the expected direction; in short, a 'better' neighbourhood tended to be associated with a lower likelihood of being overweight or obese. If we put aside the null-findings (and admittedly run the risk of over-stating and simplifying things) and use the positive evidence to devise a neighbourhood that was conducive to a healthy bodyweight then it would probably have the following characteristics:

- good street connectivity
- streets with well maintained footpaths on both sides, with low volumes of slow traffic, and that are safe and well lit at night
- a variety of land uses
- high levels of residential density
- facilitates pedestrian activity (i.e. a walkable neighbourhood)
- a variety of available and readily accessible recreation facilities and opportunity structures
- good access to public transport and a lower reliance on private motorized vehicles
- lots of greenery and interesting things to see and do
- low levels of crime and heightened perceptions of public safety
- low levels of visible incivilities, and
- high levels of social capital

This 'ideal-type' neighbourhood has a clear parallel in the findings of studies that have investigated the relationship between the neighbourhood environment and physical activity; the same neighbourhood factors listed above have often been shown to be associated with higher levels of physical activity (Saelens and Handy 2008, Wendel-Vos et al. 2007, McCormack et al. 2004, Owen et al. 2004, Humpel et al. 2002). Therefore, it seems reasonable to conclude that the evidence from this present review provides indirect support for the notion that physical activity represents an important behavioural pathway between the neighbourhood environment and bodyweight. A number of recent studies have provided more direct evidence of this by showing that the association between the neighbourhood environment and bodyweight is attenuated after adjustment for physical activity (Mujahid et al. 2008).

Irrespective of which perspective on the evidence one takes, the fact remains that the findings of studies were often inconsistent. Reconciling these differences and hence being in a position to generalize about the relationship between the neighbourhood environment and bodyweight is difficult. In part, these difficulties stem from a limited evidence base. Although 29 studies were identified (which might seem like a reasonable number in total) the number that examined any specific aspect of the physical and social environment was relatively small.

Numerous other factors contribute to the inconsistencies: the studies differed in their conceptualization and measurement of the neighbourhood, the environmental exposures, and bodyweight. For example, some studies used self-report of bodyweight whereas others used measured bodyweight: and bodyweight was variously indicated on the basis of overweight, obesity, overweight/obesity, and BMI. Differences between the studies in sample-size, analytic methods (e.g. single-level or multilevel), and the individual-level variables used as controls are also likely to have contributed to inconsistencies between studies in their findings.

Studies investigating the association between the neighbourhood environment and bodyweight have mostly been conducted in the US (23 out of 29 studies). There were no non-US studies that examined bodyweight in relation to street connectivity, land use mix, and walkability, and six of the seven studies that examined residential density originated from the US. This raises important questions about whether and to what extent the findings of US research can be generalized to other national contexts. There are many historical, cultural, political, socioeconomic, and geospatial differences between countries such as the US, New Zealand, Australia, and the UK which might play out differently vis-à-vis the neighbourhood environment and bodyweight. Interestingly, a recent article by Cummins and Macintyre (2006) reached the conclusion that most of the evidence supporting a link between food environments and obesity originates from the US, and that studies from other countries report weaker and less consistent findings. Clearly more research is needed in non-US contexts to ascertain if the seeming 'American exceptionalism' found in relation to the neighbourhood food environment also applies to the physical activity environment.

All but one of the 29 studies was based on a cross-sectional design, hence at present our interpretation of the evidence is limited to describing associations, and causal attributions cannot be made. Cross-sectional studies cannot tell us if the neighbourhood environment influences the likelihood of overweight and obesity, or if health conscious individuals in the 'normal' weight-range move to neighbourhoods that support active lifestyles. Only longitudinal studies can truly disentangle causation and selection effects, and these types of designs are urgently needed if we are going to have a stronger foundation on which to develop effective policies and interventions to address the obesity epidemic. Greater use of natural experiment studies could also add to our understanding of the directionality of the environment-bodyweight relationship. These and other technical and methodological issues relating to the challenges of establishing a casual relationship between the neighbourhood environment and obesity are discussed in Chapter 12.

This review of the evidence has suggested many other ways in which our understanding of the relationship between the neighbourhood environment and obesity could be advanced. First, most of the studies adjusted their analysis for sex; however, there were a few studies that carried out sex-specific analyses and they sometimes found a different pattern of association between the neighbourhood environment and bodyweight for men and women (Grafova et al. 2008, Mujahid et al. 2008, Boehmer et al. 2007). These latter studies point to possible important interactions between the neighbourhood environment, sex, and bodyweight (i.e. men and women may engage with their neighbourhood environments in different ways) and this should be explored further in future research. A similar reasoning could be used to justify a greater focus on different stages of the adult lifecourse. Many studies were based on all-age samples (e.g. 18 years or older) yet it is conceivable that young adults, the middle-aged, and the elderly are differentially affected by neighbourhood influences (Li et al. 2005).

Second, many of the studies defined a neighbourhood on the basis of a small-scale administrative unit such as a post-code area, census tract, or mesh-block. These choices were often made for reasons of sampling and analytic convenience rather than being underpinned by an explicit theory linking the neighbourhood environment with bodyweight. Other studies defined a neighbourhood using a catchment or buffer around each respondent's place of residence (and the size of these areas varied between studies). It is unlikely that 'these pragmatic definitions (of neighbourhood)…adequately operationalized the true spaces where people live and interact' (Black and Macinko 2008: 15). As a consequence, associations between the neighbourhood environment and bodyweight were likely to have been underestimated. Had it been possible to derive an area-unit based on people's actual reports of what in their minds constituted their local neighbourhood and what was socially and culturally meaningful in terms of their physical activity behaviours then we might have observed stronger and more consistent neighbourhood effects on bodyweight. Advancing our knowledge and understanding of how residents define and use their neighbourhoods in ways that influence their health and related behaviours will require greater use of qualitative research.

Third, the relationship between the neighbourhood environment and bodyweight might show a different pattern of association depending on other contextual factors such as socioeconomic disadvantage and rurality. It can't be assumed for example that the general relationships identified in this review will hold equally in rich and poor neighbourhoods or between urban and rural areas. Indeed, Lopez and Hynes (2006: 25) have noted an interesting paradox that illustrates this contextual heterogeneity:

> Persistent trends in overweight and obesity have resulted in a rapid research effort focused on built environment, physical activity, and overweight. Much of the focus of this research has been on the design and form of suburbs. It suggests that several features of the suburban built environment such as low densities, poor street connectivity and the lack of sidewalks are associated with decreased physical activity and an increased risk of being overweight. But compared to suburban residents, inner city populations have higher rates of obesity and inactivity despite living in neighbourhoods that are dense, have excellent street connectivity and who's streets are universally lined with sidewalks.

As challenging as these 'counterfactuals' are, our understanding of them is essential if we are to obtain a more sophisticated and realistic appreciation of the complexities that characterize the interactions between the neighbourhood environment, behaviours, and outcomes such as bodyweight.

Finally, future research on the neighbourhood environment and bodyweight would benefit from an increased multidisciplinary approach. There has been a long tradition in urban design and transport planning of investigating the links between the built environment and behaviour (Ewing 2005, Handy et al. 2002, Frank 2000). In more recent years there has been a move towards fostering greater collaborations between researchers in these disciplines and their counterparts in the health, behavioural, and social sciences (Rodriguez et al. 2006, Saelens et al. 2003a, 2003b, Sallis et al. 2002). Continued efforts in this direction will be necessary if we are to significantly advance the quality and utility of the evidence base on neighbourhood environments and bodyweight.

Conclusion

During the last few decades, rates of overweight and obesity have risen dramatically in many developed societies and the problem now constitutes a major public health epidemic. The causes of this epidemic are complex and multi-factorial and range in scale from the individual to the societal. As this review has shown, there is emerging evidence that the neighbourhood physical activity environment also has a role to play. Factors such as street connectivity, street characteristics, land-use mix, residential density, recreational facilities and opportunity structures, transport, aesthetics, crime and safety, incivilities, and social capital were each found to be associated with overweight and obesity. Moreover, physical activity appears to be an important mediator of these relationships. However, much work remains to be done: the evidence base is at a nascent stage, it is small in size and US focused; there are numerous conceptual, measurement, and analytic challenges to be addressed; and results are often mixed, inconsistent and sometimes counterintuitive. One of the great promises of the neighbourhood environment is that it has the potential to reach and influence large numbers of people simultaneously and hence make a significant measurable impact on overweight and obesity (and by extension health) at the population level. From where we sit at present, we still have a way to go before this promise is likely to be realized.

References

Bell, A.C., Ge, K. and Popkin, B. 2002. The road to obesity or the path to prevention: motorized transportation and obesity in China. *Obesity Research,* 10(4), 277-83.

Berke, E.M., Koepsell, T.D., Moudon, A.V., Hoskins, R.E. and Larson, E.B. 2007. Association of the built environment with physical activity and obesity in older persons. *American Journal of Public Health,* 97(3), 486-92.

Black, J. and Macinko, J. (2008), 'Neighborhoods and obesity' *Nutrition Reviews*, 66(1), 2-20.

Boehmer, T.K., Hoehner, C.M., Deshpande, A.D., Brennan Ramirez, L.K. and Brownson, R.C. 2007. Perceived and observed neighborhood indicators of obesity among urban adults. *International Journal of Obesity,* 31, 968-77.

Booth, K.M., Pinkston, M.M. and Poston, W.S. 2005. Obesity and the built environment. *Journal of the American Dietetic Association*, 105(5 Suppl 1), S110-7.

Brennan Ramirez, L.K., Hoehner, C.M., Brownson, R.C., et al. 2006. Indicators of activity-friendly communities: an evidence-based consensus process. *American Journal of Preventive Medicine* 31(6), 515-24.

Brownson, R.C., Baker, E.A., Housemann, R.A., Brennan, L.K. and Bacak, S.J. 2001. Environmental and policy determinants of physical activity in the United States. *American Journal of Public Health,* 91(12), 1995-2003.

Burdette, A.M. and Hill, T.D. 2008. An examination of processes linking perceived neighborhood disorder and obesity. *Social Science and Medicine,* 67(1), 38-46.

Catlin, T.K., Simoes, E.J. and Brownson, R.C. 2003. Environmental and policy factors associated with overweight among adults in Missouri. *American Journal of Health Promotion,* 17(4), 249-58.

Cerin, E., Saelens, B., Sallis, J. and Frank, L. 2006. Neighborhood Environment Walkability Scale: validity and development of a short form. *Medicine and Science in Sports and Exercise*, 38(9), 1682-91.

Cleland, V.J., Timperio, A. and Crawford, D. 2008. Are perceptions of the physical and social environment associated with mothers' walking for leisure and for transport? A longitudinal study. *Preventive Medicine,* 47(2), 188-93.

Cohen, D.A., Finch, B.K., Bower, A. and Sastry, N. 2006. Collective efficacy and obesity: the potential influence of social factors on health. *Social Science and Medicine,* 62(3), 769-78.

Cummins, S. and Macintyre, S. 2006. Food environments and obesity – neighbourhood or nation? *International Journal of Epidemiology,* 35(1), 100-4.

Davison, K.K. and Lawson, C.T. 2006. Do attributes in the physical environment influence children's physical activity? A review of the literature. *International Journal of Behavioral Nutrition and Physical Activity,* 3:19.

Diez Roux, A.V. 2004. Estimating neighborhood health effects: the challenges of causal inference in a complex world. *Social Science and Medicine,* 58(10), 1953-60.

Diez Roux, A., Evenson, K., McGinn, A., et al. 2007. Availability of recreational resources and physical activity in adults. *American Journal of Public Health,* 97(3), 493-9.

Doyle, S., Kelly-Schwatz, A., Schlossberg, M. and Stockard, J. 2006. Active community environments and health. *Journal of the American Planning Association,* 72(1), 19-31.

Duncan, M.J., Spence, J.C. and Mummery, W.K. 2005. Perceived environment and physical activity: a meta-analysis of selected environmental characteristics. *International Journal of Behavioral Nutrition and Physical Activity,* 2:11.

Egger, G. and Swinburn, B. 1997. An 'ecological' approach to the obesity pandemic. *British Medical Journal,* 315(7106), 477-80.

Ellaway, A., Macintyre, S. and Bonnefoy, X. 2005. Graffiti, greenery, and obesity in adults: secondary analysis of European cross sectional survey. *British Medical Journal* 331(7517), 611-12.

Ewing, R. 2005. Can the physical environment determine physical activity levels? *Exercise and Sport Science Reviews,* 33(2), 69-75.

Foster, S. and Giles-Corti, B. 2008. The built environment, neighborhood crime and constrained physical activity: an exploration of inconsistent findings. *Preventive Medicine,* 47(3), 241-51.

Frank, L., Andresen, M. and Schmid, T. 2004. Obesity relationships with community design, physical activity, and time spent in cars. *American Journal of Preventive Medicine,* 27(2), 87-96.

Frank, L., Saelens, B., Powell, K. and Chapman, J. 2007. Stepping towards causation: do built environments or neighborhood and travel preferences explain physical activity, driving, and obesity? *Social Science and Medicine,* 65(9), 1898-14.

Frank, L., Sallis, J., Conway, T., Chapman, J., Saelens, B. and Bachman, W. 2006. Many pathways from land use to health. *Journal of the American Planning Association,* 71(1), 75-87.

Frank, L.D. 2000. Land use and transportation interaction: Implications on public health and quality of life. *Journal of Planning Education and Research,* 20, 6-22.

Frank, L.D., Kerr, J., Sallis, J.F., Miles, R. and Chapman, J. 2008. A hierarchy of sociodemographic and environmental correlates of walking and obesity. *Preventive Medicine,* 47(2), 172-8.

Frank, L.D., Schmid, T.L., Sallis, J.F., Chapman, J. and Saelens, B.E. 2005. Linking objectively measured physical activity with objectively measured urban form: findings from SMARTRAQ. *American Journal of Preventive Medicine,* 28(2 Suppl 2), 117-25.

Gebel, K., Bauman, A.E. and Petticrew, M. 2007. The physical environment and physical activity: a critical appraisal of review articles. *American Journal of Preventive Medicine,* 32(5), 361-9.

Giles-Corti, B., Macintyre, S., Clarkson, J.P., Pikora, T. and Donovan, R.J. 2003. Environmental and lifestyle factors associated with overweight and obesity in Perth, Australia. *American Journal of Health Promotion,* 18(1), 93-102.

Glass, T.A., Rasmussen, M.D. and Schwartz, B.D. 2006. Neighborhoods and obesity in older adults: the Baltimore Memory Study. *American Journal of Preventive Medicine,* 31(6), 455-63.

Grafova, I.B., Freedman, V.A., Kumar, R. and Rogowski, J. 2008, Neighborhoods and obesity in later life. *American Journal of Public Health,* 98(11), 2065-71.

Handy, S.L., Boarnet, M.G., Ewing, R. and Killingsworth, R.E. 2002. How the built environment affects physical activity: views from urban planning. *American Journal of Preventive Medicine,* 23(2 Suppl), 64-73.

Heinrich, K.M., Lee, R.E., Suminski, R.R., et al. 2007. Associations between the built environment and physical activity in public housing residents. *International Journal of Behavioral Nutrition and Physical Activity,* 4:56.

Hendryx, M.S., Ahern, M.M., Lovrich, N.P. and McCurdy, A.H. 2002. Access to health care and community social capital. *Health Services Research,.* 37(1), 87-103.

Hill, J.O. and Peters, J.C. 1998. Environmental contributions to the obesity epidemic. *Science,* 280(5368), 1371-4.

Hill, J.O., Wyatt, H.R., Reed, G.W. and Peters, J.C. 2003. Obesity and the environment: where do we go from here? *Science,* 299(5608), 853-5.

Hong, T. and Farley, T.A. 2008. Urban residents' priorities for neighborhood features. A survey of New Orleans residents after Hurricane Katrina. *American Journal of Preventive Medicine,* 34(4), 353-6.

Humpel, N., Owen, N. and Leslie, E. 2002. Environmental factors associated with adults' participation in physical activity: a review. *American Journal of Preventive Medicine,* 22(3), 188-99.

Jeffery, R.W. and Utter, J. 2003. The changing environment and population obesity in the United States. *Obesity Research,* 11(Suppl), 12S-22S.

Joshu, C.E., Boehmer, T.K., Brownson, R.C. and Ewing, R. 2008. Personal, neighbourhood and urban factors associated with obesity in the United States. *Journal of Epidemiology and Community Health,* 62, 202-8.

Kawachi, I., Kennedy, B.P., Lochner, K. and Prothrow-Stith, D. 1997. Social capital, income inequality, and mortality. *American Journal of Public Health,* 87(9), 1491-8.

Kawachi, I., Kennedy, B.P. and Glass, R. 1999. Social capital and self-rated health: a contextual analysis. *American Journal of Public Health,* 89(8), 1187-93.

Kelleher, C.C., Lynch, J., Harper, S., Tay, J.B. and Nolan, G. 2004. Hurling alone: how social capital failed to save the Irish from cardiovascular disease in the United States. *American Journal of Public Health,* 94(12), 2162-9.

King, D. 2008. Neighborhood and individual factors in activity in older adults: results from the neighborhood and senior health study. *Journal of Aging Physical Activity,* 16(2), 144-70.

King, T., Kavanagh, A.M., Jolley, D., Turrell, G. and Crawford, D. 2006. Weight and place: a multilevel cross-sectional survey of area-level social disadvantage and overweight/obesity in Australia. *International Journal of Obesity,* 30, 281-7.

Kipke, M.D., Iverson, E., Moore, D., et al. 2007. Food and park environments: neighborhood-level risks for childhood obesity in east Los Angeles. *Journal of Adolescent Health,* 40(4), 325-33.

Lee, R., Booth, K., Reese-Smith, J., Regan, G. and Howard, H. 2005. The Physical Activity Resource Assessment (PARA) instrument: evaluating features, amenities and incivilities of physical activity resources in urban neighborhoods. *International Journal of Behavioral Nutrition and Physical Activity,* 2:13.

Leslie, E., Saelens, B., Frank, L., et al. 2005. Residents' perceptions of walkability attributes in objectively different neighbourhoods: a pilot study. *Health and Place,* 11(3), 227-36.

Leyden, K.M. 2003. Social capital and the built environment: the importance of walkable neighborhoods. *American Journal of Public Health,* 93(9), 1546-51.

Li, F., Fisher, K.J., Bauman, A., et al. 2005. Neighborhood influences on physical activity in middle-aged and older adults: a multilevel perspective. *Journal of Aging Physical Activity,* 13(1), 87-114.

Li, F., Harmer, P., Cardinal, B., et al. 2008. Built environment, adiposity, and physical activity in adults aged 50-75. *American Journal of Preventive Medicine,* 35(1), 38-46.

Lochner, K., Kawachi, I. and Kennedy, B. 1999. Social capital: a guide to its measurement. *Health and Place,* 5(4), 259-70.

Lochner, K., Kawachi, I., Brennan, R. and Buka, S. 2003. Social capital and neighborhood mortality rates in Chicago. *Social Science and Medicine,* 56(8), 1797-805.

Lopez, R.P. and Hynes, H.P. 2006. Obesity, physical activity, and the urban environment: public health research needs. *Environment and Health,* 5, 25.

Lopez, R.P. 2007. Neighborhood risk factors for obesity. *Obesity,* 15(8), 2111-9.

Lynch, J., Due, J., Muntaner, C. and Davey Smith, G. 2000. Social capital – is it a good investment strategy for public health? *Journal of Epidemiology and Community Health,* 54(6), 404-8.

McCormack, G., Giles-Corti, B., Lange, A., Smith, T., Martin, K. and Pikora, T.J. 2004. An update of recent evidence of the relationship between objective and self-report measures of the physical environment and physical activity behaviours. *Journal of Science and Medicine in Sport,* 7(1 Suppl), 81-92.

McLaren, L. 2007. Socioeconomic status and obesity. *Epidemiologic Reviews,* 29, 29-48.

McNeill. L.H., Kreuter, M.W. and Subramanian, S.V. 2006. Social environment and physical activity: a review of concepts and evidence. *Social Science and Medicine,* 63(4), 1011-22.

Mobley, L.R., Root, E.D., Finkelstein, E.A., Khavjou, O., Farris, R.P. and Will, J.C. 2006. Environment, obesity, and cardiovascular disease risk in low-income women. *American Journal of Preventive Medicine,* 30(4), 327-32.

Mujahid, M.S., Diez Roux, A.V., Shen, M., et al. 2008. Relation between neighborhood environments and obesity in the multi-ethnic study of atherosclerosis. *American Journal of Epidemiology,* 167(11), 1349-57.

Mummery, W.K., Lauder, W., Schofield, G. and Caperchione, C. 2008. Associations between physical inactivity and a measure of social capital in a sample of Queensland adults. *Journal of Science and Medicine in Sport,* 11(3), 308-15.

Muntaner, C., Lynch, J. and Davey Smith, G. 2001. Social capital, disorganized communities, and the third way: understanding the retreat from structural inequalities in epidemiology and public health. *International Journal of Health Services,* 31(2), 213-37.

Nagel, C.L., Carlson, N.E., Bosworth. M., et al. 2008. The relation between neighborhood built environment and walking activity among older adults. *American Journal of Epidemiology,* 168(4), 461-8.

Nestle, M. and Jacobson, M.F. 2000. Halting the obesity epidemic: a public health policy approach. *Public Health Reports,* 115(1), 12-24.

Ogilvie, D., Mitchell, R., Mutrie, N., et al. 2008. Perceived characteristics of the environment associated with active travel: development and testing of a new scale. *International Journal of Behavioral Nutrition and Physical Activity,* 5:32.

Owen, N., Humpel, N., Leslie, E., Bauman, A. and Sallis, J. 2004. Understanding environmental influences on walking; Review and research agenda. *American Journal of Preventive Medicine,* 27(1), 67-76.

Owen, N., Cerin, E., Leslie, E., et al. 2007. Neighborhood walkability and the walking behavior of Australian adults. *American Journal of Preventive Medicine,* 33(5), 387-95.

Papas, M.A., Alberg, A.J., Ewing, R., et al. 2007. The built environment and obesity. *Epidemiologic Reviews,* 29, 129-43.

Pearce, N. and Davey Smith, G. 2003. Is social capital the key to inequalities in health? *American Journal of Public Health,* 93(1), 122-9.

Pendola, R. and Gen, S. 2007. BMI, auto use, and the urban environment in San Francisco. *Health and Place,* 13(2), 551-6.

Poortinga, W. 2006. Perceptions of the environment, physical activity, and obesity. *Social Science and Medicine,* 63(11), 2835-46.

Poston, W.S. and Foreyt, J.P. 1999. Obesity is an environmental issue. *Atherosclerosis,* 146(2), 201-9.

Putnam, R.D. 1993. *Making democracy work: civic traditions in modern Italy.* New Jersey: Princeton University Press.

Regidor, E., Gutiérrez-Fisac, J.L., Ronda, E., et al. 2008. Impact of cumulative area-based adverse socioeconomic environment on body mass index and overweight. *Journal of Epidemiology and Community Health,* 62, 231-8.

Rodriguez, D., Khattak, A.J. and Evenson, K.R. 2006. Can *New Urbanism* encourage physical activity? *Journal of the American Planning Association,* 72(1), 43-53.

Ross, N., Tremblay, S., Khan, S., et al. 2007. Body mass index in urban Canada: neighborhood and metropolitan area effects. *American Journal of Public Health,* 97(3), 500-8.

Rundle, A., Diez Roux, A.V., Free, L.M., et al. 2007. The urban built environment and obesity in New York City: a multilevel analysis. *American Journal of Health Promotion,* 21(4 Suppl), 326-34.

Rutt, C.D. and Coleman, K.J. 2005. Examining the relationships among built environment, physical activity, and body mass index in El Paso, TX. *Preventive Medicine,* 40(6), 831-41.

Saelens, B., Sallis, J. and Frank, L. 2003. Environmental correlates of walking and cycling: findings from the transportation, urban design, and planning literatures. *Annals of Behavioral Medicine,* 25(2), 80-91.

Saelens, B., Sallis, J., Black, J. and Chen, D. 2003. Neighborhood-based differences in physical activity: an environment scale evaluation. *American Journal of Public Health,* 93(9), 1552-8.

Saelens, B.E. and Handy, S.L. 2008. Built environment correlates of walking: a review. *Medicine and Science in Sports and Exercise,* 40(7 Suppl), S550-66.

Sallis, J., Kraft, K. and Linton, L. 2002. How the environment shapes physical activity: a transdisciplinary research agenda. *American Journal of Preventive Medicine,* 22(3), 208.

Santos, R., Silva, P., Santos, P., Ribeiro, J.C. And Mota, J. 2008. Physical activity and perceived environmental attributes in a sample of Portuguese adults: results from the Azorean Physical Activity and Health study. *Preventive Medicine,* 47(1), 83-8.

Skrabski, A., Kopp, M. and Kawachi, I. 2003. Social capital in a changing society: cross sectional associations with middle aged female and male mortality rates. *Journal of Epidemiology and Community Health,* 57(2), 114-9.

Smith, K.R., Brown, B.B., Yamada. I., et al. 2008. Walkability and body mass index: density, design, and new diversity measures. *American Journal of Preventive Medicine,* 35(3), 237-44.

Strach, S., Isaacs, R. and Greenwald, M.J. 2007. Operationalizing environmental indicators for physical activity in older adults. *Journal of Aging and Physical Activity,* 15(4), 412-24.

Subramanian, S.V., Kawachi. I. and Kennedy, B.P. 2001. Does the state you live in make a difference? Multilevel analysis of self-rated health in the US. *Social Science and Medicine,* 53(1), 9-19.

Subramanian, S.V., Kim, D.J. and Kawachi, I. 2002. Social trust and self-rated health in US communities: a multilevel analysis. *Journal of Urban Health,* 79(4 Suppl 1), S21-34.

Swinburn, B., Egger, G. and Raza, F. 1999. Dissecting obesogenic environments: the development and application of a framework for identifying and prioritizing environmental interventions for obesity. *Preventive Medicine,* 29(6 Pt 1), 563-70.

Takano, T., Nakamuram, K. and Watanabe, M. 2002. Urban residential environments and senior citizens' longevity in megacity areas: the importance of walkable green spaces. *Journal of Epidemiology and Community Health,* 56(12), 913-8.

Tilt, J.H., Unfried, T.M. and Roca, B. 2007. Using objective and subjective measures of neighborhood greenness and accessible destinations for understanding walking trips and BMI in Seattle, Washington. *American Journal of Health Promotion,* 21(4 Suppl, 3), 71-9.

Turrell, G., Kavanagh. A. and Subramanian, S.V. 2006. Area variation in mortality in Tasmania (Australia): the contributions of socioeconomic disadvantage, social capital, and geographic remoteness. *Health and Place,* 12, 291-305.

van Lenthe, F. Mackenbach, J. 2002. Neighbourhood deprivation and overweight: the GLOBE study. *International Journal of Obesity,* 26, 234-240.

Wendel-Vos, W., Droomers, M., Kremers, S., et al. 2007. Potential environmental determinants of physical activity in adults: a systematic review. *Obesity Reviews,* 8(5), 425-40.

Wilson, D.K., Bowles, H. and Ainsworth, B.E. 2007. Body mass index and environmental supports for physical activity among active and inactive residents of a U.S. southeastern county. *Health Psychology,* 26(6), 710-7.

Witten, K., Hiscock, R., Pearce, J. and Blakely, T. 2008. Neighbourhood access to open spaces and the physical activity of residents: a national study. *Preventive Medicine,* 47(3), 299-303.

Wood, L., Shannon, T., Bulsara, M., et al. 2008. The anatomy of the safe and social suburb: an exploratory study of the built environment, social capital and residents' perceptions of safety. *Health and Place,* 14(1), 15-31.

Chapter 9
Childhood Obesity, Physical Activity and the Physical Environment

Melody Oliver and Grant Schofield

…a healthy community is not just one with excellent medical care; it is one with an environment that encourages physical activity and social contact and provides healthy air and landscapes. Today's cities sprawl into forest and farmland with ever widening roadways but no sidewalks or bicycle routes. With their vast asphalt parking areas and treeless streets, these cities coddle the automobile while denying children the opportunity to experience the wonder and joy of the natural world. What child can be allowed independent exploration in cities experienced as dangerous and lacking parks and sidewalks? Parents exhausted by long commutes and endless traffic can scarcely cope with their own needs after long days; consequently, they find the inevitable demands by hungry and tired children more exhausting than refreshing. In this setting, virtually every task requires a car, even a trip to school, church or the library, and the quality time for parents and their children is reduced to brief conversations in the car. The developmental and psychological effects of barren and commercialized landscapes need to be further examined.

Cummins and Jackson (2001: 1247)

Introduction

Epidemic and increasing rates of child and adolescent overweight and obesity are being observed in developed and developing countries worldwide (Wang and Lobstein 2006). Particularly striking examples of this can be observed in the United States, where a significant increase in overweight children aged 2-19 years has occurred, from 14 per cent in 1999/2000 to 18 per cent and 16 per cent in 2003/2004 for boys and girls respectively (Ogden et al. 2006). Likewise, in 2005, the proportion of obese or overweight children in the United Kingdom reached 18 per cent, reflecting more than a doubling in prevalence from 1995 (NHS Information Centre. Lifestyles Statistics 2006). Notably, obesity is not restricted to high-income countries; over three quarters of children who are overweight or obese currently reside in low- or middle-income countries (World Health Organization 2002).

The putative influence of environments on childhood obesity is via multiple mechanisms, underlying which are factors that facilitate increased energy consumption or decreased energy expenditure. Changing environments, particularly urban sprawl and modern urban developments are purported to support behaviours that tip energy balance in favour of weight gain. Physical activity is fundamental to energy balance and obesity prevention and treatment (Ness et al. 2007), as well as conferring numerous benefits to child wellbeing in its own right, including improved bone health, reduced risk factors for cardiovascular disease, and reduced risk of developing type 2 diabetes (World Health Organization 2005, Department of Health, Physical Activity, Health Improvement and Prevention 2004).

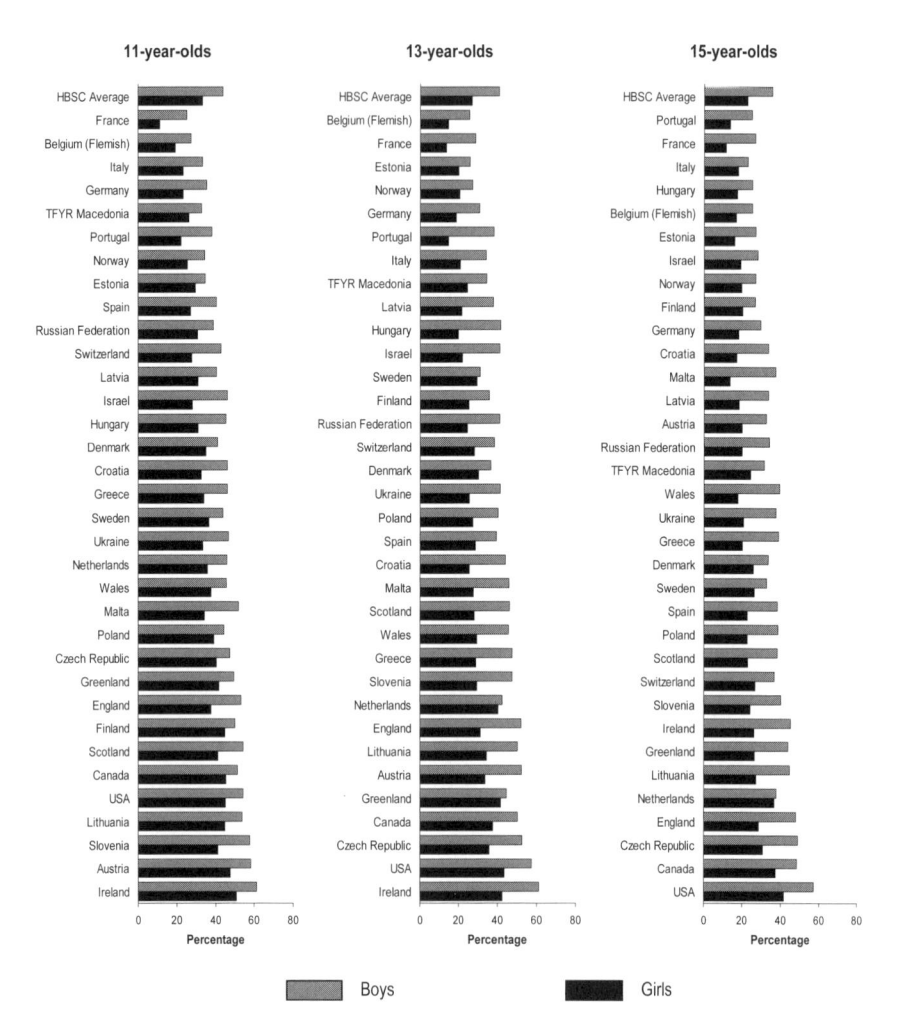

Figure 9.1 **Percentage of youth meeting guidelines for moderate-to-vigorous physical activity**

Note: Data are unavailable for Belgium (French).

Source: Data from the Health Behaviour in School-aged Children Study reprinted with permission from the World Health Organization Office for Europe (Currie et al. 2004).

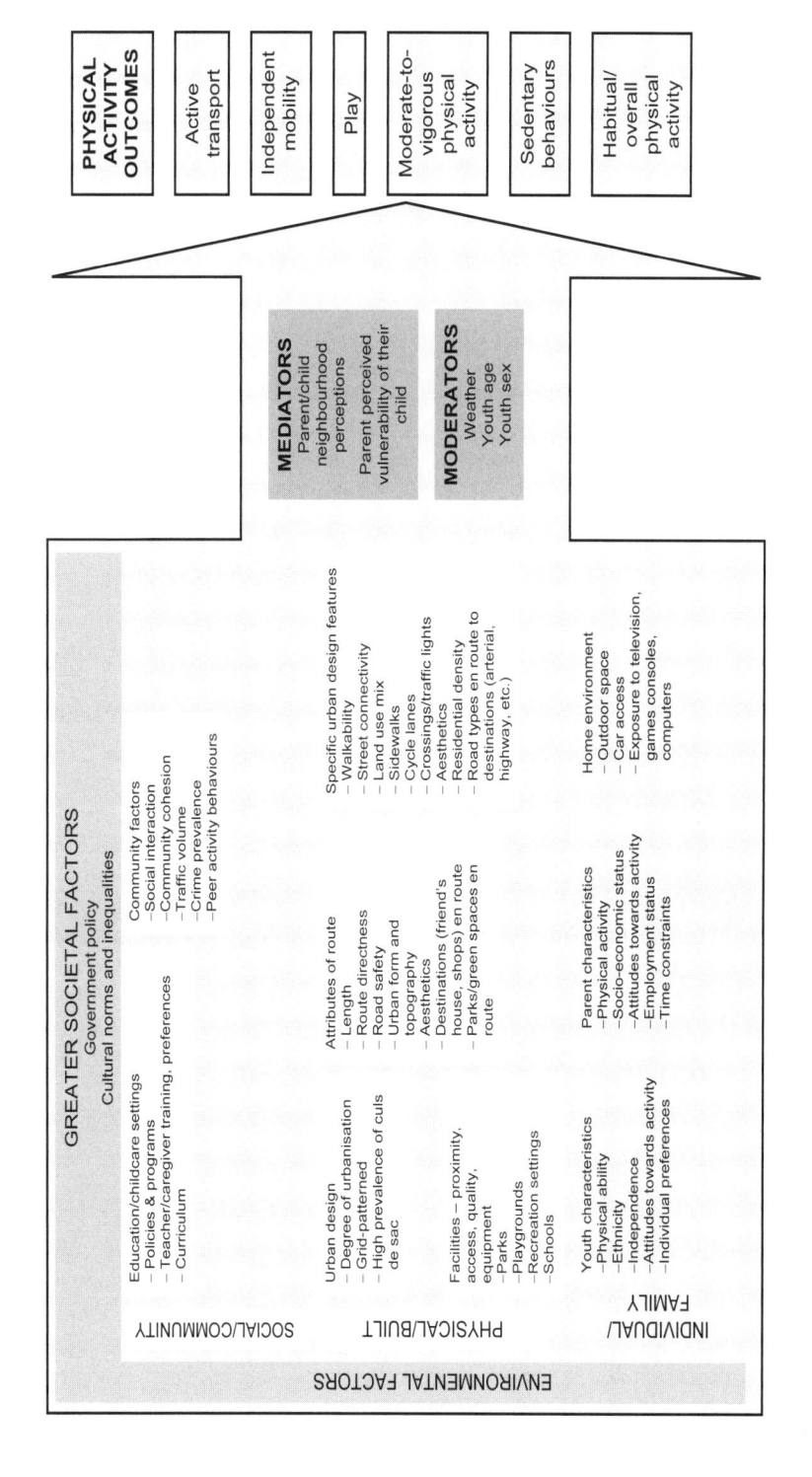

Figure 9.2 Conceptual model of environmental factors that may influence children's physical activity

Conversely, sedentary behaviours may negatively influence health separately and independently from the benefits of physical activity (Hamilton et al. 2007, Dietz 1996).

Globally, many children are insufficiently active for health gain (Figure 9.1) (Active Healthy Kids Canada 2008, Eaton et al. 2006, NHS Information Centre. Lifestyles Statistics 2006, Currie et al. 2004, Vincent et al. 2003), and children's participation in physical activity has generally decreased (Salmon and Timperio 2007). Considering activity declines with age, the importance of promoting physical activity in childhood should not be underestimated (Active Healthy Kids Canada 2008, Nader et al. 2008).

This chapter considers specific environmental aspects that may be related to physical activity and obesity-related health outcomes among children including preschoolers (3-5 years) and adolescents (13-18 years). A conceptual model of the potential environmental influences (both evidence-based and hypothesized) on children's physical activity is provided in Figure 9.2, and relevant studies have been outlined in detail in Table 9.1. Factors related specifically to children's nutrition behaviours and the physical environment are discussed in Chapter 6.

Activity, Environments and Safety

A fundamental issue within this field is the over-riding influence of parental concerns for their child's safety on the activity behaviours of their child. Children are subjected to ever-increasing restrictions on independent mobility, limiting opportunities for accumulating activity by walking, cycling, and playing in their neighbourhood (Mackett et al. 2007, Karsten 2005, O'Brien et al. 2000, Hillman and Adams 1992). Although conflicting evidence exists (McMillan et al. 2006, Schlossberg et al. 2006, Ziviani et al. 2006), a growing body of research shows parental concerns are a key inhibitor of various physical activity dimensions in youth (Ahlport et al. 2008, Cole et al. 2007, McMillan 2007, Kerr et al. 2006, Centers for Disease Control and Prevention 2002, Collins and Kearns 2001). Differing relationships have been found for perceived neighbourhood safety and activity by child age and sex and by neighbourhood type, indicating perceived vulnerability of certain population groups may confound findings and one measure of safety is also unlikely to adequately capture the range of safety concerns parents face (Carver et al. 2008a, Weir et al. 2006, Tranter and Pawson 2001).

Interestingly, crime rates have generally decreased or remained stable from the mid 1990s to present in both the United States and England (Bureau of Justice Statistics 2008, Home Office Statistics 2008) and musculoskeletal injury rates by activity exposure are relatively low with minimal severity (Spinks et al. 2006a, Spinks et al. 2006b). Sadly, however, parental concerns regarding traffic safety are often justified, with a substantially increased risk of traffic-related injuries and deaths for child pedestrians and cyclists compared with car occupants (Sonkin et al. 2006, Transportation Research Board 2002). Paradoxically, motorists are less likely to collide with pedestrians or cyclists when there is a greater prevalence of people walking or cycling (Jacobsen 2003).

Table 9.1 Environmental associates of activity in children: Studies reviewed and key findings

| Study | Country | Number of children, ages or grades | Dependent variable | Independent environmental variables | Key findings related to activity and the physical environment (after adjusting for all other covariates where available) |||
					Negative association or reduced likelihood of being active	Positive association or increased likelihood of being active	No association
Adkins (2004)	US	52 girls, 8-10y	PA^O	Parent and child neighbourhood perceptions			Child and parental perceptions of neighbourhood safety and access to facilities
Ahlport (2008)	US	37, 10y	AT^S	Focus-group derived barriers and facilitators of AT	Parental fears of child abduction were greatest barrier to AT, traffic safety, school policies, inadequate sidewalks, distance, and terrain also inhibited AT	Adequate sidewalks, school policies and facilities (bike storage, crossing guards)	
Baranowski (1993)	US	191, 3-4y	PA^D	Time spend playing indoors vs. outdoors		Outdoor location, gender, and season accounted for 75% of PA variance	
Boarnet (2005)	US	862, G3-5	AT^S	Exposure to Californian SR2S project		Exposure to Californian SR2S project	
Boldemann (2006)	Sweden	197, 4.5-6.5y	PA^O	Preschool environment^O		Trees, shrubbery, uneven play surfaces	
Braza (2004)	US	2 993, G5	AT^S	Population density, connectivity (# intersections per street mile), school size, SES (receiving public welfare), ethnicity	School size	Population density, SES, ethnicity	Street connectivity

Table 9.1 continued   Environmental associates of activity in children: Studies reviewed and key findings

Study	Country	Number of children, ages or grades	Dependent variable	Independent environmental variables	Negative association or reduced likelihood of being active	Positive association or increased likelihood of being active	No association
Bringol-Isler (2008)	Switzerland	1 031, 6-14y	MT[s]	Home and school neighbourhoods, parental neighbourhood perceptions and AT rules	Own >2 cars, distance to school (community specific), main street crossing (all ↑MT = ↓AT)		Parental neighbourhood perceptions, population density, topography, length of motorways, main streets, side streets, motorway crossing, side street crossing
Bungum (2009)	US	2 692, mean (SD) age 15.2y (1.5)	AT[s]	School neighbourhood (1.2 km radius)		Street connectivity	
Burdette (2005)	US	3 141, 3y	PA[s], TV[s]	Parent neighbourhood safety perceptions	Parents' perception of neighbourhood being unsafe (↑TV)		
Carver (2005)	Australia	347, mean (SD) age 13.0 (0.2)y	AT[s] (includes to locations other than school), RT[s]	Parent and child neighbourhood perceptions	Child concerns about roaming dogs, parent-reported traffic density	Parent-reported availability of recreation settings, child perception of road safety, proximity of convenience stores	Parent reported safety of neighbourhood for walking/cycling, take-away/fast food outlets near home
Carver (2008a)	Australia	188 children, 8-9y; 346 adolescents, 13-15y	MVPA°	Parent and adolescent neighbourhood perceptions	Girls' perception of road safety (girls' MVPA outside school hours and evenings)	Parent score of safety of child (boys' MVPA after school), parental reporting of traffic calming devices existing (girls' weekend MVPA), parental reporting of incivilities existing (boys' MVPA on weekend days)	Adolescent perceptions of personal safety and presence of incivilities

Table 9.1 continued   Environmental associates of activity in children: Studies reviewed and key findings

Study	Country	Number of children, ages or grades	Dependent variable	Independent environmental variables	Key findings related to activity and the physical environment (after adjusting for all other covariates where available)		
					Negative association or reduced likelihood of being active	Positive association or increased likelihood of being active	No association
Carver (2008b)	Australia	188 children, 8-9y; 346 adolescents 13-15y	MVPA outside school hours^O, RT^S	Objectively assessed home neighbourhood road environment features (800m radius)		2-3 traffic/pedestrian lights (adolescent girls' RT), residing on a cul-de-sac and speed humps (adolescent boys' MVPA)	All road environment variables (children)
Cole (2007)	Australia	559, G1-7	AT^S	Distance to school, parent neighbourhood perceptions	Parent-reported stranger danger, traffic, speed limits for cars, road crossings to school, distance to school ($\uparrow$ MT = $\downarrow$AT)	Parent-reported traffic calming around school entrance ($\downarrow$MT = $\uparrow$AT)	Parent-reported footpaths on the way to school
Cradock (2007)	US	248, G7-8	School PA^O	School spatial attributes		School campus area per student, play area per student, building area per student, number of days of PE class	
de Bruijn (2004)	Netherlands	3 859, mean (SD) age 14.8y (1.6)	Cycling for transport^S	School urbanisation level (>50,000 city inhabitants or not)		Urbanisation <50 000 inhabitants	
Epstein (2006)	US			Density (housing units/residential acre), design (connectivity – N of intersections/ length of street network), diversity (% of residential parcels, accessible park land)		Park area when sedentary behaviours decreased	

Table 9.1 continued Environmental associates of activity in children: Studies reviewed and key findings

Study	Country	Number of children, ages or grades	Dependent variable	Independent environmental variables	Negative association or reduced likelihood of being active	Positive association or increased likelihood of being active	No association
Ewing (2004)	US	709 school trips for children K-G12	AT^{S,M}	Distance to school, urban form	Distance to school	Proportion of arterials and collectors with sidewalks (walking only)	Proportion of arterials and collectors with sidewalks or bike lanes (biking only), school size, land use density and mix
Ewing (2005)	US	Travel demand modelling for 709 trips to 2 high schools using regional survey data	AT^{M}	Home and school neighbourhoods (density, commercial floor area ratio, sidewalk coverage, street tree coverage, regional accessibility), school age (used as indicator of related to traditional neighbourhood design, higher density, finer land use mix)		Shorter walk or bike times to school, sidewalk presence on main roads	Population density, street tree coverage in school neighbourhood, school age
Fulton (2005)	US	1 395, G4-12	AT^{S}	Neighbourhood type/size, child perceptions of neighbourhood safety, sidewalks in neighbourhood, access to motorized transportation		Children from suburb or small town more likely to AT than those from urban areas, sidewalks in neighbourhood	Child perception of neighbourhood safety, access to motorized transportation

Table 9.1 continued   Environmental associates of activity in children: Studies reviewed and key findings

Study	Country	Number of children, ages or grades	Dependent variable	Independent environmental variables	Negative association or reduced likelihood of being active	Positive association or increased likelihood of being active	No association
Grow (2008)	US	124 adolescents, mean (SD) age 14.4y (1.7) and their parents, and parents of 87 children mean (SD) age 7.6y (1.7)	AT[s]	Proximity to recreation settings, parent and adolescent neighbourhood perceptions	Crime threat (adolescents)	Perceived pedestrian infrastructure and traffic safety; use of pool, basketball court, walking/running tracks, school recreation facilities, public parks and playgrounds, trails, and public open space (adolescents); use of indoor recreation, walking/running tracks, school recreation sites, playgrounds, and public open space (children)	Proximity to indoor recreation facilities, other play fields/courts, and basketball courts, perceived neighbourhood aesthetics (adolescents)
Hart (1986)	US	40, 2-5y	Nursery PA[OD]	Traditional vs. contemporary nursery school playgrounds	Contemporary playground (increased sitting and standing)	Traditional playground (increased active climbing) and contemporary playground (increased walking)	No difference in running behaviours between playgrounds
Haug (2008)	Norway	1 347, 13y	PA during recess[s]	Perceived facilities in school and school neighbourhood (2 km radius)		Overall amount of facilities, open field space with no marking, outdoor obstacle course or activity trail, playground equipment, room with cardio/weightlifting equipment	Fenced courtyard, soccer fields, ski track, areas for boarding/skating, climbing walls, swimming/water facilities, ice skating areas, wooded areas

Table 9.1 continued   Environmental associates of activity in children: Studies reviewed and key findings

Study	Country	Number of children, ages or grades	Dependent variable	Independent environmental variables	Negative association or reduced likelihood of being active	Positive association or increased likelihood of being active	No association
Holt (2007)	Canada	68, 6–12y	AT^s, RPA^s	Neighbourhood walkability (density, connectivity, mixed land use)	Walkability (young children played significantly less in and out of home/yard environment than older children)	Walkability (AT)	
Hume (2007)	Australia	280, 10y	PA^o, walking^s	Child neighbourhood perceptions		Graffiti, easy to walk/cycle around (girls' walking; number of accessible destinations (boys' walking), lots of litter, friends living in walking/cycling distance (boys' PA)	Roads safe to cross, safe to walk/cycle to school, nice gardens; no variables were associated with girls' overall PA
Hume (2009)	Australia	957, 9–12y	MVPA^s, walking frequency^s	Perceived neighbourhood social environment		Perceived social capital, social networks (MVPA), social capital (walking frequency)	Social networks (walking frequency), physical environment (MVPA and walking frequency)
Kerr (2006)	US	259, 5–18y	AT >1 time per week^s	Neighbourhood walkability (net residential density, retail floor area ratio, intersection density, land use mix; 1 km network buffer), parent neighbourhood perceptions	Parent concerns (crime, safety, logistics, traffic, child preferences)	Neighbourhood walkability (high SES neighbourhoods only), parent perceptions of neighbourhood aesthetics, access to local stores and biking/walking facilities	Neighbourhood walkability (low SES neighbourhoods), parent perceptions of land use mix, street connectivity

Table 9.1 continued   Environmental associates of activity in children: Studies reviewed and key findings

Study	Country	Number of children, ages or grades	Dependent variable	Independent environmental variables	Key findings related to activity and the physical environment (after adjusting for all other covariates where available)		
					Negative association or reduced likelihood of being active	Positive association or increased likelihood of being active	No association
Klesges (1990)	US	222, 3-6y	Home PA^D	Time spent playing indoors vs. outdoors		Percentage of time outdoors	
Kligerman (2007)	US	98, mean 16.2y	PA^O	Neighbourhood walkability (land use mix, retail density, street connectivity, residential density) and accessibility (proximity to public and private recreation facilities)		Neighbourhood walkability	Accessibility to recreation facilities
Loucaides (2004)	Greece	256, 11-12y	PA^O	Parental neighbourhood perceptions, school location (urban, rural)		Urban school (winter), rural school (summer), parents of rural school children perceived significantly more play space in garden and neighbourhood, and safer neighbourhood, parents of urban school children reported more home exercise equipment and transportation to recreation facilities/spaces	
Martin (2007)	US	7 433, 9-15y	AT^S	Neighbourhood urbanisation level	Residing in a metro-suburban, second city, town, or rural region	Residing in an urban area	

Table 9.1 continued   Environmental associates of activity in children: Studies reviewed and key findings

Study	Country	Number of children, ages or grades	Dependent variable	Independent environmental variables	Negative association or reduced likelihood of being active	Positive association or increased likelihood of being active	No association
McDonald (2008)	US	6 508, 5-13y	AT[s]	Residential density, distance to school, estimated commute duration for walking and driving	Distance to school >1.6 km, walking time to school	Population density	
McKenzie (2008)	US	139, mean age 6.5y	Home MVPA[D]	Being indoors vs. outdoors		Being outdoors	
McMillan (2006)	US	1 244, G3-5	AT[s]	Parent neighbourhood perceptions		Distance to school <1.6 km	Perceived neighbourhood safety
McMillan (2007)	US	Caregivers of children from 16 schools, G3-5	AT[s]	School neighbourhood (800 m radius), parent neighbourhood perceptions		Windows facing street and mixed land use in school neighbourhood, distance to school <1.6 km	Proportion of street segments with complete sidewalk system
Mota (2007)	Portugal	1 561, mean (SD) age 14.7y (1.6)	RPA[s]	Adolescent neighbourhood perceptions	Girls' concerns about safety and crime	Neighbourhood aesthetics, access to facilities (girls only)	
Nelson (2006)	US	20 745, mean (SD) age 15.4y (0.1)	PA[s]	Neighbourhood type		Living in older suburban areas (moderate connectivity, moderate access to facilities, high road density) compared with newer suburbs (low access, poor street connectivity and density)	
Nelson (2008)	Scotland	4 013, 15-17y	AT[O]	Neighbourhood environment	Distance to school	Population density	

Table 9.1 continued   Environmental associates of activity in children: Studies reviewed and key findings

Study	Country	Number of children, ages or grades	Dependent variable	Independent environmental variables	Negative association or reduced likelihood of being active	Positive association or increased likelihood of being active	No association
Ridgers (2007)	England	470 elementary school children	PA[o]	Comparison of children's PA in intervention (markings and equipment provision) versus control schools		Playground intervention (playground markings for sports, multi-activity, and quiet play areas; provision of soccer goal posts, basketball hoops, fencing, seating)	
Robertson-Wilson (2008)	Canada	21 345, G9-12	AT[s], PA[s]	School location (urban, rural, suburban) and type (separate, public)	Rural school, separate school		
Roemmich (2006)	US	59, 4-7y	PA[o], TV[o]	Home neighbourhood (800 m radius)		Housing density, interaction of housing density by sex, percentage park plus recreation area	Number of TV sets in home
Roemmich (2007)	US	88, 8-12y	PA and MVPA[o], TV time[s]	Home neighbourhood (800 m radius)	TV sets in home (↑TV time)	Street connectivity (PA, MVPA), proportion park area, proportion park/recreation area (PA), Proportion park/recreation area (↓TV time)	Residential density, street width, park area, recreation area, residential land area
Salmon (2007)	Australia	720, 4-13y	AT[s]	Parent neighbourhood perceptions	Too far to walk, no direct route	Concern child may be injured in a road accident	Lack of pedestrian crossings, lights, footpaths, or car-parking; presence of speeding drivers

Table 9.1 continued   Environmental associates of activity in children: Studies reviewed and key findings

Study	Country	Number of children, ages or grades	Dependent variable	Independent environmental variables	Key findings related to activity and the physical environment (after adjusting for all other covariates where available)		
					Negative association or reduced likelihood of being active	Positive association or increased likelihood of being active	No association
Schlossberg (2006)	US	287, G6-8	AT^S	Distance to school, urban form (intersection density, dead-end density, presence of major roads and railroads, route directedness, 400 m route buffer), parent neighbourhood perceptions	Distance to school >2.4 km for walking and >4.0 km for cycling, dead-end street density (>18) for walking only	Intersection density (6 or more compared with <3 or 3<6) for walking only	Route directness, major road on route, railroad on route; intersection and dead-end density (cycling only); parent-reported dangerous traffic, traffic speed, lack of complete sidewalks
Scott (2007)	US	1 556 girls, G6	Weekend MVPA^O	Neighbourhood recreation and school weekend amenities (800 m radius)			Locked schools, unlocked schools with active facilities, no schools, parks
Shi (2006)	China	824, 12-14y	AT^S, housework^S, VPA^S	Neighbourhood urbanisation level		Residing in an urban area (girls only; overall PA score based on housework, AT, and VPA; residing in a rural area (girls only; housework)	Urbanisation level (all children: AT, VPA, boys only: housework; overall PA score based on housework, AT, and VPA)
Sirard (2005)	US	Observation in 8 school areas, total school enrolment 3,911 children	AT^D	School SES, school urbanisation level, weather conditions, temperature			School urbanisation level

Table 9.1 continued   Environmental associates of activity in children: Studies reviewed and key findings

Study	Country	Number of children, ages or grades	Dependent variable	Independent environmental variables	Negative association or reduced likelihood of being active	Positive association or increased likelihood of being active	No association
Sjolie (2002)	Norway	88, G8-9	AT^s, TV^s, RPA^s	Neighbourhood urbanisation level	Rural area (AT)		Urbanisation level (RPA, TV)
Timperio (2004a)	Australia	291 children, 5-6y; 919 children, 10-12y	AT^s	Parent and adolescent neighbourhood perceptions	Owning >2 cars (5-6y girls), limited public transport (5-6y and 10-12y girls), no lights/crossings (10-12y boys), have to cross several roads, no parks near residence (10-12y girls)	Parent perceptions of heavy traffic (5-6y boys)	Availability of sporting venues
Timperio (2006)	Australia	235 children, 5-6y, 677 aged 10-12y	AT^s	School route variables, parent and child neighbourhood perceptions, personal and family factors	Route to school: no lights or crossings, few children, busy road barrier, topography (steep incline), direct route to school	Route <800 m	Route along busy road, parental and child concerns about strangers and road safety, perceived heavy local traffic, need to cross several roads
Timperio (2008)	Australia	163 children, 8-9y; 334 adolescents, 13-15y	MVPA^o	Public open spaces (800 m radius)	Lighting along paths (weekend MVPA), number of recreational facilities (younger girls weekend and after school)	Presence of playgrounds (younger boys weekend MVPA), presence of trees providing shade and signage regarding dogs (adolescent girls after school)	Walking paths, cycling paths, water features
Tudor-Locke (2003)	Philippines	1 518, 14-16y	AT^s	Neighbourhood urbanisation level	Residing in an urban setting		
Veugelers (2008)	Canada	5 471, G5	PA^s (composite of TV and sports)	Parent neighbourhood perceptions		Safety (unsupervised sports), access to playgrounds, parks, recreational facilities (PA)	

Table 9.1 continued   Environmental associates of activity in children: Studies reviewed and key findings

Study	Country	Number of children, ages or grades	Dependent variable	Independent environmental variables	Negative association or reduced likelihood of being active	Positive association or increased likelihood of being active	No association
Weir (2006)	US	307, 5–10y	PA^S	Parent neighbourhood safety perceptions	Parental anxiety about neighbourhood safety (urban children)		No relationships for suburban children
Wen (2008b)	Australia	1 603, 9–11y	MT^S	Parent neighbourhood perceptions, distance to school	Distance to school >500 m (↑MT = ↓AT)		Parent perceived neighbourhood safety, dangerous roads
Worobey (2004)	US	40, 4–5y	PA^O	Preschool type (university, high SES or Head Start, low SES)	Head Start preschool	University preschool	
Ziviani (2004)	Australia	164, mean (SD) age 9.1y (2.0)	AT^S	Parent neighbourhood perceptions, distance to school	Parent concerns about child safety, traffic, and general concerns, distance to school		Parent concerns about pollution, outside shelter, manned crossings
Ziviani (2006)	Australia	63, G7	AT^S	Parent neighbourhood perceptions, presence of walk to school programme	Distance to school		Presence of a walk to school programme; parent-reported traffic, weather, pollution, footpath condition, manned crossings, sheltered walkways

Notes: AT = active transportation (walking or cycling), ^{D} = directly observed, G = school grade/year, K = kindergarten, km = kilometres, ^{M} = modelled activity, m = metres, MT = motorised transport, MVPA = moderate-to-vigorous physical activity, N = number, ^{O} = objectively assessed (e.g., accelerometry, pedometry, GIS), PA = physical activity, PE = physical education, RPA = recreational or leisure-time physical activity, RT = recreational transport (walking or cycling), ^{S} = self/proxy report, SD = standard deviation, SES = socioeconomic status, SR2S = Safe Routes to School project, TV = television, US = United States of America, vs. = versus, y = years of age.

Active Commuting and the Physical Environment

Studies investigating the relationships between active transportation and body size in children have produced mixed results, with no association (Fulton et al. 2005), negative associations (Gordon-Larsen et al. 2005) and a positive association (Heelan et al. 2005) observed in different studies. It is likely that the influence of active commuting on body size is mitigated by other factors, including age, sex and socioeconomic status. Even so, the increased opportunity for physical activity accumulation through active transportation warrants further exploration of this activity dimension (Tudor-Locke et al. 2003). Although conflicting reports exist (Rosenberg et al. 2006, Metcalf et al. 2004), cumulative evidence shows that compared to those who travel by motorized transport to school, actively commuting children are more physically active overall and are more likely to attain recommended activity levels (Lee et al. 2008).

Despite this, the prevalence of walking or cycling to school in developed countries is generally very low (Robertson-Wilson et al. 2008, van der Ploeg et al. 2008, McDonald 2007). Exceptions exist, however, with two European studies reporting a majority of children travelling to school by active means (Bringolf-Isler et al. 2008; Cooper et al. 2006). Environmental differences (particularly urban design) likely explain some of this difference; it is also possible that there are over-riding social and cultural influences on active transportation behaviours such as socioeconomic status and car accessibility (Yelavich et al. 2008), and cultural norms for or against car use (Lorenc et al. 2008).

There has also been a drastic decline in proportions of children actively transporting to school (van der Ploeg et al. 2008, McDonald 2007). Numerous explanations for this shift have been suggested, including parental concerns about traffic safety and harm from others, increasing distances to school, and time pressures. Encouragingly, children often report a preference for actively commuting to school irrespective of parental preferences (Lorenc et al. 2008). Considering parents are the 'gatekeepers' of commute mode choice, the ideal environments for encouraging children's active transport may still be more related to alleviating parental fears rather than child preferences.

Studies of United States, Canadian, Irish, Dutch and Norwegian children have shown greater active transportation in children residing in urbanized areas, attending urban schools, or living in high population density areas compared with those attending rural schools or living in smaller towns or rural regions (Nelson et al. 2008, Robertson-Wilson et al. 2008, Martin et al. 2007, de Bruijn et al. 2004, Sjolie and Thuen 2002). In two studies from China and the United States, however, no significant difference in active commuting was found by rural or urban areas (Shi et al. 2006, Sirard et al. 2005). A further two studies in the Philippines (Tudor-Locke et al. 2003) and United States (Fulton et al. 2005) showed that youth residing in rural or suburban areas were actually more likely to actively commute to school than those living in urban areas. Although different relationships by country or geographic region might be expected, the contrasting findings found for studies conducted within the United States exemplify a fundamental problem in this research area; the application of differing definitions of environments can result in considerably different approaches (and consequent findings) for explaining associations between geographic factors and activity behaviours (Ball et al. 2006). As may be observed below, inconsistencies exist with regard to specific environmental features and activity prevalence; as well as the measurement issues noted above, variation in sampling frames and measurement of independent variables likely contribute to this diversity.

Urban Design and Active Transportation

Large scale engineering-based changes along routes to school, such as installing sidewalks and widening cycle lanes, have shown promise for increasing children's active transportation (Boarnet et al. 2005), although conflicting evidence exists for specific urban form elements such as sidewalks (Ahlport et al. 2008, McMillan 2007, Schlossberg et al. 2006, Fulton et al. 2005, Ewing et al. 2004). It is possible that sidewalk presence on one side of the road is sufficient to encourage active transportation, with no further benefit gained for sidewalks on both street sides.

Inconsistent relationships have also been found for active commuting and presence or condition of factors such as road crossings, traffic lights, main roads, or railroads en route to school. Timperio et al. (2006) provide an example of the complexity of these relationships and measurement issues whereby no association was found between active commuting and heavy local traffic, parental concerns about road safety, or need to cross several roads; yet lack of lights/crossings, and the presence of an objectively assessed busy road barrier en route to school were negatively associated with active transport.

Of the limited work investigating street connectivity and activity in children, both positive relationships (Bungum et al. 2009, Schlossberg et al. 2006) and no associations (Braza et al. 2004) have been reported. Higher connectivity is typical of traditional urban environments and may facilitate participation in active transportation because of more direct routes, however other facets of physical activity for children may be compromised. For example, a high prevalence of culs-de-sac and low street connectivity typified in modern developments may encourage increased independent mobility and activity of children, such as playing in the streets (O'Brien et al. 2000).

A convincing body of research has shown that although findings may differ by community (Bringolf-Isler et al. 2008), distance or trip duration to school are the dominant environmental factors associated with reduced likelihood of actively transporting to school (see Table 9.1). Distinct differences can be found for walking and cycling, with the effect of distance being greater for children who walk (Nelson et al. 2008, Schlossberg et al. 2006), while increased trip duration may affect cycling more than walking (Ewing et al. 2004). This difference is likely due to the disproportionate increase in distance travelled by time for cycling when compared with walking. Using travel demand modelling and simulated durations of 10 minutes walking or 2.5 minutes biking, a near tripling in active commuting prevalence has been estimated (Ewing et al. 2006).

Localized schools nested within communities are thus likely to facilitate increased update of active transportation. Consequently, recent trends in school siting and upsizing (Centers for Disease Control and Prevention 2006) have likely created a significant barrier to children actively transporting to school, with potentially long lasting negative effects on child health.

Physical Environments and Other Physical Activity Dimensions

Physical elements of the urban environment have also been related to children's leisure time activity, overall (accumulated) physical activity, and walking and cycling for non-school trips. The concerns are discussed individually below.

Urban Design Elements and Physical Activity

A high mix of land use including retail, residential, industrial, and institutional settings can stimulate walking or cycling, as destinations are nearby and easily accessible by active means. Similarly, high retail density settings are usually pedestrian oriented, while car-parking is prioritized in low retail density environments.

Limited and conflicting research in this area exists for children; for example residential density has been associated with activity in 4-7 year old children (Roemmich et al. 2006), while in older children, street width and residential density had no relationship with activity (Roemmich et al. 2007). The influence of residential density on activity may be a function of parents being more amenable to their young children playing outside when neighbours and friends live nearby. Inconclusive findings have also been reported for street connectivity, ranging from no relationship with physical activity (Mota et al. 2007, Roemmich et al. 2006), to a significant positive association with activity in young boys but not girls (Roemmich et al. 2007). These conflicting results indicate potentially differing roles of street characteristics on various activity dimensions, which may also differ by age and sex. In contrast with active transportation, where higher connectivity facilitates easy access to destinations, other factors (e.g. perceived safety) may moderate the effect of street connectivity on other activity behaviours of children.

The combined effect of one or more of these elements (walkability) has also been positively related to moderate-to-vigorous physical activity and active transportation in children and adolescents (Kligerman et al. 2007). Differing effects by age and sex have been found (Holt et al. 2007), however, suggesting a reduced influence of the built environment for those who may be perceived as more 'vulnerable' individuals (i.e. females and younger children).

Availability and Accessibility of Recreation and Other Facilities

The existence of, and proximity to, settings that facilitate physical activity may also encourage increased activity levels in youth. Leisure-time active transportation and overall physical activity has been associated with amount of land dedicated to park and recreation areas and accessibility to parks, destinations, and settings to be active (Veugelers et al. 2008, Hume et al. 2007, Roemmich et al. 2006, Epstein et al. 2006, Carver et al. 2005, Timperio et al. 2004a).

Again, conflicting findings have been reported; access to recreation and similar facilities has not been related to girls' objectively assessed activity (Adkins et al. 2004), adolescents' self reported leisure time activity (Mota et al. 2007), or high intensity activity in adolescents (Kligerman et al. 2007, Scott et al. 2007). Contrasting effects by sex and age have also been found in children for park area, playground availability, and access to recreational facilities (Timperio et al. 2008, Roemmich et al. 2007). How neighbourhood green space in high density areas affects independent play and mobility needs investigation as this may mediate changes in activity within neighbourhoods.

Sedentary Behaviours

To date, research on environmental associates of sedentary behaviours has predominantly focused on family-level factors such as number of television sets in the home, and limited findings related to the built urban environment and sedentary behaviours have been reported (Epstein et al. 2006, Burdette and Whitaker 2005). This current dearth of information provides little guidance for urban planners to consider environmental elements that may reduce participation in sedentary activities. Simple displacement theory would suggest that manipulating environments to encourage physical activity will simultaneously reduce participation in sedentary behaviours. Although this may have some effect, children's activity profiles are complex, whereby highly active children may concurrently participate in high levels of sedentary behaviours, and vice-versa (Dollman and Ridley 2006, Proctor et al. 2003).

The Role of Schools and Early Childhood Settings

Schools are ideal settings in which physical activity may be accumulated throughout the day (Story et al. 2006). Multiple opportunities for promoting physical activity exist, including developing school wellness policies, identifying safe routes to school, and building community collaborations (Chodzko-Zajko et al. 2008). Curriculum-based interventions have included increasing physical education class time or improving activity levels within classes (Sallis et al. 1997, McKenzie et al. 1996), incorporating physical activity bouts to non-physical education classes (Stewart et al. 2004), or integrating physical activity throughout the curriculum (Oliver et al. 2006). Few such interventions have been rigorously evaluated, and results have varied greatly (van Sluijs et al. 2007, Cale and Harris 2006, Timperio et al. 2004b). School-based initiatives such as 'walk to school' days may be effective in encouraging increased active transportation (Wen et al. 2008a, Zaccari and Dirkis 2003, Tranter and Pawson 2001), however, small and intermittent activities alone will likely be insufficient to achieve sustained behaviour change. Ongoing strategies such as walking school buses and bike trains provide sustained support for active commuting, and may mitigate parental fears for their child's safety (National Center for Safe Routes to School and Pedestrian and Bicycle Information Center 2006). Interestingly, school policies have been considered as being both inhibiting (e.g. grade/age minimums for walking and cycling) and facilitating (e.g. providing crossing guards) for children walking or cycling to school (Ahlport et al. 2008). Multifaceted, whole-of-school, ecological interventions that also engage families and communities appear most worthwhile for improving children's physical activity participation (Fesperman et al. 2008, van Sluijs et al. 2007, Timperio et al. 2004b).

School design may also influence activity; the proportion of school campus, play area, and building area per student and the provision of playground markings for games and to separate play areas have been related to increased activity in children (Cradock et al. 2007, Ridgers et al. 2007, Stratton and Mullan 2005, Stratton 2000). Playground design, equipment, and quality of outdoor environments in preschools can also inhibit or encourage activity in young children (Hannon and Brown 2008, Boldemann et al. 2006, Sallis et al. 2000, Hart and Sheehan 1986). Smaller playground sizes and increased obesity prevalence have been observed in low-socioeconomic status preschools compared with their high-socioeconomic status counterparts (Worobey et al. 2004), exemplifying the potential confounding effect of wider social and demographic factors on environmental influences on activity and weight status.

Summary and Conclusions

Endeavours to stem the growing prevalence of obesity in children have extended to considering elements within the physical environment that may influence physical activity and nutrition behaviours. Inhibiting factors of walking and cycling to school are distance to school, restrictive school policies, and parental concerns about their child's safety. Much less investigation of environmental factors related to other physical activity dimensions has occurred and, in general, findings are equivocal and effects often weak. Residing close to school or in a neighbourhood with a greater proportion of land dedicated to public open space is likely to increase physical activity levels. Whether improving perceptions of the environment is sufficient to alter parental inhibitors of their child's activity is yet to be determined. The application of differing methodologies to assess activity and the physical environment further restrict understanding of environmental influences on differing physical activity dimensions. Moreover, the prevalence of insufficient activity participation makes it challenging to determine what a supportive environment for sufficient activity levels would be when conducting research with the modern child. Over-riding political and social factors may mediate or moderate the influence of the physical environment on physical activity and have not been assessed in relation to physical activity participation or obesity prevalence in children.

More credible evidence is needed to support assertions that physical environments can play a crucial role in determining how much young people move through transport, play, and within education settings. To achieve this, four fundamental developments in this research field are necessary and noted below:

1. Gather more evidence from prospective (i.e. longitudinal) designs. This evidence can serve to move the current evidence from simple associations to plausible connections and determine causality.
2. Measure associations across a wide range of possible environments. At present, relative homogeneity in measured environments means that the lack of association is often taken as being evidence of no association.
3. Objective measures of physical activity dimensions including sedentary time, standing, and sitting times are needed to gain a full comprehension of movement over a day.
4. Studies which track changes in physical activity, body size, and health after substantial redevelopments in urban design are needed to present convincing evidence to local and central governments in the domains of health and town planning; better use of 'natural experiments' will help to achieve this.

References

Active Healthy Kids Canada. 2008. *Canada's Report Card on Physical Activity for Children and Youth.* Toronto, Ontario.

Adkins, S., Sherwood, N., Story, M. and Davis, M. 2004. Physical activity among African-American girls: The role of parents and the home environment. *Obesity Research*, 12, 38S-45S.

Ahlport, K.N., Linnan, L., Vaughn, A., Evenson, K.R. and Ward, D.S. 2008. Barriers to and facilitators of walking and bicycling to school: Formative results from the non-motorized travel study. *Health Education and Behavior*, 35(2), 221-44.

Ball, K., Timperio, A. and Crawford, D. 2006. Understanding environmental influences on nutrition and physical activity behaviors: Where should we look and what should we count? *International Journal of Behavioral Nutrition and Physical Activity*, 3, 33, doi:10.1186/47/9.

Baranowski, T., Thompson, W.O., DuRant, R.H., Baranowski. J. and Puhl, J. 1993. Observations on physical activity in physical locations: Age, gender, ethnicity and month effects. *Research Quarterly for Exercise and Sport*, 64(2), 127-33.

Boarnet, M.G., Anderson, C.L., Day, K., McMillan, T. and Alfonzo, M. 2005. Evaluation of the California Safe Routes to School legislation. *American Journal of Preventive Medicine*, 28(2S2), 134-40.

Boldemann, C., Blennow, M., Dal, H., Martensson, F., Raustorp, A., Yuen, K., et al. 2006. Impact of pre-school environment upon children's physical activity and sun exposure. *Preventive Medicine*, 42(4), 301-8.

Braza, M., Shoemaker, W. and Seeley, A. 2004. Neighborhood design and rates of walking and biking to elementary school in 34 California communities. *American Journal of Health Promotion*, 19(2), 128-36.

Bringolf-Isler, B., Grize, L., Mäder, U., Ruch, N., Sennhauser, F.H., Braun-Fahrländer, C. and SCARPOL team. 2008. Personal and environmental factors associated with active commuting to school in Switzerland. *Preventive Medicine*, 46(1), 67-73.

Bungum, T.J., Lounsbery, M., Moonie, S. and Gast, J. 2009. Prevalence and correlates of walking and biking to school among adolescents. *Journal of Community Health*, 34(2):129-34.

Burdette, H.L. and Whitaker, R.C. 2005. A national study of neighborhood safety, outdoor play, television viewing, and obesity in preschool children. *Pediatrics*, 116(3), 657-62.

Bureau of Justice Statistics. 2008. *State and national level crime trend estimates (online resource).* Washington, DC: US Department of Justice, Office of Justice Programs.

Cale, L. and Harris, J. 2006. School-based physical activity interventions: Effectiveness, trends, issues, implications and recommendations for practice. *Sport, Education and Society*, 11(4), 401-20.

Carver, A., Salmon, J., Campbell, K., Baur, L., Garnett, S. and Crawford, D. 2005. How do perceptions of local neighborhood relate to adolescents' walking and cycling? *American Journal of Health Promotion*, 20(2), 139-47.

Carver, A., Timperio, A. and Crawford, D. 2008a. Perceptions of neighborhood safety and physical activity among youth: The CLAN study. *Journal of Physical Activity and Health*, 5(3), 430-44.

Carver, A., Timperio, A.F. and Crawford, D.A. 2008b. Neighborhood road environments and physical activity among youth: The CLAN study. *Journal of Urban Health*, 85(4), 532-44.

Centers for Disease Control and Prevention. 2002. Barriers to children walking and biking to school – United States, 1999. *Morbidity and Mortality Weekly Report*, 51(32), 701-4.

Centers for Disease Control and Prevention. 2006. *Kids Walk-to-school: Then and Now – Barriers and Solutions.* Atlanta, GA.

Chodzko-Zajko, W., Zhu, W., Bazzarre, T., Castelli, D., Graber, K. and Woods, A. 2008. 'We Move Kids' – The consensus report from the roundtable to examine strategies for promoting walking in the school environment. *Medicine and Science in Sports and Exercise*, 40(7S), S603-S605.

Cole, R., Leslie, E., Donald, M., Cerin, E. and Owen, N. 2007. Residential proximity to school and the active travel choices of parents. *Health Promotion Journal of Australia*, 18(2), 127-34.

Collins, D.C. and Kearns, R.A. 2001. The safe journeys of an enterprising school: Negotiating landscapes of opportunity and risk. *Health and Place*, 7(4), 293-306.

Cooper, A.R., Wedderkopp, N., Wang, H., Andersen, L.B., Froberg, K. and Page, A.S. 2006. Active travel to school and cardiovascular fitness in Danish children and adolescents. *Medicine and Science in Sports and Exercise*, 38(10), 1724-31.

Cradock, A.L., Melly, S.J., Allen, J.G., Morris, J.S. and Gortmaker, S.L. 2007. Characteristics of school campuses and physical activity among youth. *American Journal of Preventive Medicine*, 33(2), 106-13.

Cummins, S.K. and Jackson, R.J. 2001. The built environment and children's health. *Pediatric Clinics of North America*, 48(5), 1241-52.

Currie, C., Roberts, C., Morgan, A., et al. (eds.) 2004. Young people's health in context. Health Behaviour in School-aged Children (HBSC) study: International report from the 2001/2002 survey. Copenhagen: WHO Regional Office for Europe.

de Bruijn, G.J., Kremers, S.P., Schaalma, H., van Mechelen, W. and Brug, J. 2004. Determinants of adolescent bicycle use for transportation and snacking behaviour. *Preventive Medicine*, 40(6), 658-67.

Department of Health, Physical Activity, Health Improvement and Prevention. 2004. *At Least Five a Week. Evidence on the Impact of Physical Activity and its Relationship to Health. A Report from the Chief Medical Officer.* London.

Dietz, W.H. 1996. The role of lifestyle in health: The epidemiology and consequences of inactivity. *Proceedings of the Nutrition Society*, 55(3), 829-40.

Dollman, J. and Ridley, K. 2006. Differences in body fatness, fat patterning and cardio-respiratory fitness between groups of Australian children formed on the basis of physical activity and television viewing guidelines. *Journal of Physical Activity and Health*, 1, 1-14.

Eaton, D.K., Kann, L., Kinchen, S., et al. 2006. Youth risk behavior surveillance – United States, 2005. *Morbidity and Mortality Weekly Report*, 55(SS05), 1-108.

Epstein, L.H., Raja, S., Gold, S.S., et al. 2006. Reducing sedentary behavior: The relationship between park area and the physical activity of youth. *Psychological Science*, 17(8), 654-9.

Ewing, R., Schroeer, W. and Greene, W. 2004. School location and student travel: Analysis of factors affecting mode choice. *Transportation Research Record: Journal of the Transportation Research Board*, 1895, 55-63.

Ewing, R., Forinash, C.V. and Schroeer, W. 2005, Neighborhood schools and sidewalk connections. What are the impacts on travel mode choice and vehicle emissions? *Transport Research News*, 237 (March-April), 4-10.

Ewing, R., Brownson, R.C. and Berrigan, D. 2006. Relationship between urban sprawl and weight of United States youth. *American Journal of Preventive Medicine*, 31(6), 464-74.

Fesperman, C., Evenson, K., Rodriguez, D. and Salvesen, D. 2008. A comparative case study on active transport to and from school. *Preventing Chronic Disease* 5:2.

Fulton, J.E., Shisler, J.L., Yore, M.M. and Caspersen, C.J. 2005. Active transportation to school: Findings from a national survey. *Research Quarterly for Exercise and Sport*, 76(3), 352-7.

Gordon-Larsen, P., Nelson, M.C. and Beam, K. 2005. Associations among active transportation, physical activity, and weight status in young adults. *Obesity Research*, 13(5), 868-75.

Grow, H.M., Saelens, B.E., Kerr, J., Durant, N.H., Norman, G.J. and Sallis, J.F. 2008. Where are youth active? Roles of proximity, active transport, and built environment. *Medicine and Science in Sports and Exercise*, 40(12), 2071-9.

Hamilton, M.T., Hamilton, D.G. and Zderic, T.W. 2007. Role of low energy expenditure and sitting in obesity, metabolic syndrome, type 2 diabetes, and cardiovascular disease. *Diabetes*, 56(11), 2655-67.

Hannon, J.C. and Brown, B.B. 2008. Increasing preschoolers' physical activity intensities: An activity-friendly preschool playground intervention. *Preventive Medicine*, 46(6), 532-6.

Hart, C.H. and Sheehan, R. 1986. Preschoolers' play behavior in outdoor environments: Effects of traditional and contemporary playgrounds. *American Educational Research Journal*, 23(4), 668-78.

Haug, E. Torsheim, T. and Samdal, O. 2008. Physical environmental characteristics and individual interests as correlates of physical activity in Norwegian secondary schools: The Health Behaviour in School-aged Children study. *International Journal of Behavioral Nutrition and Physical Activity*, 5(1), 47.

Heelan, K.A., Donnelly, J.E., Jacobsen, D.J., Mayo, M.S., Washburn, R. and Greene, L. 2005. Active commuting to and from school and BMI in elementary school children – preliminary data. *Child: Care, Health and Development*, 31(3), 341-9.

Hillman, M. and Adams, J.G. 1992. Children's freedom and safety. *Children's Environments*, 9(2), 12-33.

Holt, N.L., Spence, J.C., Sehn, Z.L. and Cutumisu, N. 2008. Neighborhood and developmental differences in children's perceptions of opportunities for play and physical activity. *Health and Place*, 14(1), 2-14.

Home Office Statistics. 2008. *Crime in England and Wales 2007/2008*. London: Home Office.

Hume, C., Salmon, J. and Ball, K. 2007. Associations of children's perceived neighborhood environments with walking and physical activity. *American Journal of Health Promotion*, 21(3), 201-7.

Hume, C., Jorna, M., Arundell, L., Saunders, J., Crawford, D. and Salmon, J. 2009. Are children's perceptions of neighbourhood social environments associated with their walking and physical activity? *Journal of Science and Medicine in Sport*, 12(6), 637-41.

Jacobsen, P.L. 2003. Safety in numbers: more walkers and bicyclists, safer walking and bicycling. *Injury Prevention*, 9(3), 205-9.

Karsten, L. 2005. It all used to be better? Different generations on continuity and change in urban children's daily use of space. *Children's Geographies*, 3(3), 275-90.

Kerr, J., Rosenberg, D., Sallis, J.F., Saelens, B.E., Frank, L.D. and Conway, T.L. 2006. Active commuting to school: Associations with environment and parental concerns. *Medicine and Science in Sports and Exercise*, 38(4), 787-94.

Klesges, R., Hanson, C., Haddock, C.K. and Klesges, L. 1990. Effects of obesity, social interactions, and physical environment on physical activity in preschoolers. *Health Psychology*, 9(4), 435-49.

Kligerman, M., Sallis, J.F., Ryan, S., Frank, L.D. and Nader, P.R. 2007. Association of neighborhood design and recreation environment variables with physical activity and body mass index in adolescents. *American Journal of Health Promotion*, 21(4), 274-7.

Lee, M.C., Orenstein, M.R. and Richardson, M.J. 2008. Systematic review of active commuting to school and children's physical activity and weight, *Journal of Physical Activity and Health,* 5:6.

Lorenc, T., Brunton, G., Oliver, S., Oliver, K. and Oakley, A. 2008. Attitudes to walking and cycling among children, young people and parents: A systematic review. *Journal of Epidemiology and Community Health*, 62(10), 852-7.

Loucaides, C.A., Chedzoy, S.M. and Bennett, N. 2004. Differences in physical activity levels between urban and rural school children in Cyprus. *Health Education Research*, 19(2), 138-47.

Mackett, R., Brown, B., Gong, Y., Kitazawa, K. and Paskins, J. 2007. Children's independent movement in the local environment. *Built Environment*, 33(4), 454-68.

Martin, S.L., Lee, S.M. and Lowry, R. 2007. National prevalence and correlates of walking and bicycling to school. *American Journal of Preventive Medicine*, 33(2), 98-105.

McDonald, N.C. 2007. Active transportation to school: Trends among U.S. schoolchildren, 1969-2001. *American Journal of Preventive Medicine*, 32(6), 509-16.

McDonald, N.C. 2008. Children's mode choice for the school trip: The role of distance and school location in walking to school. *Transportation*, 35(1), 23-35.

McKenzie, T.L., Nader, P.R., Strikmiller, P.K., et al. 1996. School physical education: Effect of the Child and Adolescent Trial for Cardiovascular Health. *Preventive Medicine*, 25(4), 423-31.

McKenzie, T.L., Baquero, B., Noe C. Crespo, N., Arredondo, E., Campbell, N. and Elder, J. 2008. Environmental correlates of physical activity in Mexican American children at home. *Journal of Physical Activity and Health*, 5(4), 579-91.

McMillan, T., Day, K., Boarnet, M., Alfonzo, M. and Anderson, C. 2006. Johnny walks to school – does Jane? Sex differences in children's active travel to school. *Children, Youth and Environments*, 16(1), 75-89.

McMillan, T.E. 2007. The relative influence of urban form on a child's travel mode to school. *Transportation Research Part A*, 41(1), 69-79.

Metcalf, B., Voss, L., Jeffery, A., Perkins, J. and Wilkin, T. 2004. Physical activity cost of the school run: Impact on schoolchildren of being driven to school (EarlyBird 22). *British Medical Journal*, 329(7470), 832-3.

Mota, J., Gomes, H., Almeida, M., Ribeiro, J.C. and Santos, M.P. 2007. Leisure time physical activity, screen time, social background, and environmental variables in adolescents. *Pediatric Exercise Science*, 19(3), 279-90.

Nader, P.R., Bradley, R.H., Houts, R.M., McRitchie, S.L. and O'Brien, M. 2008. Moderate-to-vigorous physical activity from ages 9 to 15 years. *Journal of the American Medical Association*, 300(3), 295-305.

National Center for Safe Routes to School and Pedestrian and Bicycle Information Center. 2006. *The Walking School Bus: Combining Safety, Fun and the Walk to School*. Chapel Hill, NC: University of North Carolina Highway Safety Research Center.

Nelson, M.C., Gordon-Larsen, P., Song, Y. and Popkin, B.M. 2006. Built and social environments: Associations with adolescent overweight and activity. *American Journal of Preventive Medicine*, 31(2), 109-17.

Nelson, N.M., Foley, E., O'Gorman, D.J., Moyna, N.M. and Woods, C.B. 2008. Active commuting to school: How far is too far? *International Journal of Behavioral Nutrition and Physical Activity*, 8(5), 1.

Ness, A.R., Leary, S.D., Mattocks, C., Blair, S.N., Reilly, J.J., Wells, J., et al. 2007. Objectively measured physical activity and fat mass in a large cohort of children. *Public Library of Science Medicine*, 4(3), e97.

NHS Information Centre. Lifestyles Statistics. 2006. *Statistics on Obesity, Physical Activity and Diet: England, 2006*. Leeds, UK: Author.

O'Brien, M., Jones, D., Sloan, D. and Rustin, M. 2000. Children's independent spatial mobility in the urban public realm. *Childhood*, 7(3), 257-77.

Ogden, C.L., Carroll, M.D., Curtin, L.R., McDowell, M.A., Tabak, C.J. and Flegal, K.M. 2006. Prevalence of overweight and obesity in the United States, 1999-2004. *Journal of the American Medical Association*, 295(13), 1549-55.

Oliver, M., Schofield, G. and McEvoy, E. 2006. An integrated curriculum approach to increasing habitual physical activity in children: A feasibility study. *Journal of School Health*, 76(2), 74-9.

Panter, J.R., Jones, A.P. and van Sluijs, E.M. 2008. Environmental determinants of active travel in youth: A review and framework for future research. *International Journal of Behavioral Nutrition and Physical Activity*, 5(34), doi:10.1186/479-5868-5-34.

Proctor, M.H., Moore, L.L., Gao, D., Cupples, L.A., Bradlee, M.L., Hood, M.Y., et al. 2003. Television viewing and change in body fat from preschool to early adolescence: The Framingham Children's Study. *International Journal of Obesity and Related Metabolic Disorders*, 27(7), 827-33.

Ridgers N.D., Stratton, G., Fairclough, S.J. and Twisk, J.W. 2007. Long-term effects of playground markings and physical structures on children's recess physical activity levels. *Preventive Medicine*, 44(5), 393-7.

Robertson-Wilson, J.E., Leatherdale, S.T. and Wong, S.L. 2008. Social-ecological correlates of active commuting to school among high school students. *Journal of Adolescent Health*, 42(5), 486-95.

Roemmich, J.N., Epstein, L.H., Raja, S., Yin, L., Robinson, J. and Winiewicz, D. 2006. Association of access to parks and recreational facilities with the physical activity of young children. *Preventive Medicine*, 43(6), 437-41.

Roemmich, J.N., Epstein, L.H., Raja, S. and Yin, L. 2007. The neighborhood and home environments: Disparate relationships with physical activity and sedentary behaviors in youth. *Annals of Behavioral Medicine*, 33(1), 29-38.

Rosenberg, D.E., Sallis, J.F., Conway, T.L., Cain, K.L. and McKenzie, T.L. 2006. Active transportation to school over 2 years in relation to weight status and physical activity. *Obesity*, 14(10), 1771-6.

Sallis, J.F., McKenzie, T.L., Alcaraz, J.E., Kolody, B., Faucette, N. and Hovell, M.F. 1997. The effects of a 2-year physical education program (SPARK) on physical activity and fitness in elementary school students. *American Journal of Public Health*, 87(8), 1328-34.

Sallis, J.F., Prochaska, J.J. and Taylor, W.C. 2000. A review of correlates of physical activity of children and adolescents. *Medicine and Science in Sports and Exercise*, 32(5), 963-75.

Salmon, J., Salmon, L., Crawford, D.A., Hume, C. and Timperio, A. 2007. Associations among individual, social, and environmental barriers and children's walking or cycling to school. *American Journal of Health Promotion*, 22(2), 107-13.

Salmon, J. and Timperio, A. 2007. Prevalence, Trends and Environmental Influences on Child and Youth Physical Activity, in *Pediatric Fitness. Secular Trends and Geographic Variability,* edited by G. Tomkinson and T. Olds. Basel, Switzerland: Karger.

Schlossberg, M., Greene, J., Paulsen, P., Johnson, B. and Parker, B. 2006. School trips: Effects of urban form and distance on travel mode. *Journal of the American Planning Association*, 72(3), 337-46.

Scott, M.M., Cohen, D.A., Evenson, K.R., Elder, J., Catellier, D., Ashwood, J.S., et al. 2007. Weekend schoolyard accessibility, physical activity, and obesity: The Trial of Activity in Adolescent Girls (TAAG) study. *Preventive Medicine*, 44(5), 398-403.

Shi, Z., Lien, N., Kumar, B.N. and Holmboe-Ottesen, G. 2006. Physical activity and associated socio-demographic factors among school adolescents in Jiangsu Province, China. *Preventive Medicine*, 43(3), 218-21.

Sirard, J.R., Ainsworth, B.E., McIver, K.L. and Pate, R.R. 2005. Prevalence of active commuting at urban and suburban elementary schools in Columbia, SC. *American Journal of Public Health*, 95(2), 236-7.

Sjolie, A.N. and Thuen, F. 2002. School journeys and leisure activities in rural and urban adolescents in Norway. *Health Promotion International*, 17(1), 21-30.

Sonkin, B., Edwards, P., Roberts, I. and Green J. 2006. Walking, cycling and transport safety: An analysis of child road deaths. *Journal of the Royal Society of Medicine*, 99(8), 402-5.

Spinks, A.B., MacPherson, A.K., Bain, C.J. and McClure, J. 2006a. Injury risk from popular childhood physical activities: Results from an Australian primary school cohort. *Injury Prevention*, 12, 390-4.

Spinks, A.B., McClure, R.J., Bain, C. and Macpherson, A.K. 2006b. Quantifying the Association Between Physical Activity and Injury in Primary School-Aged Children. *Pediatrics*, 118(1), e43-e50.

Stewart, J.A., Dennison, D.A., Kohl, H.W. and Doyle, J.A. 2004. Exercise level and energy expenditure in the TAKE 10!® in-class physical activity program. *Journal of School Health*, 74(10), 397-400.

Story, M., Kaphingst, K.M. and French, S. 2006. The role of schools in obesity prevention. *The Future of Children*, 16(1), 109-42.

Stratton, G. 2000. Promoting children's physical activity in primary school: An intervention study using playground markings. *Ergonomics*, 43(10), 1538-46.

Stratton, G. and Mullan, E. 2005. The effect of multicolor playground markings on children's physical activity level during recess. *Preventive Medicine*, 41(5-6), 828-33.

Timperio, A., Ball, K., Salmon, J., Roberts, R., Giles-Corti, B., Simmons, D., et al. 2006. Personal, family, social, and environmental correlates of active commuting to school. *American Journal of Preventive Medicine*, 30(1), 45-51.

Timperio, A., Crawford, D., Telford, A. and Salmon, J. 2004a. Perceptions about the local neighborhood and walking and cycling among children. *Preventive Medicine*, 38(1), 39-47.

Timperio, A., Giles-Corti, B., Crawford, D., Andrianopoulos, N., Ball, K., Salmon, J., et al. 2008. Features of public open spaces and physical activity among children: Findings from the CLAN study. *Preventive Medicine*, 47(5), 514-18.

Timperio, A., Salmon, J. and Ball, K. 2004b. Evidence-based strategies to promote physical activity among children, adolescents and young adults: Review and update. *Journal of Science and Medicine in Sport*, 7(1 Supplement), 20-9.

Transportation Research Board. 2002. *The Relative Risks of School Travel: A Nationwide Perspective and Guidance for Local Community Risk Assessment.* Washington, DC: Transportation Research Board of the National Academies.

Tranter, P. and Pawson, E. 2001. Children's access to local environments: A case-study of Christchurch, New Zealand. *Local Environment*, 6(1), 27-48.

Tudor-Locke, C., Ainsworth, B.E., Adair, L.S. and Popkin. B.M. 2003. Objective physical activity of Filipino youth stratified for commuting mode to school. *Medicine and Science in Sports and Exercise*, 35(3), 465-71.

van der Ploeg, H.P., Merom, D., Corpuz, G. and Bauman, A.E. 2008. Trends in Australian children traveling to school 1971-2003: Burning petrol or carbohydrates? *Preventive Medicine*, 46(1), 60-2.

van Sluijs, E., McMinn, A. and Griffin, S. 2007. Effectiveness of interventions to promote physical activity in children and adolescents: Systematic review of controlled trials. *British Medical Journal*, 335(7622), 703.

Veugelers, P., Sithole, F., Zhang, S. and Muhajarine, N. 2008. Neighborhood characteristics in relation to diet, physical activity and overweight of Canadian children. *International Journal of Pediatric Obesity*, 3(3), 152-9.

Vincent, S.D., Pangrazi, R.P., Raustorp, A., Tomson, L.M. and Cuddihy, T.F. 2003. Activity levels and body mass index of children in the United States, Sweden, and Australia. *Medicine and Science in Sports and Exercise*, 35(8), 1367-73.

Wang, Y. and Lobstein, T. 2006. Worldwide trends in childhood overweight and obesity. *International Journal of Pediatric Obesity*, 1(1), 11-25.

Weir, L.A., Etelson, D. and Brand, D.A. 2006. Parents' perceptions of neighborhood safety and children's physical activity. *Preventive Medicine*, 43(3), 212-17.

Wen, L.M., Fry, D., Merom, D., Rissel, C., Dirkis, H. and Balafas, A. 2008a. Increasing active travel to school: Are we on the right track? A cluster randomised controlled trial from Sydney, Australia. *Preventive Medicine*, 47(6), 612-18.

Wen, L.M., Fry, D., Rissel, C., Dirkis, H., Balafas, A. and Merom, D. 2008b. Factors associated with children being driven to school: Implications for walk to school programs. *Health Education Research*, 23(2), 325-34.

World Health Organization. 2002. *The World Health Report 2002. Reducing Risks, Promoting Healthy Life.* Copenhagen.

World Health Organization. 2005. *The European Health Report 2005. Public Health Action for Healthier Children and Populations.* Copenhagen.

Worobey, J., Adler, A. and Worobey, H.S. 2004. Diet, activity, and risk for overweight in a sample of Head Start children. *Journal of Children's Health*, 2(2), 133-44.

Yelavich., S., Towns, C., Burt, R., Chow, K., Donohue, R., Sani, H.S., et al. 2008. Walking to school: Frequency and predictors among primary school children in Dunedin, New Zealand. *New Zealand Medical Journal*, 121(1271), 51-8.

Zaccari, V. and Dirkis, H. 2003. Walking to school in inner Sydney. *Health Promotion Journal of Australia*, 14(2), 137-40.

Ziviani, J., Scott, J. and Wadley, D. 2004. Walking to school: Incidental physical activity in the daily occupations of Australian children. *Occupational Therapy International*, 11(1), 1-11.

Ziviani, J.M., Kopeshke, R.E. and Wadley, D.A. 2006. Children walking to school: Parent perceptions of environmental and psychosocial influences. *Australian Occupational Therapy Journal*, 53(1), 27-34.

PART IV
Obesogenic Environments and Policy Responses

Chapter 10
Policy Responses and Obesogenic Food Environments

Katrina Giskes

Introduction

The rising prevalence of overweight and obesity among children and adults across countries and cultures in the developed world has raised questions about the role of environmental factors in the obesity epidemic (Schafer Elinder and Jansson 2009, Hill et al. 2008, Poston and Foreyt 1999). Technological advances and geographic movement of the population to urban areas have resulted in large changes to the food environment (Lopez 2007). As the focus of obesity research has moved to identify more upstream determinants of increasing body size the role of the food environment in contributing to people's dietary behaviours has taken on new significance (Sacks et al. 2009). In response to emerging evidence health practitioners, politicians, workers, parents and the general population are calling for changes to the food environments that promote high energy and fat intakes (Sacks et al. 2009, Swinburn 2008).

Access to an adequate supply of nutritious food is essential for human life. In modern societies and with modern lifestyles, almost everybody is dependent on the commercial food environment to obtain their nutritional needs, and the vast majority of individuals are wholly dependent on it (Morland et al. 2002a, 2002b). Food consists of multiple nutrients that are required to a greater or lesser degree (for example macronutrients, vitamins and minerals). The food environment may be a potential threat to human health if supplies of essential nutrients are insufficient. However, the greatest hazard posed by the food environment in developed countries and increasingly in developing countries is not one of insufficiency, rather one of surplus. Excess supplies of certain nutrients, namely energy, total fat and saturated fat, are the greatest risk the current food environment poses to the population (James 2008b, Branca et al. 2007, Stanton 2006).

Changes to the food environment are a necessary step to address the obesity epidemic. The food environment is driven by a myriad of forces that operate at the individual- to international-levels (James 2008b). Many of these forces, including innate human preferences for energy-dense and high-fat foods, favour obesogenic food environments (James 2008a). Policy strategies represent our best tool for tackling food environments; policies have the ability to target the many different sectors of the food environment, are relatively cost-effective, reach all segments of the population and can result in sustained change (Sacks et al. 2009). At their core, all policy responses to the food environment are geared toward making healthy food choices easy for individuals.

This chapter will identify a broad scope of policy instruments to address the food environment and will briefly review how some policies (or lack of policy action) may create an environment that contributes to weight gain, and higher prevalence of overweight and obesity. In doing so, this chapter will identify where current policy action has predominantly been concentrated, whether this activity has been synergistic in tackling obesogenic food environments and identify future policy opportunities that may address obesogenic food environments in developed countries.

Policy Responses

Policy responses may be a system of laws, regulatory measures, courses of action or funding priorities (Sacks et al. 2009). Policy instruments can be classified as 'soft' or 'hard' actions to address obesogenic food environments (Swinburn 2008). Soft policy approaches include measures that operate on informing or persuading individuals to make better dietary choices, such as social marketing, health promotion programmes and government advocacy for changes in individual and organizational behaviour. Hard policy instruments take choice and rationalisation out of the hands of the individual through the use of laws, regulations and financial implications (Swinburn 2008).

To date, governments have preferred soft policy options to tackle the obesity problem and have steered away from the 'policy battleground' of hard policy options (James 2008b, Swinburn 2008). However, the adoption of soft policy instruments has not resulted in many changes in the food environment per se since soft options only tackle how individuals engage with their environment. Fighting this uphill battle has proven to be not only challenging, but also costly and ineffective if the underlying food environments that promote overweight and obesity are not changed (Swinburn 2003, Friel et al. 2007). Additionally, there is some evidence to suggest that soft policy instruments increase health inequalities, since they are more readily adopted by higher socioeconomic groups (Swinburn 2008). In contrast, hard policy options operate across the board and target the behaviours of all socioeconomic groups equally.

However, the preference for soft policy instruments by governments is not universal across all health-related behaviours. Governments have been more progressive in adopting hard legislative and regulatory policies for other health-related behaviours, such as tobacco and alcohol consumption (Swinburn 2008, Friel et al. 2007). Examples include: the regulation of tobacco outlets; taxation of tobacco; age at which these substances can be purchased and restricting places where people can smoke. A major barrier to hard policies aimed at the food environment is resistance from the food industry (Swinburn 2008), akin to the hostile response the tobacco industry has shown for tobacco-control policies. The food industry advocates self regulation and consumer choice, and positions the obesity epidemic as largely driven by individual behaviour. The food industry also has concerns that providing more 'healthy' food choices or more healthy recipes for existing food items will result in increased food prices, and therefore less consumer demand (Swinburn 2008). The food industry is a major employer in most developed countries; federal governments are hesitant to bring about changes that may result in reduced employment and are therefore reluctant to bring about large changes in food supply or food production (see Chapter 4). Additionally, the population has a greater vested interest (for survival, and the social aspects of food consumption) in the food environment than environments that promote smoking and harmful alcohol consumption as food is consumed and enjoyed by everyone.

The Food Environment

The food environment is highly complex and results from systems interacting and operating at different levels of the food-supply chain. Components of the supply chain are: primary production (for example agriculture, fishing); food processing (for example combining, cooking and/or packaging raw ingredients); transportation (for example distribution within the country, import and export); marketing (for example endorsements, point of sale, TV advertising); retail (for example supermarkets, wholesalers) and catering/food service (for example canteens, takeaway outlets) (Figure 10.1) (Sacks et al. 2009, Friel et al. 2007). Consumers interact with the last two steps in the food-supply chain. However, the quantity and types of foods presented in the consumer's food environment are contingent on earlier steps in the food-supply chain. Comprehensive policies for the food environment need to intervene at all points of the food-supply chain to have the greatest impact on obesogenic food environments (Sacks et al. 2009, Friel et al. 2007).

Levels of Governance for Food Policy

As identified by Sacks et al. (2008, 2009), there are a number of 'levels of governance' that are responsible for the creation of policy that effects the food environment. These may vary by country, however they are relatively similar among developed countries, and include local, state and federal government, as well as organizational and international levels of governance. Recently, the World Health Organization (WHO) released a document examining the various policy options (targeting different levels of governance) to address overweight/obesity in Europe (Branca et al. 2007).

```
┌─────────────────────────────────────────────┐
│           Primary production                │
│        e.g. agriculture, fishing            │
└─────────────────────────────────────────────┘
                      ⇩
┌─────────────────────────────────────────────┐
│             Food processing                 │
│  e.g. combining, cooking or packaging raw ingredients │
└─────────────────────────────────────────────┘
                      ⇩
┌─────────────────────────────────────────────┐
│              Transportation                 │
│  e.g. distribution within the country, import, export │
└─────────────────────────────────────────────┘
                      ⇩
┌─────────────────────────────────────────────┐
│                Marketing                    │
│   e.g. endorsements, point of sale, TV advertising │
└─────────────────────────────────────────────┘
                      ⇩
┌─────────────────────────────────────────────┐
│                  Retail                     │
│        e.g. supermarkets, wholesale         │
└─────────────────────────────────────────────┘
                      ⇩
┌─────────────────────────────────────────────┐
│           Catering/food service             │
│       e.g. canteens, takeaway outlets       │
└─────────────────────────────────────────────┘
```

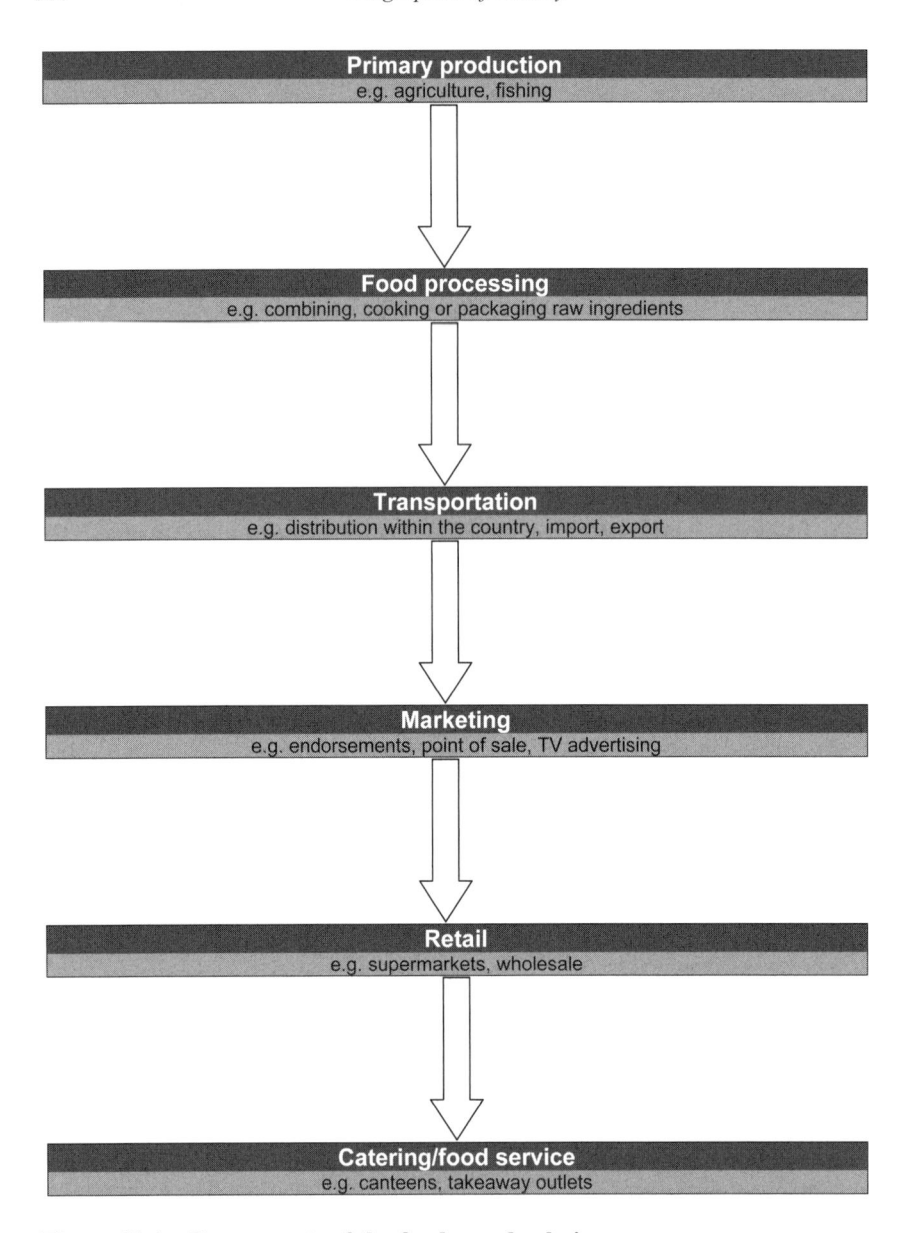

Figure 10.1 Components of the food supply chain

Local government is generally responsible for land-use management, including the number, types and locations of food outlets and the use of land for primary production. State governments are often responsible for food distribution and the types of foods sold/marketed in schools, and federal governments control food composition standards, food importation, labelling/advertising and taxes/subsidies on foods. International governance regulates trade arrangements between countries, and food labelling/advertising standards. Some policies relating to the food environment are handled across different levels of government. For example, state, federal and international governance develops fiscal policies for primary production (Sacks et al. 2009).

Organizational governance is responsible for the policies adopted within organizations, such as food companies, workplaces, schools and food stores. Policies at these levels may include: the types of foods available at schools/ workplaces; product composition standards for food catering; product placement in stores and food marketing (Sacks et al. 2009).

To identify policy actions that may influence the food environment Sacks et al. (2008, 2009) have developed a policy action framework comprising two dimensions: the food system and the level of governance. Use of this framework can identify policy opportunities that can be undertaken by each level of governance that influence the food supply and (consequently) the commercial food environment. This framework can also be used to assess the comprehensiveness and/or deficiencies of current policy actions aimed at addressing the food environment, and to map future potential policy areas (Sacks et al. 2009).

The framework shows there are policy opportunities for each level of governance to influence the food supply chain and (consequently) the commercial food environment (Sacks et al. 2009) (Table 10.1). It also highlights that, in some steps of the supply chain, concerted efforts are required across all levels of government. However, other steps are controlled more narrowly by one or two levels of governance. Some cells of the matrix are empty, meaning regulation of some aspects of the food supply are beyond the regulatory boundaries of some levels of governance. However, some of these blank cells may also represent opportunities for the development of innovative policies in the future, some of which are outlined below.

Policy Opportunities to Tackle Obesogenic Food Environments

Primary Production Subsidies and Taxes

Primary production subsidies and taxes at the state, national and international levels influence the products that food manufacturers, distributors and retailers make available to consumers and the price they are sold at. The over-supply of fats/ oils and sweeteners in national and international commodity markets has resulted in greater added fats and sugars in the food supply in recent decades, and this is thought to be a fundamental driver of the obesity epidemic seen in developed countries, and now starting in developing nations (Hill et al. 2008, James 2008b, Friel et al. 2007).

Table 10.1 Policy instruments under the jurisdiction of different levels of governance that may address the food environment

Food chain	Levels of governance			
	Local government	State government	National government	International	Organizational
Primary production	- Primary production subsidies for health-promoting foods and taxes for obesogenic foods (e.g. research into healthier food alternatives, investment in infrastructure for the production of health-promoting foods, subsidies for the production of healthy foods)			
	- Land use management			
	- Development of community gardens			
Food processing			- Food standard regulations for fat/sugar content of foods - Food labelling regulations for the healthiness of foods		
Transportation			- Importation restrictions on obesogenic foods - Subsidies for the import/transportation of health-promoting foods - Taxes for the import/transportation of foods associated with overweight/obesity - Policies promoting the accessibility and affordability of health-promoting foods to remote communities	- International trade arrangements- limits and restrictions on obesogenic foods/ ingredients - Subsidies for health-promoting foods for developing countries	

Table 10.1 continued   Policy instruments under the jurisdiction of different levels of governance that may address the food environment

| Food chain | Levels of governance |||||
	Local government	State government	National government	International	Organizational
Marketing	- Legislation and restrictions on food marketing (e.g. placement of food advertising in and around schools)	- Legislation and restrictions on food marketing (e.g. in schools)	- Legislation and restrictions on food marketing (e.g. size and content of advertising, health warning labels, advertising exposure during children's TV shows) - Taxing the marketing of obesogenic foods	- Legislation and restrictions on food marketing	- Policies on direct and indirect marketing of obesogenic foods in schools (e.g. sponsorship by food companies, advertising of food products on school premises)
Retail	- Density, type and location of supermarket and fast food outlets (town planning)				- Types of foods sold in outlets (e.g. range of health-promoting foods) - Product placement in stores (e.g. making foods less visible to children) - Point of sale product information (e.g. nutrition content information)
Catering/food service		- Nutrition requirements for accreditation - Nutrition training of childcare/teaching staff - School nutrition policies (restricting the types of foods available in schools)	- Nutrition requirements for accreditation - Nutrition training of childcare/teaching staff - School nutrition policies (restricting the types of foods available in schools)		- School nutrition policies (restricting the types of foods available in schools) - Availability and variety of obesogenic and health-promoting foods in the workplace

Source: Adapted from Sacks et al. [5].

Over the past few decades, agricultural policies in most developed countries, most notably in the US, have supported the production of grain and oilseed crops (James 2008b, Story et al. 2008). This has resulted in an over-production of these crops and a subsequent flooding of these products on to national and international food markets. Fat/oils and grain products have a relatively long shelf life and are among the food items most resistant to market or climate fluctuations (Story et al. 2008). The consequence of this over-supply has been marked reductions in the prices of fats/oils and sweeteners, with dramatic falls in the prices of these commodities over the last four decades (Hill et al. 2008, James 2008b).

Their low prices have made them attractive raw ingredients for food manufacturers, who have developed different forms for their use, such as hydrogenated fats and high fructose corn syrup. These products are thought to be even more obesogenic (Bray et al. 2004, Jacobson 2004) and artherogenic (Mozaffarian et al. 2006) than their raw ingredients, however their versatility in processed products has favoured their use among food manufacturers. In a competitive food marketplace, where consumer behaviour is largely driven by price, food manufacturers have increasingly incorporated fats/oils and refined sugars as a means to keep food prices down (Story et al. 2008). Parallel to reductions in the prices of fat/oils and sweeteners, the prices of fresh food items – particularly fruits and vegetables – have increased (James 2008b, Story et al. 2008). Consequently, energy-dense foods have become increasingly affordable (James 2008b), and therefore favoured by low-income groups, who suffer disproportionally from overweight/obesity (Friel et al. 2007) (see Chapter 4).

A policy response to combat excess fat and sugar in our food supply, primary production subsidies and taxes should be oriented toward population nutrition recommendations. Governments at all levels should support the production of fruits and vegetables, so their prices to food manufacturers and consumers are low compared to with less-healthy options. Such policy can come in many forms, such as increased dollars for research into farming practices, investment in infrastructure for their distribution and processing, farming subsidy policies that alleviate production costs and taxes placed on food items or ingredients that are high in fat and/or sugar.

Policies Aimed at Food Manufacturing

There are several key ways by which policies aimed at food manufacturing may reduce or mitigate the development of obesogenic food environments. These strategies centre on food standard regulations and food labelling, and are mainly within the jurisdictions of national governments.

National governments have food standard regulations that specify the ingredients and contents of key nutrients in different types of foods, and food manufacturers must abide by these regulations (Food and Nutrition Standards Australia and New Zealand 2008a). These standards ensure the quality of foods that are sold in the commercial food environment. However, adopting food standard regulations in line with public health nutrition recommendations may specify 'upper limits' for the fat and sugar contents of commercially-manufactured foods. Such standards, when applied throughout the country, may make the commercial food environment less obesogenic and may contribute to lower energy, fat and sugar intakes among the population. Recently, the US states of California and New York adopted food standard regulations for the trans-fatty acid contents of packaged foods and restaurant meals in an effort to target atherosclerosis (Food and Nutrition Standards Australia and New Zealand 2008b). This may set precedence for food standards regulations targeting overweight/obesity and (consequently) the fat and sugar content of foods.

Furthermore, policies aimed at food manufacturing may target food labelling as a consumer-oriented approach to changing the food environment. This may involve, for example, labelling manufactured food items as having 'low', 'medium' or 'high' contents of fat and/or sugar, so consumers can make more-educated purchasing decisions in the commercial food environment. The traffic light system of green, orange and red symbols on food packaging that denotes 'healthier' food choices, together with consumer education, has been shown to lead to consumer recognition and health-promoting changes in food purchasing behaviour (Lobstein and Davies 2009, James 2008b). To date, this strategy of enabling consumers to identify and limit their intakes of foods high in fat and other nutrients at the point of purchase, has primarily been adopted at the organizational-level (for example school/work canteens, restaurants) (Story et al. 2008). Extending this to food labelling at the national-level may provide consumers with across-the-board recognition of food choices that are more consistent with dietary recommendations. Other, more 'hard-hitting' strategies, may include health warning labels on foods that are high in fat and/or sugar, similar to those many countries have adopted for tobacco products (Story et al. 2008).

Food Distribution Policy Opportunities

Food distribution occurs at international, national and regional levels and concerted policy action at all levels may contribute to more health-promoting food environments (Sacks et al. 2009). Subsequently, lack of policy action at even one level may result in food environments that promote high energy, fat and sugar intakes.

International trade arrangements play a large role in national food supplies. The over-supply of fats/oils and sweeteners in food markets has lead to a 'dumping' of these cheap ingredients in the developing world, where some of the most rapid changes in the food environment and increases in the prevalence of overweight/obesity are being observed (see Figure 2.3 in Chapter 2) (Hill et al. 2008, James 2008b, Friel et al. 2007). Traditional diets in many of these countries have been healthy and some countries even suffered from food insufficiency. However, increased availability of cheap, energy-dense ingredients has forced many developed countries into a 'nutrition transition' (see Chapter 2) to food environments that are increasingly characterized by high energy-density choices (Astrup et al. 2008). International governing bodies, such as the World Trade Organization, are working to address these rapid changes in food environments and are exploring policy interventions to curb the nutrition transition (Chopra et al. 2002). Such policy options may restrict/limit the trade of obesogenic foods to developing countries, and may provide subsidies for developing countries for the purchase of fresh and nutritious foods from international food markets.

At the national level, there are policy opportunities that can regulate the importation of foods, subsidies/taxes to be placed on food items and policies that ensure an equitable distribution of foods within countries. Countries may choose to limit their importation of energy-dense foods and food items, in efforts to control the population prevalence of overweight/obesity and to reduce the health implications of the nutrition transition. National-level policies can also contribute to healthier food environments by subsidising nutritious foods such as fruit and vegetables. This may be particularly important in countries that grow few fruits or vegetables, where importation contributes to their high prices.

Remote communities in countries that are geographically-vast, such as Australia and Canada often report food environments that encourage the consumption of energy-dense foods. The cost of fruits and vegetables in these communities is a major barrier to their consumption, and they are often of poor quality due to long transportation times (Heading 2008, Burns et al. 2004). Consequently, energy-dense foods that are cheaper and more shelf-stable are favoured, and these communities suffer disproportionately from overweight/obesity (Heading 2008). Furthermore, these communities are often indigenous and suffer high levels of poverty, and have higher prevalence of overweight/obesity and diet-related chronic disease than their non-indigenous counterparts (Belanger-Ducharme and Tremblay 2005, Thorburn 2005). Lack of access to affordable and healthy foods may be major contributing factors to disparities in diet-related chronic disease and obesity rates (Heading 2008). Healthier food environments in these communities may be achieved if the distribution of fresh, nutritious foods is subsidized by national governments. Furthermore, these communities may benefit from the development of community gardens (where climate conditions permit) that grow produce locally, make the community more self-sufficient and may improve dietary intakes (Alaimo et al. 2008).

Food Marketing

The food environment is not only composed of the types and prices of food available to consumers. An important aspect of the food environment are the types of food messages that individuals are exposed to, as targeted media messages influence the food choices people make (Hawkes 2007, Story and French 2004). If these food marketing messages reinforce poor dietary choices, they can contribute to a food environment that encourages obesogenic dietary behaviours (Story et al. 2008).

Children, in particular, are targets for food marketing. They lack the cognitive development to understand the persuasive intent of food advertising, and food companies capitalize on their influence on household food purchasing decisions (Hawkes 2007, Story and French 2004). Food marketing aimed at children often encourages them to ask their parents to buy food products, taking advantage of the 'nag factor'. Food marketing to children can be direct marketing (for example television advertising, signage) or indirect marketing (for example competitions, product displays, sponsorship) (Samuels et al. 2003).

Insights into policy opportunities for food marketing can be obtained from tobacco and alcohol advertising (Story and French 2004). Challenges in the development of policy aimed at food marketing lie in defining the foods that are considered 'unhealthy', as these foods are not as precise a category as alcohol or tobacco (Story and French 2004). However, strategies based on this model may involve: banning all advertising of unhealthy foods in public places (especially around facilities frequented by children); placing size limits on the advertising of unhealthy food items at the point of sale; taxing the advertising of unhealthy foods to fund obesity-prevention programmes and requiring health/pictorial warnings on advertisements for unhealthy foods (Story and French 2004). To date, no country has adopted such a comprehensive policy approach to the promotion of unhealthy food choices.

While some policy action is better than none, policies targeting the advertising/promotion of unhealthy foods will be most effective if policy is adopted across the various environments with which individuals interact (Samuels et al. 2003). For example, school-level policies banning the marketing of unhealthy foods are likely to be most successful if marketing is restricted simultaneously around schools, on television programmes and in food stores. To date, in many countries and regions there have not been concerted activities across the different environments with which individuals interact (Story et al. 2008, Story and French 2004).

The majority of research examining food advertising among children has examined the influence of television advertising. This research has shown that much of the advertising during children's television programmes focuses on food products, especially foods and beverages that are high in fat/sugar, energy-dense and low in fibre (Story and French 2004). Children who are exposed to greater television food advertising have shown higher rates of overweight/obesity compared to their counterparts who are exposed to less advertising (Nestle 2006). A number of countries such as Sweden, Norway and Italy have developed national policies restricting or banning the television advertising of processed food products during children's television viewing times (Samuels et al. 2003).

Other policy actions at the national, state, local and organizational levels have aimed at decreasing children's exposure to indirect marketing strategies. For example, local governments in Brazil have restricted the advertising of 'junk foods' on billboards and bus shelters within buffer zones around schools (Story and French 2004). Other policies, generally adopted by lower levels of governance or at the organizational (school) levels, have focussed on restricting the promotion and sponsorship of high-sugar beverages (for example soft drinks) and companies that produce these beverages and/or fast food companies in schools (Story and French 2004, Samuels et al. 2003). However, schools are often coerced into the promotion/marketing of these foods through the provision of resources (such as sporting equipment and much-needed infrastructure) provided by such companies in exchange for access to a captive audience with which they can foster brand loyalty (Swinburn et al. 2008, Samuels et al. 2003). Policies at the national, state, local and/or school-level can restrict the indirect marketing strategies of food companies in schools. Some countries, such as Australia, are also targeting *how* unhealthy foods are advertised to children, restricting the use of licensed or cartoon characters to endorse or advertise products (Parliament of Australia 2008).

Food Retail

The food retail sector encompasses a broad range of activities and establishments, presenting many potential policy opportunities to influence the food environment and, ultimately, contribute to food choices more consistent with a lower prevalence of overweight and obesity. Food retail includes: the number, types and placement of food outlets; the types of products sold in outlets; point of sale product information and product placement in stores (Sacks et al. 2009).

The food retail environment at the community level may be an important determinant of an individual's dietary choices. Among the range of food outlet stores, supermarkets offer the greatest variety of foods and at the lowest prices (Morland et al. 2002a). There is inconsistent evidence that access to supermarkets is associated with healthier food intakes (Pearce et al. 2008, Morland et al. 2002a), and some studies have shown that access to a supermarket is associated with lower BMI (Lopez 2007, Powell et al. 2007).

There is also some evidence that access to supermarkets differs between socioeconomically-advantaged and deprived areas (see Chapter 5). Evidence that socioeconomically-deprived areas have poorer access to supermarkets is more consistently seen in the US (Cummins and Macintyre 2006). Studies in the US and Canada have found that deprived and minority areas are less serviced by supermarkets, but have more independent grocery stores (that stock a limited range of items at higher prices) (Cummins and Macintyre 2006). However, the opposite trend was found for retail provision in the UK; a greater number of supermarkets/discounters were located in deprived areas, and the price and availability of a range of foods did not show much variation by area deprivation (Cummins and Macintyre 2002). An Australian study did not find any association between access to food retail outlets (assessed in terms of distance, numbers of local shops and their opening hours) and area deprivation (Winkler et al. 2006). The provision of supermarkets in communities represents an important potential policy intervention in urban/town planning, whereby the food environments of communities can be improved with access to fresh, nutritious foods at relatively low prices. Other potential strategies aimed at food retail in communities and that provide increased opportunity for healthy dietary intakes are: farmer's markets; community gardens; and incorporating healthy produce and fresh foods into corner stores and convenience stores (Story et al. 2008).

Another important aspect of the food retail environment is fast food and takeaway outlets. There is an increasing trend of consuming foods from takeaway establishments (Bowman et al. 2004, Bowman and Vinyard 2004). The number of takeaway and fast food establishments has increased dramatically over the past two decades (Wang et al. 2008). Furthermore, the proportion of household food expenditure on foods consumed away from home has also increased (Greenwood and Stanford 2008, Bowman et al. 2004). Foods sold at takeaway and fast food establishments tend to be more energy-dense and of poorer nutritional value than foods prepared at home (Greenwood and Stanford 2008). Furthermore, more frequent takeaway food consumption has been associated with higher prevalence of overweight/obesity (Branca et al. 2007). Another body of research has examined accessibility to fast-food/takeaway outlets, dietary intakes and overweight/obesity. US research by Morland et al. (2002a, 2002b) found no association between the presence of fast food outlets and residents' reported intakes of recommended foods and nutrients (Morland et al. 2002a). Similarly, Burdette and Whitaker (2004) and Pearce et al. (2009) found no association between overweight and proximity to fast food outlets in the US and New Zealand, respectively. By contrast, Maddock (2004) observed a strong correlation between the number of residents per fast-food outlet and state-level obesity rates across 50 US states, and Alter and Eny (2005) reported that across 380 administrative regions in Ontario Canada, per capita rate of fast food outlets was positively related with all-cause mortality and hospitalisation for acute coronary syndromes after adjustment for age, sex, and socioeconomic characteristics.

There is mixed evidence of the accessibility of takeaway/fast-food outlets in socioeconomically-deprived areas. An Australian study examining the prevalence of franchised fast-food outlets in Melbourne found that density of stores was significantly related with area-disadvantage, with deprived areas having more outlets per head of population (Reidpath et al. 2002). However, another Australian study found that access to both franchised and un-franchised takeaway stores (ascertained as number of takeaway shops and road distance to takeaway shops) was similar in advantaged and deprived areas (Turrell and Giskes 2008). In the US, Morland et al. (2002b) found no association between the prevalence of fast-food outlets and neighbourhood wealth, whereas later work by Block et al. (2004) showed that low-income neighbourhoods had a higher density of fast-food outlets in New Orleans. The evidence from Britain is similarly inconsistent, with Cummins et al. (2005) and Macdonald et al. (2007) finding that more deprived areas in England and Scotland had a larger number of fast-food outlets per head of population, whereas Macintyre et al. (2005) found no association between the prevalence of fast-food chain restaurants and neighbourhood disadvantage in Glasgow, Scotland. Despite there being mixed evidence about accessibility to takeaway/fast food outlets and their role in overweight/obesity (and socioeconomic inequalities in these), policy regulating takeaway outlets in areas can still be justified in terms of making healthy food choices easy choices for the population. There are policy opportunities for local and state governments (through town-planning regulations) to control the location, density and types of takeaway/fast food outlets in areas. Increasingly, such town planning principles are being taken up in the 'healthy cities' movement (O'Neill and Simard 2006), which acknowledges that individual's health behaviours are influenced by the physical environments with which they interact.

Town planning regulations are not the only way that the food retail environment can promote healthier food choices. There are a number of potential policy strategies that can take place within food stores. Point-of-sale nutrition information may also help consumers to identify healthier food choices (for example nutrient-content information at restaurants/takeaway outlets). Some interventions, such as identifying healthier choices (for example the traffic-light system to identify more 'healthy' choices) and promoting the nutrient contents of fresh produce have been trialled in supermarkets with some success (James 2008b). Modest evidence has also shown that interventions to increase the availability, variety, pricing and promotional strategies of healthy foods result in their increased purchase (Story et al. 2008). Some major franchise takeaway outlets are increasing their range of healthier takeaway options and providing nutrition information about their products, however this is not a legislated requirement in most countries. Organizational policies, such as those adopted by some supermarkets in Australia, are creating environments where consumers can restrict children's exposure to the advertising and marketing of unhealthy foods. These policies regulate the product placement of food in stores, making these foods less visible to children, restricting the promotion of unhealthy choices and creating check-out lanes that are free from sweets.

Catering/Food Service

The childcare and school food environments can have a large impact on children's dietary intakes as a large proportion of the daily meals may be consumed at these locations. Similarly, worksites are important food environments for adults (see Chapter 6). There are a number of opportunities for policy action in childcare, schools and workplaces to curb the obesity epidemic by encouraging individuals to make healthier food choices (Story et al. 2008).

Childcare and school settings provide a valuable opportunity for the promotion of healthy dietary behaviours that can help prevent the development of overweight and obesity in adulthood. There is an increasing momentum around the development of policy/regulations for the types of foods available in schools and childcare settings (Story et al. 2008, Cooke et al. 2007, Pollard et al. 1999). Policy about the types of foods available in schools/childcare settings has been largely managed by each school/childcare centre separately, however higher levels of governance (for example state governments) are becoming increasingly involved (Cooke et al. 2007, Pollard et al. 1999). In many developed countries, such as Australia, there are nutrition requirements that childcare centres must meet in order to be accredited. However, these requirements have traditionally focussed on the nutritional adequacy of foods provided in childcare (in terms of meeting dietary intake recommendations), rather than promoting dietary intakes that result in reduced prevalence of overweight/obesity (Story et al. 2008, Nutrition Australia 2002). Policy opportunities exist for higher levels of governance to incorporate the prevention of overweight and obesity in the nutrition-related requirements for the accreditation of childcare centres. The health and nutrition training of childcare workers also represents a policy area that has received little attention and one that has the potential to help ensure healthier food environments in childcare settings (Story et al. 2008).

Nutrition policy at schools that stipulate the types of foods that may be available have, until recently, been adopted by schools on an ad-hoc basis. In countries that serve school meals, such as the US and UK, there is a federal-level policy that these meals must meet national dietary guideline standards (Story et al. 2008, Nestle 2006). In Australia and the US, there has been increased activity in the development of state-based policy about what foods school canteens can sell. However, the development of such policy is often fragmented in these countries, and further action needs to unify policy across states. There has also been an increased movement by individual schools in Australia toward policies that restrict the types of foods children are permitted to bring to school, banning foods that are high in fat and sugar. Furthermore, policies that limit the availability of energy-dense foods for children at schools should be complemented with classroom-based nutrition education to increase children's skills for developing non-obesogenic dietary habits.

The worksite food environment is a less regulated food environment compared to childcare centres and schools (Story et al. 2008). A number of worksite nutrition programs have shown that nutrition behaviours can be positively influenced by changes to the worksite food environment (French 2005). Dietary intakes have been shown to be positively influenced by strategies such as increasing the availability and variety of healthy food options, reducing the price of healthy options in canteens and vending machines, and by tailored nutrition-education messages (Story et al. 2008, French 2005). Further activities may identify the barriers (both social and contextual) to consuming healthy foods faced by employees.

Concerted Policy Action is Required

The food environment is complex and it is influenced by a number of different factors. Furthermore, individuals interact with numerous food environments on a daily basis. While singular aspects of the food environment can be targeted by specific policies outlined in the previous sections, these efforts may only bring about limited change if other parts of the food environment remain unaddressed

(Sacks et al. 2009, Swinburn 2008). For example, improving the availability and price of healthy food options in schools or workplace canteens may not be effective in reducing overweight/obesity if other changes (such as the promotion of unhealthy food choices) are not made. A commitment to reducing overweight/obesity is required from all levels of governance, and this needs to be implemented in all sectors of the food chain if the food environment is to become less obesogenic (Sacks et al. 2009). Once there are changes in the food environment, the translation of these to improvements in the population prevalence of overweight/obesity needs to be bridged by 'soft' policy strategies (such as health promotion/social marketing) at the national-, state- and local-levels that inform individuals and influence social norms about the importance of healthy dietary choices in controlling overweight/obesity (Swinburn 2008). Therefore, governments need to make a conscious effort to address the food environments and commit to developing and adopting policies across all levels of governance and all sectors of the food chain to make the most impact on reducing obesogenic food environments (see Chapter 4).

Conclusions

The food environment is increasingly being seen as an important determinant of overweight/obesity, consequently tackling food environments has become a fertile policy space and battleground for addressing the obesity epidemic. However, regulating the food environment is complex and requires coordinated action from different sectors spanning from primary production to food retail and foodservice establishments and harmonising policy development at all levels of governance, ranging from the international level to the local government level. A challenge for policy makers is that it is not only the food environment that needs hard policy options to tackle the obesity epidemic, but also the way in which people interact with their food environments needs to be addressed through soft policy options so that people have the necessary personal resources to make healthy food choices. At some levels of governance, and in some sectors, there is resistance to addressing obesogenic food environments with issues such as encroaching on personal choice, maintaining profit margins in the food industry and retaining jobs in the food sector being major stumbling blocks to addressing them. Although this chapter has outlined extensive policy opportunities to address obesogenic food environments, we are often powerless to implement changes at all levels. However, there is still much more to be gained from taking small steps to address only one aspect of the food environment, such as food labelling regulations, or to tackling smaller food environments such as our local, school or workplace food environments.

References

Alaimo, K., Packnett, E., Miles, R.A. and Kruger, D.J. 2008, Fruit and vegetable intake among urban community gardeners. *Journal of Nutrition Education and Behavior,* 40, 94-101.

Alter, D. and Eny, K. 2005. The relationship between the supply of fast-food chains and cardiovascular outcomes. *Canadian Journal of Public Health,* 96, 173-7.

Astrup, A., Dyerberg, J., Selleck, M. and Stender, S. 2008. Nutrition transition and its relationship to the development of obesity and related chronic diseases. *Obesity Reviews*, 9(Suppl 10), 48-52.

Belanger-Ducharme, F. and Tremblay, A. 2005. Prevalence of obesity in Canada. *Obesity Reviews,* 6, 183-6.

Block, J.P., Scribner, R.A. and DeSalvo, K.B. 2004. Fast food, race/ethnicity, and income: a geographic analysis. *American Journal of Preventive Medicine*, 27, 211-17.

Bowman, S.A., Gortmaker, S.L., Ebbeling, C.B., Pereira, M.A. and Ludwig, D.S. 2004. Effects of fast-food consumption on energy intake and diet quality among children in a national household survey. *Pediatrics*, 113, 112-18.

Bowman, S. and Vinyard, B. 2004. Fast food consumption of U.S. adults: impact on energy and nutrient intakes and overweight status. *Journal of the American College of Nutrition*, 23, 163-8.

Branca, F., Haik Nikogosian, H. and Lobstein, T. (eds.) 2007. Dietary determinants of obesity, in *The challenge of obesity in the WHO European region and the strategies for response.* Copenhagen: WHO Regional Office for Europe.

Bray, G.A., Nielsen, S.J. and Popkin, B.M. 2004. Consumption of high-fructose corn syrup in beverages may play a role in the epidemic of obesity. *American Journal of Clinical Nutrition*, 79, 537-43.

Burdette, H. and Whitaker, R. 2004. Neighborhood playgrounds, fast food restaurants, and crime: relationships to overweight in low-income preschool children. *Preventive Medicine*, 38, 57-63.

Burns, C.M., Gibbon, P., Boak, R., Baudinette, S. and Dunbar, J.A. Food cost and availability in a rural setting in Australia. *Rural and Remote Health*, 4, 311.

Chopra, M., Galbraith, S. and Darnton-Hill, I. 2002. A global response to a global problem: the epidemic of overnutrition. *Bulletin of the World Health Organization*, 80, 952-8.

Cooke, L., Sangster, J. and Eccleston, P. 2007. Improving the food provided and food safety practices in out-of-school-hours services. *Health Promotion Journal of Australia*, 18, 33-8.

Cummins, S. and Macintyre, S. 2002. 'Food deserts' – evidence and assumption in health policy making. *British Medical Journal*, 325, 436-8.

Cummins, S. and Macintyre, S. 2006. Food environments and obesity – neighbourhood or nation? *International Journal of Epidemiology*, 35, 100-4.

Cummins, S.C., McKay, L. and MacIntyre, S. 2005. McDonald's restaurants and neighborhood deprivation in Scotland and England. *American Journal of Preventive Medicine*, 29, 308-10.

Food and Nutrition Standards Australia and New Zealand. 2008a. *About FSANZ*. Available at: http://www.foodstandards.gov.au/aboutfsanz [accessed: 9 January 2009].

Food and Nutrition Standards Australia and New Zealand. 2008b. Trans Fatty Acids Available at: http://www.foodstandards.gov.au/newsroom/factsheets/factsheets2008/transfattyacidsaugus3973.cfm [accessed: 9 January 2009].

French, S. 2005. Public health strategies for dietary change: schools and workplaces. *Journal of Nutrition*, 135, 910-12.

Friel, S., Chopra, M. and Satcher, D. 2007. Unequal weight: equity oriented policy responses to the global obesity epidemic. *British Medical Journal*, 335, 1241-3.

Greenwood, J. and Stanford, J. 2008. Preventing or improving obesity by addressing specific eating patterns. *Journal of the American Board of Family Medicine*, 21, 135-40.

Hawkes, C. 2007. Regulating and litigating in the public interest: regulating food marketing to young people worldwide: trends and policy drivers. *American Journal of Public Health*, 97, 1962-73.

Heading, G. 2008. Rural obesity, healthy weight and perceptions of risk: struggles, strategies and motivation for change. *Australian Journal of Rural Health*, 16, 86-91.

Hill, J.O., Peters, J.C., Catenacci, V.A. and Wyatt, H.R. 2008. International strategies to address obesity. *Obesity Reviews,* 9(Suppl 1), 41-7.

Jacobson, M. 2004. High-fructose corn syrup and the obesity epidemic. *American Journal of Clinical Nutrition*, 80, 1081.

James, W. 2008a. The epidemiology of obesity: the size of the problem. *Journal of Internal Medicine*, 263, 336-52.

James, W. 2008b. The fundamental drivers of the obesity epidemic. *Obesity Reviews*, 9(Suppl 1), 6-13.

Lobstein, T. and Davies, S. 2009. Defining and labelling 'healthy' and 'unhealthy' food. *Public Health Nutrition,* 12(3):331-40.

Lopez, R. 2007. Neighborhood risk factors for obesity. *Obesity*, 15, 2111-19.

Macdonald, L., Cummins, S. and Macintyre S. 2007. Neighbourhood fast food environment and area deprivation-substitution or concentration? *Appetite*, 49, 251-4.

Macintyre, S., McKay, L., Cummins, S. and Burns, C. 2005. Out-of-home food outlets and area deprivation: case study in Glasgow, UK. *International Journal of Behavioural Nutrition and Physical Activity*, 2, 16.

Maddock, J. 2004. The relationship between obesity and the prevalence of fast food restaurants: state-level analysis. *American Journal of Health Promotion*, 19, 137-43.

Morland, K., Wing, S. and Diez Roux, A. 2002a. The contextual effect of the local food environment on residents' diets: the atherosclerosis risk in communities study. *American Journal of Public Health*, 92, 1761-7.

Morland, K., Wing, S., Diez Roux, A. and Poole, C. 2002b. Neighborhood characteristics associated with the location of food stores and food service places. *American Journal of Preventive Medicine*, 22, 23-9.

Mozaffarian, D., Katan, M.B., Ascherio, A., Stampfer, M.J. and Willett, W.C. 2006. Trans fatty acids and cardiovascular disease. *New England Journal of Medicine*, 354, 1601-13.

Nestle, M. 2006. Food marketing and childhood obesity – a matter of policy. *New England Journal of Medicine*, 354, 2527-9.

Nutrition Australia. 2002. *Food and Nutrition Accreditation Guidelines for Child Care Centres.* Brisbane: Nutrition Australia and Queensland Health.

O'Neill, M. and Simard, P. 2006. Choosing indicators to evaluate Healthy Cities projects: a political task? *Health Promotion International*, 21, 145-52.

Parliament of Australia (Senate). 2008. Inquiry into protecting children from junk food advertising (broadcasting amendment) Bill 2008 Available at: http://www.aph.gov.au/SENATE/committee/clac_ctte/protecting_children_junk_food_advert/index.htm [accessed 9 January 2009].

Pearce, J., Hiscock, R., Blakely, T. and Witten, K. 2008. The contextual effects of neighbourhood access to supermarkets and convenience stores on individual fruit and vegetable consumption. *Journal of Epidemiology and Community Health*, 62, 198-201.

Pearce, J., Hiscock, R., Blakely, T. and Witten, K. 2009. A national study of the association between neighbourhood access to fast-food outlets and the diet and weight of local residents. *Health and Place,* 15, 193-7.

Pollard, C., Lewis, J. and Miller, M. 1999. Food service in long day care centres – an opportunity for public health intervention. *Australian and New Zealand Journal of Public Health*, 23, 606-10.

Poston, W. and Foreyt, J. 1999. Obesity is an environmental issue. *Atherosclerosis*, 146, 201-9.

Powell, L.M., Auld, M.C., Chaloupka, F.J., O'Malley, P.M. and Johnston, L.D. 2007. Associations between access to food stores and adolescent body mass index. *American Journal of Preventive Medicine*, 33, S301-7.

Reidpath, D.D., Burns, C., Garrard, J., Mahoney, M. and Townsend, M. 2002. An ecological study of the relationship between social and environmental determinants of obesity. *Health and Place*, 8, 141-5.

Sacks, G., Swinburn, B. and Lawrence, M. 2008. A systematic policy approach to changing the food system and physical activity environments to prevent obesity. *Australian and New Zealand Health Policy*, 5, 13.

Sacks, G., Swinburn, B. and Lawrence, M. 2009. Obesity Policy Action framework and analysis grids for a comprehensive policy approach to reducing obesity. *Obesity Reviews,* 10, 76-86.

Samuels, S., et al. 2003. *Food industry practices aimed at children: strategies preventing overweight and obesity*. San Francisco: California Endowment.

Schafer Elinder, L. and Jansson, M. 2009. Obesogenic environments – aspects on measurement and indicators. *Public Health Nutrition,* 12(3), 307-15.

Stanton, R. 2006. Nutrition problems in an obesogenic environment. *Medical Journal of Australia*, 184, 76-9.

Story, M. and French, S. 2004. Food Advertising and Marketing Directed at Children and Adolescents in the US. *International Journal of Behavioural Nutrition and Physical Activity*, 1, 3.

Story, M., Kaphingst, K.M., Robinson-O'Brien, R. and Glanz, K. 2008. Creating healthy food and eating environments: policy and environmental approaches. *Annual Review of Public Health*, 29, 253-72.

Swinburn, B. 2003. The obesity epidemic in Australia: can public health interventions work? *Asia Pacific Journal of Clinical Nutrition*, 12(Suppl), S7.

Swinburn, B. 2008. Obesity prevention: the role of policies, laws and regulations. *Australian and New Zealand Health Policy*, 5, 12.

Swinburn, B., Sacks, G., Lobstein, T., Rigby, N., Baur, L.A., Brownell, K.D., et al. 2008. The 'Sydney Principles' for reducing the commercial promotion of foods and beverages to children. *Public Health Nutrition*, 11, 881-6.

Thorburn, A. 2005. Prevalence of obesity in Australia. *Obesity Reviews*, 6, 187-9.

Turrell, G. and Giskes, K. 2008. Socioeconomic disadvantage and the purchase of takeaway food: a multilevel analysis. *Appetite*, 51, 69-81.

Wang, M.C., Cubbin, C., Ahn, D. and Winkleby, M.A. 2008. Changes in neighbourhood food store environment, food behaviour and body mass index, 1981-1990. *Public Health Nutrition*, 11, 963-70.

Winkler, E., Turrell, G. and Patterson, C. 2006. Does living in a disadvantaged area mean fewer opportunities to purchase fresh fruit and vegetables in the area? Findings from the Brisbane food study. *Health and Place*, 12, 306-19.

Chapter 11
Policy Responses and the Physical Environment

Mylène Riva and Sarah Curtis

During the past 50 years, policy and planning in regional development have favoured strategies supporting segregated land uses and transportation by car (Frank et al. 2003); as a result of this type of development, fewer trips are now made by walking and cycling. In parallel, increasing sedentary work, use of labour-saving devices and sedentary pursuits during leisure time are contributing to declining levels of physical activity (PA). Some argue that physical activity has been 'engineered out' of daily activities (Sallis et al. 1998).

This chapter considers the policy responses to the public health priority of preventing obesity and promoting healthy weights. Based on a social-ecological model linking individuals and their environment (Sallis et al. 2006), the focus is on the potential for policy interventions to modify aspects of the physical environment at different geographical scales in order to increase population levels of physical activity through both active recreation (leisure-time PA) and active living, i.e. integrating physical activity, such as walking and cycling, into daily routines (Active Living Research 2006). With respect to environmental influence on obesity levels, recent reviews support links between the physical and social conditions of local environments and overweight and obesity (Papas et al. 2007, Booth et al. 2005). These associations may be partly explained by physical activity, which is involved in the aetiology of obesity. Throughout the chapter, many international policies are presented, with a particular emphasis on examples from the United Kingdom (UK) and Canada.

Physical Activity and Physical Environment

In developed countries, overall levels of PA are declining; people are less physically active at work, at home, in their daily commute and are increasingly sedentary. These trends are also evident in other parts of the world, such as China (Ng et al. 2009), so that this is becoming a truly global issue. The physical environment, comprising both natural and built aspects, is an important determinant of PA and obesity as it creates a climate which promotes increased energy consumption and reduced energy expenditure (Papas et al. 2007). Natural environments encompass elements such as weather, topography (for example beaches and river banks, water bodies) and other natural elements (for example flora and fauna in green spaces). These may present opportunities for physical activity, but barriers may reduce this potential, for example where natural areas are hard to reach, hilly topography difficult to climb and weather conditions such as ice, snow, heat and humidity impede on people's mobility. The built environment is defined by some authors to include three components: the transportation system, land use patterns and design features (Transportation Research Board and Institute of Medicine of the National Academies [TRB and IOM] 2005, Frank et al. 2003, Handy et al. 2002); they are described in Box 11.1. The built environment may impose barriers to physical activity, for example, by favouring car use over active transportation such as walking and cycling. Variation in individual capacities and other characteristics can also cause differences in how physical environments relate to PA opportunities and barriers, as discussed below.

Box 11.1   Three components of the built environment

Transportation system
Physical infrastructures and services that provide connections between activities and determines how well they are connected.

Land use patterns
Spatial distribution and arrangement of structures on the landscape; it is linked to the ideas of density and compactness both in terms of people and structures, and of land use mix i.e. different types of uses are located in proximity to one another. Central to these is the proximity and accessibility of services necessary for daily living and located within walking distance from where people live.

Design features
Physical and functional qualities of the built environment, such as the design of buildings, streetscapes, sidewalks, traffic calming and public places, their safety and attractiveness.

Source: Adapted from Frank et al. 2003 and Handy et al. 2002.

From a geographical perspective, the conditions of community settings are important correlates of physical activity, both recreational (undertaken for discretionary reasons in one's leisure time) and utilitarian (undertaken to accomplish other purposes, such as commuting to work, running errands and so on) (Frank et al. 2003). Indeed, several studies report significant and consistent associations between physical activity levels and conditions of local environments such as residential and population density, proximity of resources and services, land use mix, street connectivity, presence and accessibility of parks and public open spaces, and safety (see Chapters 7 and 8; Saelens and Handy 2008, TRB and IOM 2005). Furthermore

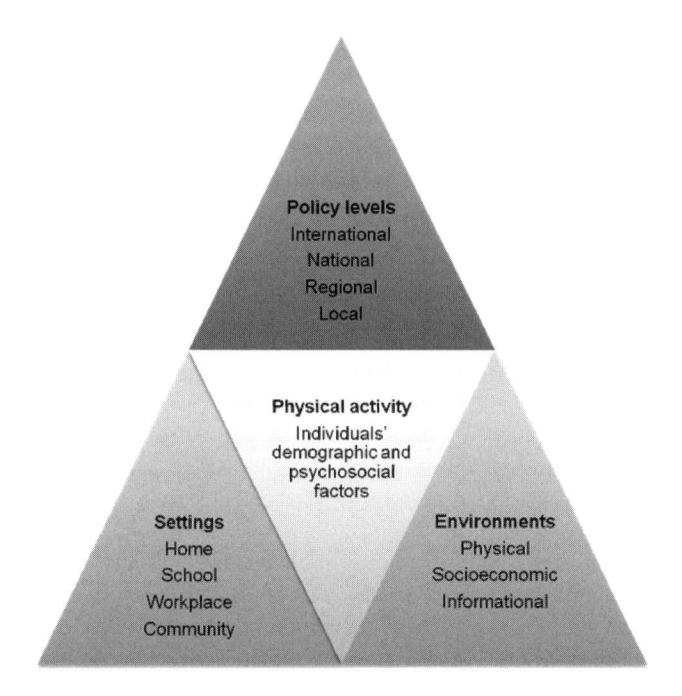

Figure 11.1   Policy levels, environments and settings influencing physical activity

Source: Adapted from Sallis et al. 2006 and Schmid et al. 2006.

it is important to consider other settings such as homes, schools, workplaces, socioeconomic and informational environments, and the extent of individuals' 'activity space' i.e. the space delimited by travels between the residence and the locations of daily activities e.g. work, shopping, recreational activities, etc. (Papas et al. 2007). Figure 11.1 illustrates the multiple settings where PA may unfold and the types of environments influencing physical activity levels. The figure further highlights the need to consider the connections between the 'dimensions' represented in terms of policy level, settings and types of environment.

Promoting Physical Activity through Public Policies

The goal of increasing levels of physical activity in the population presents complex challenges (Baranowski 2006, Resnicow and Vaughan 2006). Traditional approaches to health promotion were mainly based on theories and models focussing on psychosocial influences on individual behaviours, e.g. attitudes, intentions, and self-efficacy. The success of these individual-level interventions in increasing PA levels has been limited, with modest recruitment rates, small to moderate effect sizes and poor maintenance of PA following interventions (Sallis et al. 2006). Furthermore, interventions targeting small numbers of people deemed to be 'at-risk' may be seen as victim blaming, placing too much responsibility on individuals for changing their behaviours and may contribute to increasing social inequalities in health because the structural constraints on behavioural change are socially uneven. Also, these approaches give little attention to the contextual dimensions of the settings in which people live, learn, work, play and socialize. Yet, individuals are unlikely to change their behaviours when their environment does not support or encourage such change. Intervening (solely) at the individual or small-group level is unlikely to translate into wider population change in physical activity patterns and levels. In contrast, public policy interventions targeting environments are designed to influence large groups of people, even entire populations.

Public policies are guides to action, at any level of government, to achieve intended goals and function to set priorities and to guide resource allocation (Milio 2001). By targeting physical and organizational structures, policy interventions aim to increase physical activity by creating environments that provide opportunities for adopting and maintaining active lifestyles, thereby making physical activity an 'easier choice'. Examples of policy interventions include modifying/formulating regulations, organizational norms and social practices, creating pedestrian and cycle paths, increasing access to resources and facilities, building and improving facilities for leisure physical activity, zoning and land use strategies that support active transport, building construction encouraging PA, and incentives to promote physical activity in the workplace (Heath et al. 2006). Public policy interventions therefore have the potential to shift rates of physical activity at the population level, rather than modifying behaviour of smaller numbers of 'high-risk' individuals (Rose 1994) and are more likely to trigger and sustain long-term population-wide change in PA levels (Kumanyika et al. 2008, World Health Organization 2004).

As Bull and colleagues observe: 'governments at all levels have a key role to play in initiating, coordinating and implementing public policies that promote PA, enhance environments and provide a better access to opportunities for PA for the whole population' (Bull et al. 2004: 93). Increasing population levels of PA will require strategic, concerted and intersectoral efforts involving the health sector and sectors whose practices are not driven primarily by public health concerns, e.g. transportation, urban planning, housing, employment, education, services provision, environmental protection and the private sector. These efforts will have to unfold from decisions made at various and interrelated geographical levels, from setting national targets and regulations to establishing regional and local action plans.

Policy Responses: Designing Supportive Physical Environments for Physical Activity

In recent years, a number of systematic reviews have identified policy strategies and interventions targeting physical environments as effective approaches for promoting physical activity (see for example Heath et al. 2006, Bull et al. 2004, Foster and Hillsdon 2004, Task Force on Community Preventive Services 2002). Results of these reviews have informed recommendations to increase PA levels, for example the US Preventive Services Task Force (Task Force on Community Preventive Services 2002), the UK National Institute for Health and Clinical Excellence (NICE) Public Health Guidance *Promoting and Creating Built or Natural Environments That Encourage and Support Physical Activity* (National Institute for Health and Clinical Excellence 2008), and the proposition for an 'Obesity Policy Action Framework' (Sacks et al. 2008, World Health Organization 2006).

In this section we consider the *potential* for policy interventions to modify the conditions of physical environments by illustrating the many interrelated geographical scales at which different policies can be directed to promote and increase physical activity levels. This discussion is guided by Figure 11.1 suggesting articulation of different levels of policy and illustrated with examples of policy interventions at the local level. We present selected international examples of policy guidance and interventions and discuss the links between global/international, national, regional and local scales of actions to increase PA and the different settings for interventions. Clearly, policy levels, settings and environmental dimensions are cross-cutting and are not confined to the categories presented; their structure may also vary between countries.

An International Call for Action

In response to the global public health burden of obesity, the World Health Organization (WHO) adopted in 2004 the *Global Strategy on Diet, Physical Activity and Health* with the goal to 'promote and protect health by guiding the development of an enabling environment for sustainable actions at individual, community, national and global levels that, when taken together, will lead to reduced disease and death rates related to unhealthy diet and physical inactivity' (World Health Organization 2004: 3). The Global Strategy serves as a comprehensive guidance tool and calls upon member states to develop, implement and evaluate sustainable and integrated policy action plans at the national, regional and local levels to increase (among other goals) population proportions of physical activity, by engaging all sectors including civil society. Endorsed by Member States, the Global Strategy acts as an international commitment to strengthen existing national efforts to prevent and control chronic diseases and their common risk factors.

Subsequently, the WHO published *A Guide for Population-Based Approaches to Increasing Levels of Physical Activity: Implementation of the Who Global Strategy on Diet, Physical Activity and Health* (World Health Organization 2005). The purpose of this document was to provide Member States with guidance on policy options for effectively promoting physical activity at the national and subnational levels and to assist with the development and implementation of national physical activity plans. Since its formulation, several countries (but not all) have implemented or plan to implement policy interventions in response to the request made by the Global Strategy, either independently or as part of existing programmes (Branca et al. 2007).

This 'top-down' approach, where Member States jointly pursue agreed actions, is supplemented by 'bottom-up' local initiatives such as those under the WHO *Healthy Cities* network. The *Healthy Cities* network puts health on the agenda of local decision-makers and promotes comprehensive and systematic policy and planning frameworks by 'engaging local governments in health development through a process of political commitment, institutional change, capacity building, partnership-based planning and innovative projects [and] strives to include health considerations in urban economic, regeneration and development efforts' (Edwards and Tsouros 2008: 3). Physical activity is one of the core themes as illustrated by the WHO guide *A Healthy City is an Active City: A Physical Activity Planning Guide* which is designed to help city leaders create a plan for physical activity, active living and sport in their city or community (Edwards and Tsouros 2008). Different approaches for promoting physical activity in line with the *Healthy City* network and other related local frameworks have been applied in many countries, showing the breadth and scope of possible interventions; some examples are presented in Box 11.2.

Strategic planning targeting transport systems, land use planning and design is necessary to initiate and maintain environmental changes that support PA and active living. Planning movements such as *New Urbanism* and *Smart Growth* have emerged, especially in North America, in response to problems arising from low-density, auto-dependent urban and suburban types of developments. The *New Urbanism* movement is a coalition of architects and planners embracing the design of the compact traditional neighbourhood (built prior to World War II) characterized by higher population and building densities and mixed land uses (Frank et al. 2003). In parallel *Smart Growth* programmes are designed to counter urban sprawl through 'the planning, design, development and revitalization of cities, towns, suburbs and rural areas to create and promote social equity, a sense of place and community, and to preserve natural as well as cultural resources' (American Planning Association 2002: 21). Principles of *Smart Growth* appear in Box 11.3. *Smart Growth* programmes aim to bring residents closer to destinations and provide viable alternatives to driving, therefore helping to reduce urban sprawl and automobile use; examples of policies used include mixed-use zoning, transit-oriented development and bicycle and pedestrian infrastructures (Handy et al. 2005).

Box 11.2   Selected examples of local applications of healthy cities and active living/healthy by design frameworks

---

Healthy cities (examples are from Edwards and Tsouros 2008; for more examples, please refer to this document and to the Healthy Cities website http://www.euro.who.int/healthy-cities)
- Brighton and Hove, UK: Demonstration project for cycling; aim to increase the proportion of cyclists by 5 per cent yearly through building and extending cycle infrastructures in the local area. Implementation for 2006-2012 of a comprehensive action plan to promote and increase population's participation in PA (http://www.brighton-hove.gov.uk/downloads/bhcc/713_Sports_ Strategy.v4.pdf).
- Kvasice, Czech Republic: Half-day weekend walks for children and their parents, including visits to local sites of historical interest, amusing activities such as treasure hunts, competitions and challenges, and linking health and educational objectives.
- Rijeka, Croatia: Free, supervised and adapted exercise programmes for older adults.
- Sunderland, UK: Community wellness programme for disadvantaged populations which provides opportunities for physical activity in the heart of the community with trained staff and volunteers.

Active Living/Healthy by Design
Frameworks for action designed in response to local governments' requests for practical guidance in designing walkable, activity-friendly and more liveable communities:
- Healthy by Design, Australia (http://www.goforyourlife.vic.gov.au/hav/articles.nsf/pracpages/Healthy_by_Design?OpenDocument).
- Coalition for Active Living, Canada (http://www.activeliving.ca/).
- Active Living by Design, Unites States of America (http://www.activelivingbydesign.org)
  - Seattle, WA: Create walkable neighbourhoods in five communities within the city envisioning walking to school as the norm and safe streets for pedestrians of all ages. Activities include implementation of Safe Routes to School programmes, walking school buses, improvements to streets and sidewalks, provision of active living education to physicians in low-income health clinics through a series of guidelines for medical providers.

It is notable that many of the examples in Box 11.3 invoke solutions which are intended to address social inequalities in opportunities for healthy living, with measures to achieve a mix of socioeconomic groups and people at different stages of the lifecourse. These measures require action beyond adjustments to the built infrastructure, such as targets for mixed land use and adjustments to social and financial processes which influence affordability and inclusion of different social groups within communities.

Box 11.3   Principles of *Smart Growth*

• **Mix land uses**. Each neighbourhood has a mixture of homes, retail, business and recreational opportunities. • **Build well-designed compact neighbourhoods**. Residents can choose to live, work, shop and play in close proximity. People can easily access daily activities, transit is viable and local businesses are supported. • **Provide a variety of transportation choices**. Neighbourhoods are attractive and have safe infrastructure for walking, cycling and transit (in addition to driving). • **Create diverse housing opportunities**. People in different family types, life stages and income levels can afford a home in the neighbourhood of their choice. • **Encourage growth in existing communities**. Investments in infrastructures (such as roads and schools) are used efficiently and developments do not take up new land.	• **Foster a unique neighbourhood identity**. Each community is unique, vibrant, diverse and inclusive. • **Preserve open spaces, natural beauty and environmentally sensitive areas**. Development is to respect natural landscape features and to have high aesthetic, environmental and financial value. • **Utilize smarter and cheaper infrastructure and green buildings**. Green buildings and other systems can save both money and the environment in the long run. • **Protect and enhance agricultural lands**. A secure and productive land base provides food security, employment, and habitat, and is maintained as an urban containment boundary. • **Nurture engaged citizens**. Places belong to those who live, work and play there. Engaged citizens participate in community life and decision-making.

Source: Downs 2005, SmartGrowthBC http://www.smartgrowth.bc.ca/, no date.

Setting National Targets and Policy Guidance

Many countries have identified increasing physical activity levels as an important component or target of their national public health policy. Targets are established in order to monitor progress and inspire actions; national examples are presented in Box 11.4. Several of the national targets reflect the principles promulgated by international initiatives; yet targets vary between countries in terms of their scope (i.e. groups targeted) and their precision (specification of quantifiable goals). Whereas international initiatives generally rely on diplomacy and knowledge exchange, at a national level (or at the regional/provincial level in countries with strong regional tiers of government) it may be more feasible for societies to introduce institutional, legal or fiscal means to promote physical activity. However some of the targets may be unrealistic if they do not equate with the resources made available to achieve them.

Box 11.4   Physical activity targets by country

Québec, Canada (2003) *National program of public health 2003-2012*
By 2012, decrease the prevalence of sedentary behaviours among three groups: young people aged between 10-19 and adults aged between 25-44 and those over 55 years; increase regular involvement in physical activity among young people (younger than 15 years of age); increase by 5 per cent the proportion of people aged 15 years and older meeting public health recommendations in physical activity.

New Zealand (2007) *Health targets: Moving towards healthier futures 2007-2008*
Improving nutrition, increasing physical activity, reducing obesity, with a focus on children and youth.

United Kingdom (2002) *Game Plan: A strategy for delivering Government's sport and physical activity objectives*
By 2020, increase to 70 per cent the proportion of the population undertaking 30 minutes of physical activity five days a week with an interim target of 50 per cent by 2011.

By 2008, increase by 3 per cent the number of adults and young people aged 16 and above from priority groups engaged in at least 30 minutes of moderate-intensity-level sport, at least three times a week; priority groups defined as those with a physical or mental disability, black or minority ethnic groups, those from lower socioeconomic groups and women.

Enhance the take-up of sporting opportunities by 5 to 16 year olds so that the percentage of school children in England who spend a minimum of two hours each week on high quality physical education and school sport within and beyond the curriculum increases from 25 per cent in 2002 to 75 per cent by 2006 and to 85 per cent by 2008.

United States of America (2000) *Healthy people 2010*
Through 15 different targets, the objectives of Healthy People 2010 are to increase the amount of moderate and vigorous physical activity performed by adults, adolescents and children, reduce sedentary behaviour among adolescents and increase opportunities for physical activity through creating and enhancing access to places and facilities where people can be physically active.

National policies represent a commitment by governments to make environments more supportive of PA. Examples of national policies include the UK *Choosing Activity: A Physical Activity Action Plan* (Department of Health 2005) and the *Program 2005-2008 Action Priorities* of *Kino-Quebec*, a cross ministerial programme in the province of Quebec, Canada (Le May 2005). Both policy documents set guidelines and recommendations to make environments more conducive to active living and call upon local governments and organizations to play a leadership role in changing environments while providing them with information support and tools for doing so.

In 2008, the UK National Institute for Health and Clinical Excellence (NICE) published a public health guidance entitled *Promoting and Creating Built or Natural Environments that Encourage and Support Physical Activity*, setting out evidence-based recommendations on interventions to improve physical environments to encourage and support physical activity (National Institute for Health and Clinical Excellence 2008). The recommendations are presented in Box 11.5. The guidance is aimed at many settings, sectors and actors with direct and indirect roles and responsibilities for the physical environment, e.g. planning and transport agencies, regional and local authorities, architects, designers, developers, employers, facility managers, school authorities, local communities and voluntary and private sectors. In conjunction with other national documents such as *Choosing Activity: A Physical Activity Action Plan* (Department of Health 2005) and *Healthy Weight, Healthy Lives: A Cross-Government Strategy for England* (Cross-Government Obesity Unit Department of Health and Department of Children Schools and Families 2008), the NICE guidance provides information and tools to support organizations to identify priorities for actions and ways to create supportive environments.

Box 11.5   NICE's guidance on promoting and creating built or natural environments encouraging and supporting physical activity

- Any planning applications for new developments prioritize the need for people to be physically active in their daily routine;
- Public open spaces and public paths can be reached on foot or by bicycle and are maintained to a high standard;
- Any new workplaces are linked to walking and cycling networks; different parts of the worksite are linked by pedestrian and cycle paths;
- Pedestrians, cyclists and users of other active modes of transport are given the highest priority when developing or maintaining roads;
- During building design or refurbishment, staircases are designed, positioned and clearly signposted to encourage their use;
- School playgrounds are designed to encourage varied and physically active play.

Source: National Institute for Health and Clinical Excellence 2008.

Kino-Quebec (Canada) has developed and widely distributed a provincial document which targets municipal and regional decision-makers, providing information about the concept of 'environments conducive to physical activity' and inviting decision-makers to consider aspects of environments that are supportive of active living, formulate laws and regulations and develop urban and regional planning strategies (Otis et al. 2005). The document identifies seven domains for action:

- Land use mix and population density;
- Proximity and accessibility of services;
- Safety from traffic and crime;
- Planning/maintenance of sidewalks, walking and cycling networks;
- Access to green spaces and water surfaces;
- Building and public spaces supportive of active lifestyle;
- Planning and policies.

Both the *Kino-Quebec* and the NICE guidance considered above focus mainly on urban areas, with little consideration directed to rural environments. Indeed most research and reviews of policy interventions are set in urban and suburban settings of developed countries. It is thus unclear whether and how physical activity is influenced by the components of physical environments of rural settings (Duncan et al. 2009). Similarly, it is not clear that the guidance would translate readily to urban settings located in developing countries and especially in countries such as China that are experiencing rapid industrialization and urbanization (Bauman et al. 2008, *Obesity Reviews* 2008). Also, in initiatives already underway in high income countries, there appears to be an emphasis on urban design and planning 'solutions and scenarios' for new urban developments rather than modifying existing environments. These barriers and other challenges are discussed below.

Translating Policies into Actions: Changing Environments at the Local Level

To implement the types of broad scale policies, guidance and actions mentioned above, local government is arguably the administrative level best placed for action. Theoretically, it has been established for some time that the local municipality is a level at which resources can be most effectively organized to meet collective goals for health and well-being (Castells 1977) and where interventions can be designed to fit well with local conditions and to meet local needs. This is probably why, for example, the *Healthy Cities* movement has increasingly developed a degree of local diversity and geographical diffusion which was perhaps not anticipated at its inception. Indeed, current trends in public health are showing that 'health and well-being are being reclaimed as a legitimate area of activity for local governments' (Curtis 2008: 296). 'City leaders and other decision-makers can provide leadership, legitimacy and an enabling environment for developing and implementing policies that support active living for all' (Edwards and Tsouros 2006: ix) through interventions targeting transport infrastructures, land use planning, building design and providing access to open natural spaces. It is important to note that these are interventions that fall outside the health care sector in most countries and can be seen as part of a general process of 'demedicalisation' of the public health agenda. Box 11.6 presents examples of potential policy interventions in terms of their likely effectiveness; interventions identified with an asterisk are those for which there is strong and sufficient evidence of their effectiveness for increasing physical activity (as identified by the Task Force on Community Preventive Services 2002

Box 11.6 Examples of policy interventions for changing the physical environment of local communities

Transportation systems
- Creating/extending cycle networks
- Providing access to active means of transportation

Land use planning and design
- Regulations and practices encouraging mixed land uses*
- Creation or enhanced access to places and opportunities for physical activity combined with informational approaches*
- Reducing traffic speed; ease/safety of street intersection; traffic calming measures*
- Improved street lighting*
- Sidewalk quality and connectivity*
- Availability of safe and attractive playgrounds

Note: * Interventions for which there are strong and sufficient evidence of their effectiveness in increasing physical activity levels (National Institute for Health and Clinical Excellence 2008, Task Force on Community Preventive Services 2002).

National Institute for Health and Clinical Excellence 2008). We must consider, however, that applicability of evidence concerning effectiveness of interventions may vary for different countries depending on local conditions and needs.

The following points summarize the current evidence on the influence of the physical environment on physical activity (Frank et al. 2002):

- Different land uses (e.g. residential, retail, employment, schools, parks, open space) should be integrated rather than segregated and resources and services should be located near the home so that people can easily fulfil basic utilitarian needs by walking or cycling;
- High density and compact development puts more people within walking distance of parks, schools, transit, shops and services, and provides the vital market for those services;
- Streets and buildings that are built to prioritize pedestrians create safe, vibrant and interesting places for walkers, cyclists and transit users;
- Highly interconnected street and trail networks reduce the time and distance needed for pedestrians and cyclists to get from one place to another.

Policy interventions aiming to implement the above mentioned principles have to:

- Target multiple settings such as local communities, schools and workplaces;
- Rely on multiple strategies for changing physical, socioeconomic and informational environments;

- Be inclusive in reaching populations across the lifespan, i.e. children, adolescents, adults and older adults, across different socioeconomic and cultural groups, e.g. ethnic minorities, low-income individuals, and people with disabilities;
- Be intersectoral and involve various stakeholders and citizens.

Promoting active transportation and safe commuting In Europe 50 per cent of trips made by car are less than 5 km, whereas 30 per cent of trips are less than 3 km; such distances could be covered by cycling in 15-20 minutes or by walking in 30-50 minutes (Edwards and Tsouros 2006). Access to public transport promotes physical activity as many trips involve walking or cycling to get to and from a transit stop. Examples of interventions for promoting PA through active commuting include those aiming at improving pedestrian, cycling and transit access, increasing pedestrian and cyclist activity and safety and reducing car use.

Increasing the number of trips made by walking and cycling requires regional and local planning and infrastructure development that encourages and enables the use of active transportation. This could be achieved by creating/enhancing cycle paths and sidewalks or by providing access to active means of transportation. Although evidence about the effectiveness of transportation interventions is currently insufficient due to scarce evaluation studies (Heath et al. 2006), several studies report significant relationships between features of transportation systems and the adoption of active modes of transportation such as walking and cycling (for a review of reviews, see Saelens and Handy 2008). Cycling levels are higher in countries having a long tradition of cycling-friendly environments and dedicated policies, e.g. the Scandinavian countries and the Netherlands (Branca et al. 2007). Initiatives to develop and extend cycle networks, such as the UK National Cycle Network (http://www.sustrans.org.uk) and public bicycle rental schemes implemented in many European cities, for example Velo'v in Lyon (http://www.velov.grandlyon.com), Velib in Paris (http://www.velib.paris.fr) and Call a Bike in Germany (http://www.callabike-interaktiv.de) encourage people to cycle to cover short distances by providing access to infrastructure and means of transportation.

Changing transport behaviours will not happen solely in response to changes in the physical environment. Intersectoral collaboration is needed to exploit the full potential of human-powered mobility and long-term investment and commitment are required to change social attitudes, perceptions and norms to increase the traditionally low levels of active commuting. These will need to be paralleled and supported by economic incentives to increase active transportation including measures such as subsidising public transit, providing incentives to car or van pool, increasing the cost of parking and congestion charge schemes. For example, although the key objective of the London congestion charge (first implemented in 2003 and extended westwards in 2007) was to reduce traffic problems, other outcomes have included an increase in cycling mileage and potential increase in trips made by walking (Cavell 2007).

Creating activity-friendly local communities through land use planning and design Land use and design policies, regulations and practices aimed at creating more activity-friendly local communities encompass a broad set of actions providing opportunities for physical activity and active living, for example zoning regulations and building codes, street layouts, density of development and the location of services, schools, workplaces and leisure facilities in proximity to where people live. Additionally, the quality, safety and appearance of local environments influence residents' opportunity and willingness to actively use common spaces (Heath et al. 2006).

Environments characterized by mixed land uses, higher densities and geographic proximity of services have been consistently associated with greater levels of walking in studies from different country settings (Saelens and Handy 2008, TRB and IOM 2005). There is sufficient evidence to support municipal regulations and policies in the realization of development plans where street design includes pedestrian and bicycle facilities and where the location of places such as retail and employment are in close proximity to the home and are accessible via safe, connected and attractive pathways (National Institute for Health and Clinical Excellence 2008, Raine et al. 2008, Task Force on Community Preventive Services 2002). Other initiatives include designing mixed use and car-free developments such as the Slateford Green development in Edinburgh (http://www.scotland.gov.uk/Topics/SustainableDevelopment/CaseStudies/SlatefordGreen) and in the city of Freiburg in Germany (http://www.freiburg.de/servlet/PB/show/1199617_l2/GreenCity_E.pdf). Such policies are likely to encourage utilitarian forms of PA.

The creation and improvement of access to green spaces, family-oriented parks and places for physical activity, such as new cycle and walking paths, have been shown to be associated with increased frequencies of physical activity (Raine et al. 2008, Task Force on Community Preventive Services 2002). Planning policies could, for example, encompass frameworks for the provision and maintenance of recreational facilities and parks (Department for Communities and Local Government 2002). Policy interventions targeting natural environments have the potential to promote leisure-time PA through providing better access to seashores, rivers, lakes and forests within and around the city, and initiatives to develop and preserve green and public spaces. With respect to building design, public buildings should provide convenient and visible stairs as well as signage to encourage their use (Brownson et al. 2006).

At a smaller geographic scale, streets are central to local life as purposive and meaningful locations for movement and transport as well as social exchange and interaction. Street scale and urban design characteristics have been shown to be associated with greater levels of walking and cycling and sufficient evidence supports the effectiveness of policy strategies aiming at changing the physical environment at the street level (Heath et al. 2006). Examples of interventions include improving the connectivity of streets and sidewalks, increasing the ease and safety of street crossing, introducing traffic calming measures, improving street lighting and enhancing the aesthetics of the street through landscaping and building design. The UK Department for Transport (DfT) published in 2007 a *Manual for Streets* which provides guidance for designing streets that are people-oriented, encouraging for people to walk and cycle to local destinations and inclusive for everyone regardless of age or ability (Department for Transport 2007). Recommendations include layout and connectivity of streets, aesthetics of places, traffic signs and markings, street furniture and street lighting. These recommendations are embedded within existing policy, technical and legal frameworks and as such, their implementation does not require major changes in government regulations.

Changing the conditions of local environments has the potential to benefit different groups of the population, for example older adults and people living on low incomes (Frank et al. 2008). Designing safe and healthy physical environments with safe places where children can play and be active can contribute to the prevention of obesity in children, one group of the population where the prevalence of obesity is rapidly increasing and identified as requiring priority action (Wendel et al. 2008, World Health Organization 2003).

Schools and Workplaces

In addition to local communities, two other settings are important to consider in an overall strategy to increase PA levels: schools and workplaces.

Schools have an important role to play in promoting physical activity among pupils as they are the settings where they are likely to learn, develop and practice many of the attitudes, values and skills related to physical activity and active living that may last them throughout their life (see Chapter 9). Active commuting to and from school is important for increasing active living because it happens at least twice a day on all school days; yet, many pupils are now driven to school (McMillan 2007). In the UK, the Department for Transport's *Travelling to School: An Action Plan* outlines a series of measures for national and local governments and for schools to promote walking, cycling and bus use (Department for Transport and Department for Education and Skills 2003). School travel plans engage parents, pupils, schools and local authorities to define measures to make walking, cycling and bus use safe and attractive alternatives for the journey to school. Since 1998, more than 2,000 schools have completed travel plans, many with the help of the DfT's funding scheme. In collaboration, several local authorities have introduced lower speed zones to manage traffic speeds in residential areas whereas others have created extensive cycle routes (Department for Transport 2005). The DfT aim is that by 2010, every school in England will have implemented a travel plan. With declining national trends in walking or cycling to schools, promoting active transport to and from school is a potential strategy for increasing PA among children.

In addition, providing active and *safe routes to school* has proven to be an effective strategy for increasing activity levels in many countries (see for example Canada http://www.saferoutestoschool.ca, UK http://www.sustrans.org.uk,

US http://www.saferoutesinfo.org and Australia http://www.travelsmart.gov.au/schools/schools2.html). Other interventions in the school settings would be to develop partnership between local government and school authorities so that children, young people and other community members have after-hour access to school facilities, allowing them to engage in physical activity and sports.

In workplace settings (private and public), employers and trade unions have a role to play in establishing activity friendly environments. Workplace strategies to improve PA levels and encourage active living could include building design that promotes active choices by providing cycle racks and showers. New employment sites should be located in mixed use developments, linked to walking and cycling networks and close to public transit stations (National Institute for Health and Clinical Excellence 2008). As noted above the ergonomic design of buildings can be adjusted to encourage use of stairs rather than elevators and prevent workers being immobile at their workstations for prolonged periods.

Socioeconomic and Informational Environments

In this chapter, policy responses to promote PA have been presented with a focus on physical environments. However, it is important to also consider the role of the socioeconomic environment in influencing PA behaviours as evidence suggests that policies and interventions are more likely to be successful if they modify both the physical and socioeconomic environments (Ball 2006).

There are social inequalities in the distribution of physical in/activity in the population, with women, older adults, selected ethnic groups, people with lower income and education, people living with mental or physical disabilities and rural dwellers showing more sedentary lifestyles. In addition, sub-groups of the population, e.g. children and the elderly, may be differently affected by the context of their residential environment (Frank et al. 2008). Promoting physical activity should be linked to overall strategies to reduce social inequalities in health. For example, neighbourhood renewal schemes have the potential to reduce inequalities in access and choices for physical activity by including the provision of facilities and infrastructures for active living and team sports, safe routes to school, active transportation and the design of safe public open spaces in their strategies (Social Exclusion Unit 2001). Also, policies improving land use mix and the 'walkability' of a local area may increase overall community cohesion, due to urban design that helps increase personal security and encourages neighbours to watch out for and help each other (Raine et al. 2008). For school children and their carers, the *Walking School Bus* where a group of children make their way to and from school on foot supervised by adults (see for example Canada http://www.saferoutestoschool.ca/walkingschoolbus.asp and in the UK http://www.dft. gov.uk) is a model of an intervention with a social 'character'.

The social and cultural context of implementation is essential, as an intervention which is successful in one community may not do as well elsewhere. Interventions have to be locally defined to meet the needs of a local population. In a study set in the US, Kelly and colleagues examined how the creation of a walking path, a recommendation to increase physical activity formulated by the American Task Force on Community Preventive Services, was interpreted and valued by a group of African-American adults (higher and lower income) living in an urban setting in the Midwest (Kelly et al. 2007). Participants argued that simply creating a walking path would not encourage more physical activity in their community if other social aspects were not considered. For example, women argued that the creation of the path would need to be combined with other interventions ensuring a safe environment where they can be active. For men and lower income adults, safety and law enforcement were identified as two factors to be considered prior to improving access to the path, or to other facilities. This illustrates that prior to creating walking trails or improving the condition of parks, policies or programmes need to be adapted to local social norms and culture.

Other types of interventions are also likely to contribute to changing levels of PA: informational approaches including 'point-of-decision' prompts encouraging people to use stairs instead of escalators or elevators; community-wide campaigns involving multi-component approaches to promote physical activity such as media/informational channels (television, radio, billboards, newspapers and so on), self-help and support groups, community events, and risk factor screening and education in multiple settings; social support interventions which build, strengthen and maintain social networks in the community; and economic incentives encouraging densification and active commuting and discouraging urban sprawl, e.g. tax incentives for employers who provide fitness facilities to their employees, congestion charges and fiscal policies (Sallis et al. 2006, Pratt et al. 2004, Task Force on Community Preventive Services 2002).

Conclusion

Changing physical environments is not an easy endeavour and the following obstacles are likely to pose barriers or challenge the implementation of policy interventions (Heath et al. 2006):

- Changing the urban landscape is a costly and gradual process;
- Zoning regulations may prohibit mixed-use neighbourhoods;
- Changing social norms regarding modes of transportation, lifestyle and physical activity patterns requires a degree of socio-political consensus and concerted action at all political and geographical levels;
- Establishing effective communication between different sectors i.e. urban planning, architecture, transportation engineering and public health requires new ways of institutional and professional working.

Inclusive, intersectoral collaborations across all government levels are necessary if the aim is to sustain long-time change in community design and trends in physical activity. For example, a significant criticism of the *New Urbanism* and *Smart Growth* concepts is that often these new developments are located in suburban areas on the outskirts of large cities; although walking and cycling trips may increase locally, these types of developments are unlikely to change the use of the automobile to cover longer journeys, e.g. commuting to work (Frank et al. 2003).

Intersectoral collaborations in formulating and implementing policies are often not evidence based, as policy initiatives are not always evaluated. More systematic evaluations of the impacts of policies, interventions and natural experiments on physical activity and obesity are needed (Brownson et al. 2008, Schmid et al. 2006, Cummins et al. 2005). To date, most of the evidence is cross-sectional and therefore does not allow us to impute causality between environmental change and increased levels of PA (see Chapter 12).

Although physical inactivity is a public health issue, responsibilities and actions for changing environments extends beyond the public health sector. This sector must however provide a common ground for setting goals and guidelines and reach across governmental departments and ministries, encouraging them to assess the potential impacts on PA of their policies. Furthermore, we have emphasized that while effective action is often best implemented at the local level, such actions must find support and incentives from national and international levels in order to be initiated and sustained. Creating supportive environments will need to encompass other strategies addressing the individual and social environmental determinants of physical activity and healthy body weights.

References

Active Living Research. 2006. *What is active living?* Available at: http://www.activelivingresearch.org/about/whatisactiveliving [accessed 10 February 2009].

American Planning Association. 2002. *Planning for Smart Growth: 2002 State of the States*. Chicago, IL.

Ball, K. 2006. People, places and other people? Integrating understanding of intrapersonal, social and environmental determinants of physical activity. *Journal of Science and Medicine in Sport,* 9(5), 367-70.

Baranowski, T. 2006. Crisis and chaos in behavioral nutrition and physical activity. *International Journal of Behavioral Nutrition and Physical Activity,* 3:27.

Bauman, A., Allman-Farinelli, M., Huxley, R. and James, W.P. 2008. Leisure-time physical activity alone may not be a sufficient public health approach to prevent obesity – a focus on China. *Obesity Reviews,* 9(Suppl 1), 119-26.

Booth, K.M., Pinkston, M.M. and Carlos Poston, W.S. 2005. Obesity and the built environment. *Journal of the American Dietetic Association,* 105(5 Suppl 1), S110-S117.

Branca, F., Nikogosian, H. and Lobstein, T. 2007. *The challenge of obesity in the WHO European Region and the strategies for response.* Copenhagen: WHO Regional Office for Europe.

Brownson, R.C., Haire-Joshu, D. and Luke, D.A. 2006. Shaping the context of health: a review of environmental and policy approaches in the prevention of chronic diseases. *Annual Review of Public Health,* 27, 341-70.

Brownson, R.C., Kelly, C.M., Eyler, A.A., Carnoske, C., Grost, L., Handy, S.L., Maddock, J.E., Pluto, D., Ritacco, B.A., Sallis, J.F. and Schmid, T.L. 2008. Environmental and policy approaches for promoting physical activity in the United States: A research agenda. *Journal of Physical Activity and Health,* 5(4), 488-503.

Bull, F.C., Bellew, B., Schoppe, S. and Bauman, A.E. 2004. Developments in National Physical Activity Policy: an international review and recommendations towards better practice. *Journal of Science and Medicine in Sport,* 7(Suppl 1), 93-104.

Castells, M. 1977. *The urban question: A Marxist approach,* Cambridge, MA: MIT Press.

Cavell, N. (ed.) 2007. *Building Health: Creating and enhancing places for healthy, active lives. What needs to be done?* London: National Heart Forum, in partnership with Living Streets and CABE.

Cross-Government Obesity Unit, Department of Health and Department Of Children Schools And Families. 2008. *Healthy weight, healthy lives: A cross-government strategy for England.* London: Department of Health.

Cummins, S., Pettigrew, M., Higgins, C., Findlay, A. and Sparks, L. 2005. Large-scale food retailing as health intervention: quasi-experimental evaluation of a natural experiment. *Journal of Epidemiology and Community Health* 59(12), 1035-40.

Curtis, S. 2008. How can we address health inequality through healthy public policy in Europe? *European Urban and Regional Studies,* 15(4), 293-305.

Department for Communities and Local Government. 2002. *Planning Policy Guidance 17: Planning for open space, sport and recreation.* London.

Department for Culture Media and Sport. 2002. *Game Plan: A strategy for delivering Government's sport and physical activity objectives.* London: Strategy Unit, Cabinet Office.

Department of Health. 2005. *Choosing Activity: a physical activity action plan.* London.

Department for Transport. 2005. *Transport statistics bulletin: National Travel Survey: 2004*, London: The Stationery Office.

Department for Transport and Department for Education and Skills. 2003. *Travelling to school: an action plan.* London.

Direction générale de la santé publique. 2003. *Programme national de santé publique 2003-2012 (National programme of public health 2003-2012).* Gouvernement du Québec 2003. Québec: Ministère de la Santé et des Services sociaux.

Downs, A. 2005. *Smart Growth*: Why we discuss it more than we do it. *Journal of the American Planning Association*, 71(4), 367-78.

Duncan, M.J., Mummery, W.K., Steele, R.M., Caperchione, C. and Schofield, G. 2009. Geographic location, physical activity and perceptions of the environment in Queensland adults. *Health and Place*, 15(1), 204-9.

Edwards, P. and Tsouros, A.D. 2006. *Promoting physical activity and active living in urban environments: The role of local governments*. Copenhagen: WHO Regional Office for Europe.

Edwards, P. and Tsouros, A.D. 2008. *A healthy city is an active city: a physical activity planning guide*. Copenhagen: WHO Regional Office for Europe.

Foster, C. and Hillsdon, M. 2004. Changing the environment to promote health-enhancing physical activity. *Journal of Sports Sciences*, 22(8), 755-69.

Frank, L.D., Kavage, S. and Litman, T. 2002. *Promoting public health through Smart Growth: Building healthier communities through transportation and land use policies and practices*. [Online: SmartGrowthBC]. Available from: http://www.vtpi.org/sgbc_health.pdf [accessed 10 January 2009].

Frank, L.D., Engelke, P. and Schmid, T.L. 2003. *Health and community design: The impact of the built environment on physical activity*. Washington, DC: Island Press.

Frank, L.D., Kerr, J., Sallis, J.F., Miles, R. and Chapman, J. 2008. A hierarchy of sociodemographic and environmental correlates of walking and obesity. *Preventive Medicine*, 47(2) 172-8.

Handy, S.L., Boarnet, M.G., Ewing, R. and Killingsworth, R.E. 2002. How the Built Environment Affects Physical Activity. Views from Urban Planning. *American Journal of Preventive Medicine*, 23(2S), 64-73.

Handy, S.L., Cao, X. and Mokhtarian, P. 2005. Correlation or causality between the built environment and travel behavior? Evidence from Northern California. *Transportation Research Part D*, 10(6), 427-44.

Heath, G.W., Brownson, R.C., Kruger, J., Miles, R., Powell, K.E., Ramsey, L.T. and the Task Force on Community Preventive Services. 2006. The effectiveness of urban design and land use and transport policies and practices to increase physical activity: a systematic review. *Journal of Physical Activity and Health*, 3(Suppl 1), S55-S76.

Kelly, C.H., Baker, E.A., Brownson, R.C. and Schootman, M. 2007. Translating research into practice: Using concept mapping to determine locally relevant intervention strategies to increase physical activity. *Evaluation and Program Planning*, 30(3), 282-93.

Kumanyika, S.K., Obarzanek, E., Stettler, N., Bell, R., Field, A.E., Fortmann, S.P., et al. 2008. Population-Based Prevention of Obesity: The Need for Comprehensive Promotion of Healthful Eating, Physical Activity, and Energy Balance: A Scientific Statement From American Heart Association Council on Epidemiology and Prevention, Interdisciplinary Committee for Prevention Formerly the Expert Panel on Population and Prevention Science. *Circulation*, 118(4 July), 428-64.

Le May, D. 2005. *The 2005-2008 Action Plan of Kino-Quebec [Les cibles d'action 2005-2008 du programme Kino-Québec]*. Kino-Québec: Ministère de l'Éducation, du Loisir et du Sport; Gouvernement du Québec; Bibliothèque nationale du Québec.

McMillan, T.E. 2007. The relative influence of urban form on a child's travel mode to school. *Transportation Research Part A*, 41(1), 69-79.

Milio, N. 2001. Glossary: healthy public policy. *Journal of Epidemiology and Community Health*, 55(9), 622-3.

Minister of Health. 2007. *Health Targets: Moving towards healthier futures 2007/08*. Wellington: Ministry of Health, New Zealand.

National Institute for health and Clinical Excellence. 2008. *NICE public health guidance 8. Promoting and creating built or natural environments that encourage and support physical activity*. London.

Ng, S.W., Norton, E. and Popkin, B. 2009. Why have physical activity levels declined among Chinese adults? Findings from the 1991-2006 China Health and Nutrition Surveys. *Social Science and Medicine*, 68(7), 1305-14.

Obesity Reviews. 2008. Special Issue: Obesity in China. *Obesity Reviews*, 9(s1), 1-161.

Otis, L., Clements, C., Boudreault, D. and Manfredi, S. 2005. *Aménageons nos milieux de vie pour nous donner le goût de bouger pour une meilleure qualité de vie*. Kino-Québec : Ministère de l'Éducation, du Loisir et du Sport.

Papas, M.A., Alberg, A.J., Ewing, R., Helzlsouer, K.J., Gary, T.L. and Klassen, A.C. 2007. The Built Environment and Obesity. *Epidemiologic Reviews*, 29(1), 129-43.

Pratt, M., Macera, C.A., Sallis, J.F., O'Donnell, M. and Frank, L.D. 2004. Economic Interventions to Promote Physical Activity Application of the SLOTH Model. *American Journal of Preventive Medicine*, 27(3S), 136-45.

Raine, K., Spence, J.C., Church, J., Boulé, N., Slater, L., Marko, J., Gibbons, K. and Hemphill, E. 2008. *State of the evidence review on urban health and healthy weights*. Ottawa: CIHI.

Resnicow, K. and Vaughan, R. 2006. A chaotic view of behavior change: a quantum leap for health promotion. *International Journal of Behavioral Nutrition and Physical Activity*, 3:25.

Rose, G. 1994. *The strategy of preventive medicine*. Oxford: Oxford Medical Publications.

Sacks, G., Swinburn, B. and Lawrence, M. 2008. Obesity Policy Action framework and analysis grids for a comprehensive policy approach to reducing obesity. *Obesity Reviews*, 10.1111/j.1467-789X.2008.00524.x.

Saelens, B.E. and Handy, S.L. 2008. Built environment correlates of walking: A review. *Medicine and Science in Sports and Exercise*, 40, S550-S566.

Sallis, J.F., Bauman, A. and Pratt, M. 1998. Environmental and policy – Interventions to promote physical activity. *American Journal of Preventive Medicine*, 15(4), 379-97.

Sallis, J.F., Cervero, R.B., Ascher, W., Henderson, K.A., Kraft, M.K. and Kerr, J. 2006. An ecological approach to creating active living communities. *Annual Review of Public Health,* 27, 297-322.

Schmid, T.L., Pratt, M. and Witmer, L. 2006. A framework for physical activitypolicy research. *Journal of Physical Activity and Health,* 3(Suppl 1), S20-S29.

SmartGrowthBC. No date. *10 Smart Growth Principles*. Available at: http://www.smartgrowth.bc.ca/Default.aspx?tabid=133 [accessed: 10 February 2009].

Social Exclusion Unit 2001. *A new commitment to neighbourhood renewal: a national strategy action plan*. London, Cabinet Office.

Task Force on Community Preventive Services. 2002. Recommendations to increase physical activity in communities. *American Journal of Preventive Medicine,* 22(4S), 67-72.

Transportation Research Board and Institute of Medicine of the National Academies [TRB and IOM] 2005. *Does the built environment influence physical activity? Examining the evidence*. Special Report 282. *Transportation Research Board*. Washington, DC.

US Department of Health and Human Services 2000. *Healthy People 2010*. 2nd Edition. Washington, DC.

Wendel, A.M., Dannenberg, A.L. and Frumkin, H. 2008. Designing and building healthy places for children. *International Journal of Environment and Health,* 2(2/4), 338-55.

World Health Organization. 2003. *Diet, Nutrition and the Prevention of Chronic Diseases*. Technical Report Series No 916. Geneva: World Health Organization.

World Health Organization. 2004. *Global Strategy on Diet, Physical Activity and Health*. Geneva: World Health Organization.

World Health Organization. 2005. *A guide for population-based approaches to increasing levels of physical activity: implementation of the WHO global strategy on diet, physical activity and health*. Geneva: World Health Organization.

World Health Organization. 2006. *Global Strategy on Diet, Physical Activity and Health: A Framework to Monitor and Evaluate Implementation*. Geneva: World Health Organization.

PART V
Future Research Challenges

Chapter 12
Residential Environments and Obesity – Estimating Causal Effects

Graham Moon

Introduction

Earlier chapters in this book have considered the causes of obesity in terms of the standard 'energy balance' model. In this model, obesity results from an imbalance between energy input, captured by measures of diet and nutrition (Chapter 5), and energy output, understood as exercise, fitness and associated activity constructs (Chapter 8). The additional emphasis and unique selling point of the book has been to highlight further relationships associated with the environmental settings within which people live and work. Thus, certain environments have been highlighted as obesogenic and others as leptogenic. The factors implicated in these environmental constructs may link back to the energy balance model – food deserts or walkable neighbourhoods – or they may relate to area-based interventions designed to reduce obesity.

It is not the purpose of this chapter to rehearse again the research evidence linking obesity to inequalities in energy balance or to remake the case for environmental 'explanations' of variations in obesity prevalence. Rather, our focus here is methodological. We probe the ways in which researchers have sought to understand variations in obesity, examining the various techniques and approaches that have been employed. Our central concerns are with issues of causality and research design – how have researchers linked obesity as an outcome to particular exposures and how have they endeavoured to assess the relative importance of environmental effects? More importantly, what do these linkages and assessments tell us about the role of place, environment or area in the causation of obesity?

To address these questions, the chapter is organized in two sections. We commence with a discussion of causality in epidemiology, drawing out the implications that established theoretical and methodological positions have for research on obesity. We extend this discussion of causality into the area of research design, looking not so much at the way particular studies of obesity have been designed, but focussing more on the epistemological claims of different research designs and their strengths and weaknesses with regard to causal analysis. In the second section we exemplify the theoretical and methodological issues under discussion. The chapter concludes by setting out future directions in the study of the causal impact of residential environments on obesity.

On Causality

Research in social epidemiology generally considers causality in relation to criteria set out by Austin Bradford Hill (1965). Bradford Hill built on philosophical arguments developed by Mill (1843) and by Hume (1978). He was centrally concerned to separate association, a non-causal relationship, from a 'truly causal' linkage. In our case an example would be the relationship between obesity, an 'outcome' variable, and dietary intake or the extent of exercise, two candidate causal or 'predictor' variables. More specifically for the purposes of this book, we extend this example to incorporate an investigation of area effects: whether predictor variables measured at an area level are causally related to levels of obesity. Though they are well-known, Hill's criteria merit discussion to link them to our arguments about causality, area effects and obesity.

The Bradford Hill Criteria

Hill proposed nine criteria for consideration when investigating claims of causality (see Table 12.1). Importantly, he argued that his criteria should not be adopted in a mechanistic all-or-nothing way; interpretation and judgement were seen as essential elements in adjudging causality:

> None of my nine viewpoints can bring indisputable evidence for or against the cause-and-effect hypothesis and none can be required as a sine qua non. What they can do, with greater or less strength, is to help us to make up our minds on the fundamental question – is there any other way of explaining the set of facts before us, is there any other answer equally, or more, likely than cause and effect? (Bradford Hill 1965: 300).

Furthermore, he additionally recognized that statistical methods, what he referred to as 'formal tests of significance', were also only an element in understanding causality.

Hill's approach to causality is not without problems. Few if any causal relationships would embody all nine criteria. The desire for strong associations should not, for example, rule out the possibility that weak associations may be causal. Indeed many outcomes have multiple causes, each in themselves weakly associated with the outcome but crucial to the causation of that outcome. Interaction and synergy between possible causal variables are important concepts that can render weak individual relationships stronger when placed alongside others. Equally, confounding or intervening variables can create artefactual relationships that appear causal but in fact are not. Consistency is not essential as the form of causal relationships can change with context; some factors may only be causal in particular circumstances. Nor need causal relationships necessarily exhibit monotonic dose-response relationships. Threshold effects, for example, may give rise to clear non-linearities. To these caveats we may add

Table 12.1 The Bradford Hill criteria for causation

Criterion	Evidence	Example
Association	Strong correlation coefficient; significant and large regression coefficient; control for other influences.	Comparisons of body mass index between people with differing levels of calorific intake; areal obesity rates and levels of provision for physical exercise.
Consistency	No variation by setting or over time in magnitude and direction of relationship; need for replication studies.	Higher levels of obesity should always be associated in the same way with higher levels of calorific intake.
Specificity	A (single) cause should result in a single specific outcome. Need to recognize importance of interactions.	Eating too much of the wrong sort of food leads to obesity but uni-causal specificity is negated by exercise.
Temporality	Cause must precede effect.	Does low exercise frequency cause obesity, or is it the other way round?
Dose-response relationship	Change in causal factor should result in change in outcome variable.	As food intake increases so body-mass should also rise.
Biological plausibility	The presence of a biological mechanism linking a cause to an effect or outcome. This criterion privileges biology. In the social world it needs further extension.	What are the mechanisms within the body that convert calorific intake to body size? Area effects on obesity prevalence should reflect social processes likely to impact distally upon diet or exercise.
Coherence	Causal interpretations should confirm existing knowledge or at least not conflict with that knowledge.	-
Design	Experimental evidence is strongest.	See Table 12.2.
Analogy	Analogous situations bolster claims for causality.	-

concerns about the direction of causality – is lack of exercise a cause or a result of obesity – and the importance of recognizing the impact of sampling variation on quantitative assessments of cause.

Rothman's Criteria

A more recent and more flexible approach to causality is provided by Kenneth Rothman (Rothman and Greenland 2005, Rothman 1976). This approach begins with the contention that the baseline criterion for causality is precedence. Simplistically, and obviously, causes have to precede effects. Having identified antecedent factors, it is then important to recognize that, in most social-world contexts, these factors work together. Several factors may be present, each of which is necessary for an effect to eventuate. Rothman visualizes these 'necessary' causes as slices of a pie or cake. Each slice has to be present to make up the pie; each component cause must be in place for causation to occur.

Two ideas follow from Rothman's conception of causality. First, interaction between component causes is crucial. Indeed it may be interaction that is causal rather than the component causes themselves. Causality results from different factors coming together prior to an outcome event. This is particularly the case in complex socio-epidemiological situations. Area effects on obesity prevalence are thus a constellation of many interacting factors reflecting diet, exercise and the factors that underpin them, including socioeconomic status. Second, if there are component causes to a phenomenon, it must be possible to identify attributable fractions – the proportion of cases that result from each of the component causes. Crucially, as a consequence of interactions, the sum of component attributable fractions will inevitably add up to more than 100 per cent. In this way we can recognize that the impact of diet on obesity and the impact of exercise have a synergistic effect due to the interaction of the two component causes.

Rothman's perspective on causality takes us ultimately to the idea of the web of interacting and mutually dependent causes. While causal webs are often conceptualized as path diagrams or nomograms as used in latent variable analysis and structural equation models (Loehlin 2004, Land 1969), they are perhaps best seen as an actual spider's web. Bhopal makes this link in his vision of the casual web as a series of interlinked concentric circles (Bhopal 2002). His model places broad determinants on the edge of the circular web and successively more individualized causes towards the middle. Viewing the web from different angles can reveal alternative perspectives on causality. In the case of obesity, Bhopal's perspective allows us to place diet and exercise closer to the centre of the web with socioeconomic status nearer the outer edge. A genetic predisposition to obesity would be nearest to the centre while area or environmental effects would form an outer ring. Critically each ring interacts with its neighbours.

In summary, Bradford Hill can be said to represent the traditional perspective on causality in epidemiology. His criteria are, arguably, fairly clinical and biomedical in their emphasis and difficult to apply in population health settings. Rothman and Bhopal are more attuned to the complexity of the social contexts within which area effects on obesity need to be considered. Mevyn Susser, a leading scholar of causality in epidemiology has reflected at length on the ontological and epistemological background to these attempts to define what is understood by the idea of causality (Susser 1973, 1991, 2001), Susser's work helpfully emphasizes the key importance of direction in understanding the presence (or absence) of causation in social epidemiology. This fundamental position, that causes must of necessity precede effects in time, takes us to the matters of research design and the extent to which different research designs allow the identification of causation.

Causality and Research Design

The notion of direction in causality suggests the existence of a causal pathway between cause(s) and effect. It has been suggested that some research designs are more effective than others at isolating causal pathways and uncovering causation. This hierarchy of research designs has a direct bearing on the degree to which epidemiological studies of obesity are able to ascribe causality to environmental factors.

The Hierarchy of Research Designs

The classic hierarchy of research designs (Table 12.2) evolved alongside the rise of evidence-based medicine and focusses on the strength of evidence regarding clinical (or public health) interventions (Glasziou et al. 2004, Rychetnik et al. 2002, Harbour and Miller 2001). Only in third place and below do we find the observational designs that characterize most social epidemiology and much of the research on environmental effects on obesity. Within observational studies there is a clear internal hierarchy with cohort studies occupying 'top' position, followed by case control studies, then cross-sectional studies and finally ecological studies; the latter are arguably better seen as a particular form of cross-sectional study.

It is worth dwelling briefly on why these conclusions have been drawn. As experimental studies, randomized controlled trials in epidemiology would envisage a study population being randomly distributed between a control and an intervention group. Randomization would ensure that any confounding variables are present equally in the two groups so the intervention (the supposed causal factor) should only result in change in the intervention group. The design has forward directionality so the key criterion for causality is satisfied and the potential for confounding is minimized. Replications of well-designed randomized controlled trials provide very strong evidence for causation. As we have noted above

Table 12.2   The hierarchy of epidemiological evidence

Rank	Design	
1 (strong)	Meta-analysis of randomized controlled trials	Experimental Studies
2	Randomized controlled trials	
3	Non-randomized controlled trials	
4	Cohort studies	Observational Studies
5	Case-control studies	
6	Cross-sectional studies	
7	Ecological studies	
8	Case reports and series	Opinion
9	Expert opinion	
10 (weak)	Anecdote	

however, these designs are not easy to implement in social epidemiology; not least there is the ethical difficulty of exposing people to something that may do harm. Cohort studies share forward directionality but lack randomization. Populations are monitored through time to see if outcomes eventuate in groups that are pre-exposed to a hypothesized causal factor. The absence of randomization means that causality can be confounded by extraneous factors. Following a group of people through time to see if obesity succeeds exposure to limited exercise opportunities cannot deny the possibility that some other (unmeasured) factor may impact upon the outcome. Case-control studies look backwards in time to compare people who have an outcome with those who do not. The comparison focuses on the historic presence of a hypothesized cause. The absence of randomization is further compromised by recall bias among other issues. Cross-sectional studies essentially report responses to surveys conducted at a particular point in time; these responses may be for individuals or, in the case of ecological studies, they may be aggregated to geographical areas. Cross-sectional studies lack directionality and are thus of limited use in identifying causal relationships. With these strictures in mind it is important to note that the majority of study designs in epidemiology, and work on obesity is no exception, are cross-sectional (Swinburn et al. 2005).

The absence of randomization in observational epidemiology and, to a lesser extent, the issue of directionality are problematic because of their implications for confounding. With weaker designs it becomes difficult to separate cause from effect or to assess the magnitude of a causal effect in the face of other variables that may reduce, distort or even remove an observed effect (Moon et al. 2000, Jones and Moon 1992). At issue is the identification of a correct causal sequence and the appropriate assessment of all candidate causal factors to validate their causality and identify their role in the sequence. Only with effective control for relevant factors can a true assessment of the causal impact of one variable upon another be identified.

Strutural Equation Modelling

A range of techniques exist for achieving this all-important control and enabling observational epidemiology to say something about causality (Little and Rubin 2000). Traditionally observational epidemiology relied on cross-tabulations but for some time now linear models have been the tool of choice. Other criteria for causality being satisfied (which is a considerable assumption) these models enable the researcher to assess the impact of a range of causal factors on an outcome with each causal effect being an independent coefficient, controlling for the other measured variables with a residual term capturing 'unexplained' variation. Identifying the correct sequencing of the effects found in these models can be achieved through graphical approaches (Pearl 2001) and structural equation models (Hoyle 1995). Causal models are represented as directed acyclic graphs where nodes are the implicated variables and the nodes are linked by arrows showing the direction and magnitude of the relationship.

Examples of structural equation modelling in obesity research include McKeown-Eyssen (2006) and Stafford et al. (2007). The latter is particularly interesting as it considers environmental effects. The study is presented as an attempt to 'unpack the "black box" of area deprivation to show which specific local social and physical environmental characteristics impact upon health' (p. 1882). Using data from the English and Scottish national surveys of health, they tested models of potential causal pathways for both body mass index and waist-hip ratio. Essentially similar results were found. Participation in sport influenced obesity but was itself influenced by neighbourhood disorder with more disorderly neighbourhoods being less conducive to active participation in sport. Independent effects were evident for access to local services. These environmental effects remained after controlling for individual age, sex and socioeconomic status.

Multilevel Modelling

The Stafford et al. paper also extends the debate about the need to uncover the sequencing of potential causal effects. It shows how such effects can also operate at different spatial scales. Until recently much epidemiological research did not take account of this issue and analyses proceeded on a single scale. Multilevel analysis has overcome this problem and is now an established method (Kawachi and Subramanian 2007). Outcomes are seen as a function of causes that work within a nested hierarchy of individual determinants located within areas within regions and so on. Using a multilevel approach, studies of obesity are able to take account of both individual determinants, such as dietary intake, exercise propensity and the confounding effects of socioeconomic status, and, simultaneously, the effects of area-level determinants. The latter, following Macintyre et al. (2002), may be either aggregational or ecological, that is they may represent the local summation of individual effects, or they may be genuinely area-based. An example of an aggregational effect would be the impact of living in an area where most residents take little exercise; there would be effects for both the individual taking little exercise and, at the same time, an additional effect stemming from location within an area where lack of exercise is the norm. In contrast an ecological effect might reflect, for example, the presence or absence of a zoning policy limiting fast food restaurants within the higher level units of analysis.

Multilevel models with obesity as an outcome are now capable of considerable sophistication. They enable the identification of complex neighbourhood or residential effects. These can operate both on their own (main effects) and also take the form of interactions with individual-level determinants of obesity. Moreover, area effects can themselves be at different levels, for example though in no way presenting a causal model, Moon and colleagues discuss a three-level model of obesity prevalence in England (Moon et al. 2007) with individuals nested in both local and larger areas. The area effects that emerge from such multilevel perspectives are of course replete with limitations (Stafford et al. 2008a, Mitchell 2001, Stafford et al. 2001). Galster (2008) has described a number of challenges that are faced when seeking to identify area effects. These include identifying the appropriate scale at which such effects operate and ensuring appropriate measurement of both neighbourhood and individual variables. Critically for our purposes he also notes the need to identify the causal mechanisms that link neighbourhoods to outcomes and the risks posed by endogeneity.

As Cummins (2007) has argued in relation to causal processes: '...the challenge in understanding neighbourhood effects in epidemiology has moved on from simply describing that "place" matters independently of the "individual" to identifying the plausible causal pathways by which neighbourhood social and material environments may affect health' (p. 355). Cummins cautions against the standard deprivation-amplification hypothesis in which neighbourhood socioeconomic disadvantage cumulates and is compounded by resource poverty. He argues that area effects need to be understood at multiple scales, going beyond the constraints of the administrative areas that are usually used as surrogates for neighbourhoods in multilevel studies. Multilevel analysis can accommodate this possibility (Chaix et al. 2005) but it does not do so with ease. It may be necessary to consider contagion effects from adjacent or more distant neighbourhoods. What is clear is the need to incorporate 'multiple levels of determination' in the study of health outcomes (Diez-Roux 1998). Individual-level models on their own neglect the potential importance of context in framing individual action; they also deny the possibility that context matters in its own right. Thus, in the case of obesity, an individual may have a high BMI as a result of individual excess calorific intake, the lifestyle modelling associated with living in a community where large intakes are a norm, and the interaction between the two levels. Moreover, the 'area effect' may be influenced by adjacent areas and, indeed, ultimately by macro-level upstream processes such as globalisation and urbanisation (Cummins and Macintyre 2006).

Endogeneity

Endogeneity poses significant risks to the interpretation of area effects. Put simply, while it may be the case that people who live in 'walkable' neighbourhoods are less likely to be obese (Saelens et al. 2003b), this putative area effect may lack causal significance. The less obese individuals may self-select into more walkable neighbourhoods. Equally, as Kawachi and Subramanian (2007) argue: '...it is commonly supposed that the presence of fast-food outlets in a neighbourhood increases the risk of obesity for local residents. However, it is equally plausible that the decision of fast food franchises to open their businesses in particular locations occurs in response to the [appetites] of local residents' (p. 3). These possibilities take us back to research design. Cross-sectional designs cannot address the endogeneity problem. To do that we require longitudinal designs, cohort studies that can take account of the complex timing relationships between obesity onset, population mobility and individual and contextual determinants.

Area Effects and Obesity

The summary position suggested by the previous sections is that effective study of causality in relation to area/residential/environmental effects on obesity should entail a multilevel design, preferably with a longitudinal element within a clear causal framework. In this section we consider the extent to which current research has approached this goal. The section reports separately on the impact on obesity of diet/food environments, exercise/activity environments, and neighbourhood socioeconomic status, focusing on recent studies.

Central to the idea of the area effect is the notion of the obesogenic environment. This is encapsulated by neighbourhood-level measures of access to healthy food and exercise opportunities and the interaction of these with socioeconomic status (Procter et al. 2008, Townshend and Lake 2008). Obesogenic environments have increasingly been proposed as key triggers for individual obesity (Harrington and Elliott 2009, Lebel et al. 2009, Smith and Cummins 2009, Black and Macinko 2008). The environment is seen as a constraint on individual behaviour and an important factor in determining the uptake of health promotion measures directed at obesity. It also has implications for health inequalities in that obesogenic environments are generally to be found in areas of lower social status. Understanding more about area effects as determinants or causes of obesity raises the possibility of area-based interventions to reduce obesity prevalence (Michael and Yen 2009).

The Dietary Environment and Obesity

Cummins and Macintyre suggest that there are two aspects to the dietary environment (Cummins and Macintyre 2006). On the one hand consideration can be given to the retail environment associated with the purchase of foods for cooking. Supermarkets or larger outlets may promote choice and offer healthier products (albeit alongside less healthy lines). Access to independent grocers and other outlets such as farmers' markets or healthy food cooperatives may also assist healthy eating. Conversely, food retail opportunities may restrict choice and involve poorer quality products, often with high fat or sugar content. A second aspect of the dietary environment relates to the availability of fast-food outlets and assumes that such ready-cooked food is seldom nutritious and often obesogenic. Areas with lots of restaurants or fast food outlets may have high levels of obesity.

Weak but consistent evidence from ecological studies in North America suggests that obesity levels are lower in areas with more supermarkets and higher when areas are characterized by small groceries. Morland and Evenson (2009) provide a recent example of such work. Sturm and Datar (2005) suggest that, for children, there are area effects for fruit and vegetable prices that drive down obesity while outlet density and area dairy prices are less important. More complex studies have employed longitudinal and multilevel approaches (Inagami et al. 2006, Morland et al. 2006). Wang et al. (2008), using a repeated cross-sectional design, traced the temporal relationship between the food retail environment and obesity outcomes among the adult population of four cities in California, US. The longitudinal element to this study, though weak in research design terms, enabled them to link a growth in small neighbourhood foodstores to an increase in obesity while also confirming the negative relationship between supermarket provision and obesity. A multilevel study (Veugelers et al. 2008) extended this generalization both to Canada and to the impact of supermarkets on child obesity suggesting that child obesity is less in neighbourhoods where parents perceived that access to healthy food shopping was good.

Two review papers (Larson et al. 2009, Ford and Dzewaltowski 2008) synthesize the available North American evidence on the neighbourhood-level relationship between obesity levels and the types of retail outlets selling food for cooking. The first authors highlight the interaction between the quality of the retail food environment, the socioeconomic status of an area, and individual disadvantage, noting particular implications for obesity levels among minority ethnic groups. They reviewed 54 studies in the US concluding that neighbourhoods with more supermarkets and fewer convenience stores tend to have reduced obesity prevalences. They note that low-income, minority, and rural neighbourhoods are those most often affected by this association.

Larson et al. (2009) also indicate that studies of the link between obesity and neighbourhood measures of the accessibility and availability of restaurants and fast food outlets are less consistent in their conclusions. High levels of obesity among black New Yorkers have been linked to neighbourhood fast food density in an ecological study, and shown to be independent of income (Kwate et al. 2009). In a small-scale single level study of adults in the rural south of the US regular fast food consumption and restaurant dining was associated with high rates of self-reported obesity having controlled for age, education and gender (Casey et al. 2008). Similar findings were evident in Southern California among a child population (Ayala et al. 2008). Not only were restaurants and fast food outlets implicated in this finding, eating away from the home with relatives, neighbours or friends was also found to raise obesity levels. Type of restaurant was not found to be important. Maddock (2004) offers a state-level confirmation of the association between obesity and fast food outlet density.

Better evidence for area effects is provided in a small multilevel study in the urban setting of Portland Oregon, US using geographic information systems to calculate fast food outlet density (Li et al. 2009). This study indicated that the individual-level association between fast food consumption and obesity also manifests as an area effect in which individuals living in a neighbourhood with a high density of fast food restaurants are approximately twice as likely to be obese. A much larger multilevel study (Mehta and Chang 2008) using individual data from the US Behavioural Risk Factor Surveillance System provides still better evidence of causal association. It linked in data on restaurant density and restaurant mix (the ratio of fast-food to full-service restaurants) and confirmed that neighbourhood fast-food restaurant density is an important determinant of obesity. In contrast to the small Ayala et al. (2008) study, the type of outlet was found to matter: the density of full-service restaurants was associated with a lower likelihood of being obese.

Area effects linking fast food to obesity are challenged by small studies in Minnesota and in Cincinnati (Jeffery et al. 2006, Burdette and Whitaker 2004). Outside the US, evidence from Australia and New Zealand using GIS offers further contradiction of the hypothesis that areal concentrations of fast food provision are associated with obesity (Pearce et al. 2009, Crawford et al. 2008). The Australian study covered both children and adults and was based in Melbourne. Its results contradicted the evidence from the US: obesity was reduced for people with fast food outlets within two kilometres of their residence. It was based on a rather small sample of respondents to a cross-sectional survey. The New Zealand study, in contrast, used a national survey with linked data on proximity to fast food outlets (Pearce et al. 2009). As in Australia however, residents living furthest from a fast food outlet were more likely to be overweight. Distance was also positively associated with consuming a recommended intake of vegetables.

These ideas about the relationship between obesity and the dietary environment come together in the concept of the food desert (Wrigley et al. 2003, Cummins and Macintyre 2002, Wrigley 2002). Food deserts can be defined as locales where there is poor provision of retail outlets selling affordable and healthy foods (Black and Macinko 2008). They are generally associated with socioeconomic deprivation. On each of the issues identified in this section there is plentiful evidence that, stepping back from obesity as an outcome, deprivation relates closely to the patterns of food retail both for cooking and for direct consumption. Research on the city of Edmonton, Canada serves as an example (Hemphill et al. 2008, Smoyer-Tomic et al. 2008). Fast food outlets and supermarket provision are both shown to exhibit clear spatial biases in their location.

The Exercise Environment and Obesity

A link with socioeconomic status is also evident for exercise environments (Saelens et al. 2003b, King et al. 2000). The onward linkage to obesity has been the subject of rather less research effort but has received somewhat more attention than dietary environments. The review paper by Black and Macinko (2008) notes that environments that are not conducive to physical activity are consistently associated with obesity. This finding is substantiated for both adults and for children (Goran et al. 1999). While much of the research, as in the case of dietary environments, is based in the US, Townshend and Lake (2008) argue for similar conclusions in the UK. They note however that disputed causal mechanisms and interactions with other factors make the association of a poor exercise environment with obesity a complex matter.

Ecological studies of this relationship are of course weak in causal terms. Nonetheless they offer pointers to the complexity of the link between exercise/ activity and obesity. In part this complexity results from differences in the measurement of activity and exercise. One study considered vehicle miles of travel and commuting time as indicators of activity levels hypothesizing a positive relation with obesity and drawing on ideas about inactivity in a care-centred society (Lopez-Zetina et al. 2006). The study, based in California, controlled self-reported walking, cycling and activity levels and found significant county-level correlations between obesity and physical inactivity, vehicle miles travelled and commuting time. A larger national study of state-level variations in obesity (Vandegrift and Yoked 2004) focused on suburban sprawl. Using a measure of the growth of developed land and controlling for population size, it was found that states where city sprawl had increased had also experienced large rises in obesity levels.

Slightly stronger designs offer more complex conclusions. A comparison of two Black neighbourhoods with differing levels of deprivation but similar overall urban design did not reveal any great differences in levels of obesity (Miles et al. 2008). It was suggested that detailed analysis is needed focussing on particular barriers to physical activity. In contrast to these US findings Ellaway et al. (1997), using the West of Scotland 2007 study, studied adults aged 40 and 60 who were long-standing residents of four socially contrasting urban neighbourhoods in Glasgow, Scotland. They found that neighbourhood deprivation had a significant effect on a range of obesity measures, even after controlling for individual respondent circumstances such as age, sex, class and personal deprivation.

Addressing barriers to physical activity has been part of the urban design philosophy known as New Urbanism. A study in the US has compared a new urbanist neighbourhood, designed to maximize physical activity and place-centredness, with traditional suburbs (Brown et al. 2008). In a form of case-control design, the areas were matched for age and value of housing and accessibility. While there was variation in obesity with respect to individual respondent activity patterns, aggregate levels of obesity did not vary between the new urbanist and other neighbourhoods. This finding is substantiated by Eid et al. (2008) who argue, in a longitudinal study, that urban sprawl cannot be linked to obesity in the US and that endogeneity needs greater consideration: do obese people choose to live in sprawling neighbourhoods? Individual level cross-sectional studies investigating this question have noted that that individual perceptions of how pleasant an area is for physical activity can impact on obesity after controlling for age, sex and education status (Casey et al. 2008). Cross-sectional studies have also found significant negative correlations with street connectivity, exercise facility provision, quality and physical accessibility in deprived populations in the mid-western US (Heinrich et al. 2008) and significant associations between neighbourhood walkability and low obesity in men and high obesity in women and non-whites in Atlanta, US (Frank et al. 2004, 2008). High walkability neighbourhoods have a general association with lower levels of obesity (Saelens et al. 2003a) but this may not extend to older people (Berke et al. 2007). Child obesity appears to be lower when children undertake physical activity but this may or may not be constrained by neighbourhood availability of resources for physical activity (Franzini et al. 2009, Gordon-Larsen et al. 2006), indeed it may not be the case at all (Burdette and Whitaker 2004). Spence et al. (2008) rightly note that studies of child obesity have been inconsistent. Their own study in Edmonton Canada suggests that, contrary to the US adult evidence, girls from walkable neighbourhoods were more likely to be obese.

Inconsistency is perhaps the best conclusion that can be drawn from recent ecological and cross-sectional studies of obesity in relation to the activity environment. Walkability may or may not matter; there may or may not be particular links with gender. In theory multilevel studies and work using allied techniques, though still largely cross-sectional, should help to unpick this confusion. A study of child obesity in the US (Singh et al. 2008) controlled for individual measures including physical activity, and also for area deprivation. Area variations in child obesity remained evident after this adjustment process with highest levels in south-central states and individual physical activity was a significant predictor. A study of a child and adolescent population in Canada (Veugelers et al. 2008) reveals that children in areas with good access to activity provision – parks, playgrounds – are less likely to be obese. For older North American populations, there are significant relationships between individual obesity and area measures of land-use mix (more mixed land use is associated with lower obesity) (Li et al. 2008) and older women are less likely to be obese if they live in neighbourhoods with high street connectivity (Grafova et al. 2008). For adult populations, Australian evidence suggests that urban sprawl has a significant independent effect on obesity after controlling for individual and other area factors (Garden and Jalaludin 2009). This may also be the case in Canada (Ross et al. 2007) while some work in the US now argues that personal and neighbourhood barriers to physical activity confound the direct effect of spawl (Joshu et al. 2008). Such neighbourhood barriers may include land-use mix and access to public transport (Lovasi et al. 2009) and low quality walking environments (Mujahid et al. 2008). Although these associate with low levels of obesity among non-deprived groups, there is no consistent converse association. Rundle et al. (2008) explore this matter, arguing that neighbourhood walkability is best seen as an intervening variable in the relationship between deprivation and obesity, having a particular impact on obesity among women. Some support for this view from outside the US is offered in recent studies from Portugal (Santana et al. 2009) and the UK (Stafford et al. 2007).

A further theme, connected with the idea of the activity environment, is the relationship between obesity and the 'greenness' of an environment. Green environments might be expected to offer more informal activity opportunities and act as an encouragement to outdoor activity. A small longitudinal study of changes in body mass among children in one US county used satellite measures of neighbourhood 'greenness', controlling for baseline body mass index, residential density and socioeconomic status (Bell et al. 2008). Work from Sweden with a more substantial empirical base (Bjork et al. 2008) also suggests that perceived greenness within residential neighbourhoods is associated with low body mass.

Socioeconomic Status and Obesity

Our third area of concern regarding area effects on obesity is the role of socioeconomic status. Here we focus on area socioeconomic status rather than the contested association between individual status and obesity. A recent review of work on individual-level studies of child populations (Shrewsbury and Wardle 2008) has remarked on the inconsistent nature of the conclusions that have been drawn to date. A similar conclusion could be drawn about adult populations and about research on area effects.

Multiple measures of socioeconomic status have been associated with obesity levels (Harrington and Elliott 2009, Moon et al. 2007, Monden et al. 2006, Nelson et al. 2006, Van Lenthe and Mackenbach 2002, Sundquist et al. 1999). Such relationships include crime (Mobley et al. 2006), neighbourhood wealth (Ackerson et al. 2008), household instability and area poverty (Chambers et al. 2009). The latter study suggested that mothers in unstable households in low poverty areas were more likely to be obese. The ecological study of Drewnowski et al. (2009) linked the prevalence of overweight children with neighbourhood poverty and suggested that ethnicity was not a mediating factor. In Glasgow, people in more deprived neighbourhoods were more likely to be obese (Ellaway et al. 1997). Grafova et al. (2008) suggested that both men and women are less likely to be obese in economically advantaged neighbourhoods, and male obesity is more common in areas with high levels of immigrants. Ross et al. (2007) found against the latter point in the Canadian context, suggesting instead that it applies to women rather than men, and went on to link obesity to area measures of low education attainment.

Another Canadian study suggests that women are more likely to be obese if they live in poor areas and men if they live in affluent areas, raising the possibility that obesity reflects gender-specific responses to relative disadvantage (Matheson et al. 2008, Pickett et al. 2005, Robert and Reither 2004, Diez-Roux et al. 2000, Kahn et al. 1998). In further thinking through the pathways by which socioeconomic status may impact upon obesity, a small number of researchers are now also focussing on the idea of neighbourhood disorder, hypothesizing that the stresses and strains of living in disordered neighbourhoods promote inappropriate eating and exercise behaviour (Burdette and Hill 2008, Mujahid et al. 2008, Stafford et al. 2008b, Boehmer et al. 2007, Glass et al. 2006, Lumeng et al. 2006). Others are exploring links with social capital (Poortinga 2006) as the basis for a hypothesis of causation by community stress while, as noted above, Cummins and Macintyre (2006) suggest a theory of deprivation amplification. In the case of neighbourhood disorder, findings suggest that individual obesogenic behaviour and individual stress tend to be more important than the area effects of neighbourhood disorder and social capital though there are important cross-level interactions with both notions, for example women living in areas with little residential mobility tend to be more likely to be obese (Grafova et al. 2008).

While multilevel designs are relatively common among these studies, there are few studies that offer the greater insight into causality that is possible with longitudinal research. One such study (Oliver and Hayes 2008) controlled for a range of individual circumstances among a sample of Canadian children and found that residents in the poorest neighbourhoods were more likely to have an increasing body mass index. Another (Park et al. 2008), has explored the links between obesity and acculturation. The hypothesis here is that immigrants to the US are more likely to become obese as their period of time in a new country increases. Park et al. considered the confounding impact on this theory of living in an immigrant community, noting the generally higher levels of obesity found in immigrant areas (Chang 2006). After controlling for a range of relevant individual variables they concluded that relationship between obesity and the proportion of local residents born outside the US declined to insignificance for all migrant communities except Hispanics.

Conclusions and Future Directions

The studies that have been reviewed above add much to our knowledge of area effects on obesity, but do they add significantly to understanding the causes of obesity? In many ways the cases that are presented fall short of the ideals discussed in the earlier parts of this chapter. Most studies are cross-sectional and, as such, fall foul of arguments about directionality and causal precedence; they present snapshot associations rather than clear causal pathways. Most studies are also either ecological or individual. They commit the ecological or atomistic fallacies; we cannot read off environmental causation from either design. A minority of studies with longitudinal designs avoid the cross-sectional problem but with few exceptions focus on the individual rather than on area effects. A growing number of multilevel studies offer simultaneous consideration of individuals within areas and thus the possibility of unravelling the confounding of individual and area. While this represents progress, such studies only infrequently incorporate a longitudinal element. 'Strong' studies as conceptualized in the traditional hierarchy of study designs are thus rare. With respect to the Bradford Hill criteria, there is some consistency but frequent exceptions. Further shortcomings are detailed in Black and Macinko (2008). Health geographers can at least draw comfort from the fact that there is a great deal of geographic variation in the multitude of marginally different studies that make up the current pattern of evidence.

Recent studies are now starting to resolve this complex picture. In part this complexity is intriguing, interesting and potentially useful; it may be important to know, if indeed it is the case, that area effects differ between men and women or that they may vary between nations. Resolving the complexity demands a move away from looking at simple association towards the theorisation of causal pathways. Do area effects operate through individual diet and exercise to 'cause' obesity? Or is the pathway one in which area effects modify the impact of individual behaviour?

Does area socioeconomic status 'cause' areal variations in diet and exercise, or is the causal pathway reversed? These questions take us simultaneously back to Rothman (1976), Loehlin (2004), Bhopal (2002) and the web of causation and forward to recent work that has sought to frame obesity outcomes in terms of causal pathways (Stafford et al. 2007, 2008b).

Operationalizing the causal pathway approach requires sophisticated analysis using structural equation modelling, ideally within a multilevel framework to address the ecological and atomistic fallacies, and incorporating a longitudinal element to address endogeneity (Kawachi and Subramanian 2007). Fortunately large (and complex) data sets are now available that enable this task. Adequate theorising is an essential precursor of the empirical testing of such pathway analysis. The approach, if done well, has promise as a means of linking aetiological studies to the development of interventions to reduce levels of obesity (Foresight 2007, Egger and Swinburn 1997, Swinburn et al. 1999, 2005).

Beyond the intersection of multilevel modelling, longitudinal analysis and structural equation modelling, the wider literature on area effects suggests further methodological developments that may improve our understanding of the causal possibilities associated with areas (Galster 2008). With regard to obesity, two other themes deserve mention as additional future directions for study. First, there is a case for more sophisticated thinking about obesity as a learned behaviour; proximity may confer an area effect but it may not be proximity per se that is the issue. Rather, there may be a case for considering the importance of social networks (Cohen-Cole and Fletcher 2008, Christakis and Fowler 2007). To take it to an extreme, obesity may be contagious. More realistically the social habits of eating and exercising may be shared within a social network that may or may not include elements of geographical proximity. The connectivities of networks are likely to interact in complex and potentially confounding ways with area effects. Christakis and Fowler suggest that networks based on proximity are relatively less important. It is possible to operationalize a consideration of social networks within a multilevel framework.

A second future direction for study concerns experimental and quasi-experimental design. Such approaches, in a traditional sense, produce stronger evidence of causation. As we have noted however, they do not fit easily with the research questions that are posed in social epidemiology and they can pose ethical challenges. Nevertheless, there is a future for experimental approaches to the study of area effects on obesity. This future relates particularly to natural experiments and experiments of opportunity that eventuate when new food retail developments are planned, land uses change or area based policies to reduce obesity are initiated (Veugelers et al. 2008). There are also methodological approaches, such as instrumental variable estimation and propensity scoring that can be used to mimic experimental designs (Kawachi and Subramanian 2007). Cummins et al. (2005) report an example of an experimental study in Glasgow, Scotland. A prospective quasi-experimental design compared baseline and follow up data in an 'intervention' community with a matched 'comparison' community. The intervention was the introduction of a large-scale food retail outlet. Outcome measurement focussed on food consumption however rather than obesity and suggested that the intervention had little effect on consumption. Similarly limited evidence arose from a cluster community trial in five cities in the US that examined the impact of poverty on a range of physical health outcomes including obesity (Kling et al. 2007).

Looking still further and moving some way from area effects, a final further direction for research on the causation of obesity is genomics (McKeown-Eyssen 2006). There is some logic in extending the study of individual determinants of obesity to 'go within the individual' and emergent research is also considering environment-gene interaction (Bouchard 2008). Particular genotypes may be associated with metabolic responses to diet or exercise (Galgani and Ravussin 2008) and twin studies suggest that genetic inheritance may be linked to up to two thirds of the population variation in individual body mass (Ravussin and Bogardus 2000). Genetics, alongside upbringing may be implicated in the established relationship between infant feeding, maternal nutrition and obesity among children (Von Kries et al. 1999, Whitaker and Dietz 1998).

In conclusion, there is some way to go before we can definitively ascribe causality to the relationship between area effects and obesity. Much has been achieved but, in common with other areas of social epidemiology and public health, many of the observed area effects are better seen as risk factors or health determinants (McDowell 2008, Pearce 1996). Some may be true causes but many will simply be markers. The evidence to date is reasonably positive but by no means conclusive. Future research will need to unpack pathways and causal mechanisms if we are to clarify fully the role of area effects.

References

Ackerson, L.K., Kawachi, I., Barbeau, E.M. and Subramanian, S.V. 2008. Geography of underweight and overweight among women in India: a multilevel analysis of 3,204 neighborhoods in 26 states. *Economics and Human Biology* 6, 264-80.
Ayala, G.X., Rogers, M., Arredondo, E.M., Campbell, N.R., Baquero, B., Duerksen, S.C. and Elder, J.P. 2008. Away-from-home food intake and risk for obesity: examining the influence of context. *Obesity* 16, 1002-8.
Bell, J.F., Wilson, J.S. and Liu, G.C. 2008. Neighborhood greenness and 2-year changes in body mass index of children and youth. *American Journal of Preventive Medicine* 35, 547-53.
Berke, E.M., Koepsell, T.D., Moudon, A.V., Hoskins, R.E. and Larson, E.B. 2007. Association of the built environment with physical activity and obesity in older persons. *American Journal of Public Health* 97, 486-92.
Bhopal, R. 2002. Cause and effect: the epidemiological approach, in *Concepts of Epidemiology: an integrated introduction to the ideas, theories, principles and methods of epidemiology,* edited by R. Bhopal. Oxford: Oxford University Press.

Bjork, J., Albin, M., Grahn, P., Jacobsson, H., Ardo, J., Wadbro, J., Ostergren, P.O. and Skarback, E. 2008. Recreational values of the natural environment in relation to neighbourhood satisfaction, physical activity, obesity and wellbeing. *Journal of Epidemiology and Community Health* 62(4), e2.

Black, J.L. and Macinko, J. 2008. Neighborhoods and obesity. *Nutrition Reviews* 66, 2-20.

Boehmer, T.K., Hoehner, C.M., Deshpande, A.D., Brennan Ramirez, L.K. and Brownson, R.C. 2007. Perceived and observed neighborhood indicators of obesity among urban adults. *International Journal of Obesity* 31, 968-77.

Bouchard, C. 2008. Gene-environment interactions in the etiology of obesity: Defining the fundamentals. *Obesity* 16(Suppl 3), S5-S10.

Bradford Hill, A. 1965. The environment and disease: association or causation. *Proceedings of the Royal Society for Medicine* 58, 295-300.

Brown, A.L., Khattak, A.J. and Rodriguez, D.A. 2008. Neighbourhood types, travel and body mass: a study of new urbanist and suburban neighbourhoods in the US. *Urban Studies* 45, 963-88.

Burdette, A.M. and Hill, T.D. 2008. An examination of processes linking perceived neighborhood disorder and obesity. *Social Science and Medicine* 67, 38-46.

Burdette, H.L. and Whitaker, R.C. 2004. Neighborhood playgrounds, fast food restaurants, and crime: relationships to overweight in low-income preschool children. *Preventive Medicine* 38, 57-63.

Casey, A.A., Elliott, M., Glanz, K., Haire-Joshu, D., Lovegreen, S.L., Saelens, B.E., Sallis, J.F. and Brownson, R.C. 2008. Impact of the food environment and physical activity environment on behaviors and weight status in rural U.S. communities. *Preventive Medicine* 47, 600-4.

Chaix, B., Merlo, J. and Chauvin, P. 2005. Comparison of a spatial approach with the multilevel approach for investigating place effects on health: the example of healthcare utilisation in France. *Journal of Epidemiology and Community Health* 59, 517-26.

Chambers, E.C., Duarte, C.S. and Yang, F.M. 2009. Household instability, area poverty, and obesity in urban mothers and their children. *Journal of Health Care for the Poor and Underserved* 20, 122-33.

Chang, V.W. 2006. Racial residential segregation and weight status among US adults. *Social Science and Medicine* 63, 1289-303.

Christakis, N.A. and Fowler, J.H. 2007. The spread of obesity in a large social network over 32 years. *New England Journal of Medicine* 357, 370-9.

Cohen-Cole, E. and Fletcher, J.M. 2008. Is obesity contagious? Social networks vs. environmental factors in the obesity epidemic. *Journal of Health Economics* 27, 1382-7.

Crawford, D.A., Timperio, A.F., Salmon, J.A., Baur, L., Giles-Corti, B., Roberts, R.J., Jackson, M.L., Andrianopoulos, N. and Ball, K. 2008. Neighbourhood fast food outlets and obesity in children and adults: the CLAN Study. *International Journal of Pediatric Obesity* 3, 249-56.

Cummins, S. 2007. Commentary: Investigating neighbourhood effects on health – avoiding the 'Local Trap'. *International Journal of Epidemiology* 36, 355-7.

Cummins, S. and Macintyre, S. 2002. 'Food deserts' – evidence and assumption in health policy making. *British Medical Journal* 325, 436-8.

Cummins, S. and Macintyre, S. 2006. Food environments and obesity – neighbourhood or nation? *International Journal of Epidemiology* 35, 100-4.

Cummins, S., Petticrew, M., Higgins, C., Findlay, A. and Sparks, L. 2005. Large scale food retailing as an intervention for diet and health: quasi-experimental evaluation of a natural experiment. *Journal of Epidemiology and Community Health* 59, 1035-40.

Diez-Roux, A.V. 1998. Bringing context back into epidemiology: variables and fallacies in multilevel analysis. *American Journal of Public Health* 88, 216-22.

Diez-Roux, A.V., Link, B.G. and Northridge, M.E. 2000. A multilevel analysis of income inequality and cardiovascular disease risk factors. *Social Science and Medicine* 50, 673-87.

Drewnowski, A., Rehm, C., Kao, C. and Goldstein, H. 2009. Poverty and childhood overweight in California Assembly districts. *Health and Place* 15, 631-5.

Egger, G. and Swinburn, B. 1997. An 'ecological' approach to the obesity pandemic. *British Medical Journal* 315, 477-80.

Eid, J., Overman, H.G., Puga, D. and Turner, M.A. 2008. Fat city: questioning the relationship between urban sprawl and obesity. *Journal of Urban Economics* 63, 385-404.

Ellaway, A., Anderson, A. and Macintyre, S. 1997. Does area of residence affect body size and shape? *International Journal of Obesity* 21, 304-8.

Ford, P.B. and Dzewaltowski, D.A. 2008. Disparities in obesity prevalence due to variation in the retail food environment: three testable hypotheses. *Nutrition Reviews* 66, 216-28.

Foresight. 2007. *Tackling Obesities: Future Choices.* London: Government Office for Science.

Frank, L.D., Andresen, M.A. and Schmid, T.L. 2004. Obesity relationships with community design, physical activity, and time spent in cars. *American Journal of Preventive Medicine* 27, 87-96.

Frank, L.D., Kerr, J., Sallis, J.F., Miles, R. and Chapman, J. 2008. A hierarchy of sociodemographic and environmental correlates of walking and obesity. *Preventive Medicine* 47, 172-8.

Franzini, L., Elliott, M.N., Cuccaro, P., Schuster, M., Gilliland, M.J., Grunbaum, J. A., Franklin, F. and Tortolero, S.R. 2009. Influences of physical and social neighborhood environments on children's physical activity and obesity. *American Journal of Public Health* 99, 271-278.

Galgani, J. and Ravussin, E. 2008. Energy metabolism, fuel selection and body weight regulation. *International Journal of Obesity* 32, S109-S119.

Galster, G. 2008. Quantifying the effect of neighbourhood on individuals: challenges, alternative approaches, and promising directions. *Schmollers Jahrbuch* 128, 7-48.

Garden, F.L. and Jalaludin, B.B. 2009. Impact of urban sprawl on overweight, obesity and physical activity in Sydney, Australia. *Journal of Urban Health* 86, 19-30.

Glass, T.A., Rasmussen, M.D. and Schwartz, B.S. 2006. Neighborhoods and obesity in older adults. The Baltimore memory study. *American Journal of Preventive Medicine* 31, 455-63.

Glasziou, P., Vandenbroucke, J. and Chalmers, I. 2004. Assessing the quality of research. *British Medical Journal* 328, 39-41.

Goran, M.I., Reynolds, K.D. and Lindquist, C.H. 1999. Role of physical activity in the prevention of obesity in children. *International Journal of Obesity* 23, 18-33.

Gordon-Larsen, P., Nelson, M.C., Page, P. and Popkin, B.M. 2006. Inequality in the built environment underlies key health disparities in physical activity and obesity. *Pediatrics* 117, 417-24.

Grafova, I.B., Freedman, V.A., Kumar, R. and Rogowski, J. 2008. Neighborhoods and Obesity in Later Life. *American Journal of Public Health* 98, 2065-71.

Harbour, R. and Miller, J. 2001. A new system for grading recommendations in evidence based guidelines. *British Medical Journal* 323, 334-6.

Harrington, D.W. and Elliott, S.J. 2009. Weighing the importance of neighbourhood: a multilevel exploration of the determinants of overweight and obesity. *Social Science and Medicine* 68, 593-600.

Heinrich, K.M., Lee, R.E., Regan, G.R., Reese-Smith, J.Y., Howard, H.H., Haddock, C.K., Poston, W.S. and Ahluwalia, J.S. 2008. How does the built environment relate to body mass index and obesity prevalence among public housing residents? *American Journal of Health Promotion* 22, 187-94.

Hemphill, E., Raine, K., Spence, J.C. and Smoyer-Tomic, K.E. 2008. Exploring obesogenic food environments in Edmonton, Canada: the association between socioeconomic factors and fast-food outlet access. *American Journal of Health Promotion* 22, 426-32.

Hoyle, R.H. 1995. *Structural Equation Modeling: Concepts, Issues, and Applications.* London: Sage.

Hume, D. 1978. *A Treatise of Human Nature*, edited with an analytical index by L.A. Selby-Bigge. 2nd Edition (with text revisions and variant readings by P.H. Nidditch). Oxford: Clarendon Press.

Inagami, S., Cohen, D.A., Finch, B.K. and Asch, S.M. 2006. You are where you shop: grocery store locations, weight, and neighborhoods. *American Journal of Preventive Medicine* 31, 10-17.

Jeffery, R.W., Baxter, J., McGuire, M. and Linde, J. 2006. Are fast food restaurants an environmental risk factor for obesity? *International Journal of Behavioural Nutrition and Physical Activity* 3(1), 2.

Jones, K. and Moon, G. 1992. *Health, Disease and Society.* London: Routledge.

Joshu, C.E., Boehmer, T.K., Brownson, R.C. and Ewing, R. 2008. Personal, neighbourhood and urban factors associated with obesity in the United States. *Journal of Epidemiology and Community Health* 62, 202-8.

Kahn, H.S., Tatham, L.M., Pamuk, E.R. and Heath Jr, C.W. 1998. Are geographic regions with high income inequality associated with risk of abdominal weight gain? *Social Science and Medicine* 47, 1-6.

Kawachi, I. and Subramanian, S.V. 2007. Neighbourhood influences on health. *Journal of Epidemiology and Community Health* 61, 3-4.

King, A.C., Castro, C., Wilcox, S., Eyler, A.A., Sallis, J.F. and Brownson, R.C. 2000. Personal and environmental factors associated with physical inactivity among different racial-ethnic groups of U.S. middle-aged and older-aged women. *Health Psychology* 19, 354-64.

Kling, J.R., Liebman, J.B. and Katz, L.F. 2007. Experimental analysis of neighborhood effects. *Econometrica* 75, 83-119.

Kwate, N.O., Yau, C.Y., Loh, J.M. and Williams, D. 2009. Inequality in obesigenic environments: fast food density in New York City. *Health and Place* 15, 364-373.

Land, K.C. 1969. Principles of path analysis. *Sociological Methodology* 1, 3-37.

Larson, N.I., Story, M.T. and Nelson, M.C. 2009. Neighborhood environments: disparities in access to healthy foods in the US. *American Journal of Preventive Medicine* 36, 74-81.

Lebel, A., Pampalon, R., Hamel, D. and Theriault, M. 2009. The geography of overweight in Quebec: a multilevel perspective. *Canadian Journal of Public Health* 100, 18-23.

Li, F.Z., Harmer, P.A., Cardinal, B.J., Bosworth, M., Acock, A., Johnson-Shelton, D. and Moore, J.M. 2008. Built environment, adiposity, and physical activity in adults aged 50-75. *American Journal of Preventive Medicine* 35, 38-46.

Li, F.Z., Harmer, P., Cardinal, B.J., Bosworth, M. and Johnson-Shelton, D. 2009. Obesity and the built environment: does the density of neighborhood fast-food outlets matter? *American Journal of Health Promotion* 23, 203-9.

Little, R.J. and Rubin, D.B. 2000. Causal effects in clinical and epidemiological studies via potential outcomes: concepts and analytical approaches. *Annual Review of Public Health* 21, 121-45.

Loehlin, J.C. 2004. *Latent variable models: an introduction to factor, path, and structural equation analysis.* New Jersey: Guilford.

Lopez-Zetina, J., Lee, H. and Friis, R. 2006. The link between obesity and the built environment. Evidence from an ecological analysis of obesity and vehicle miles of travel in California. *Health and Place* 12, 656-64.

Lovasi, G.S., Neckerman, K.M., Quinn, J.W., Weiss, C.C. and Rundle, A. 2009. Effect of individual or neighborhood disadvantage on the association between neighborhood walkability and body mass index. *American Journal of Public Health* 99, 279-84.

Lumeng, J.C., Appugliese, D., Cabral, H.J., Bradley, R.H. and Zuckerman, B. 2006. Neighborhood safety and overweight status in children. *Archives of Pediatrics and Adolescent Medicine* 160, 25-31.

Macintyre, S., Ellaway, A. and Cummins, S. 2002. Place effects on health: how can we conceptualise, operationalise and measure them? *Social Science and Medicine* 55, 125-39.

Maddock, J. 2004. The relationship between obesity and the prevalence of fast food restaurants: state-level analysis. *American Journal of Health Promotion* 19, 137-43.

Matheson, F.I., Moineddin, R. and Glazier, R.H. 2008. The weight of place: a multilevel analysis of gender, neighborhood material deprivation, and body mass index among Canadian adults. *Social Science and Medicine* 66, 675-90.

McDowell, I. 2008. From risk factors to explanation in public health. *Journal of Public Health* 30, 219-223.

McKcown-Eysscn, G. 2006. Mcthodologic issucs for thc study of obesity. *Epidemiology* 17, 134.

Mehta, N.K. and Chang, V.W. 2008. Weight status and restaurant availability – a multilevel analysis. *American Journal of Preventive Medicine* 34, 127-33.

Michael, Y.L. and Yen, I.H. 2009. Built environment and obesity among older adults – can neighborhood-level policy interventions make a difference? *American Journal of Epidemiology* 169, 409-12.

Miles, R., Panton, L.B., Jang, M. and Haymes, E.M. 2008. Residential context, walking and obesity: two African-American neighborhoods compared. *Health and Place* 14, 275-86.

Mill, J.S. 1843. *A system of logic, ratiocinative and inductive: being a connected view of the principles of evidence and the methods of scientific evidence.* New York: Harper.

Mitchell, R. 2001. Multilevel modeling might not be the answer. *Environment and Planning A*, 33, 1357-1360.

Mobley, L.R., Root, E.D., Finkelstein, E.A., Khavjou, O., Farris, R.P. and Will, J.C. 2006. Environment, obesity, and cardiovascular disease risk in low-income women. *American Journal of Preventive Medicine* 30(4), 327-32.

Monden, C.W., van Lenthe, F.J. and Mackenbach, J.P. 2006. A simultaneous analysis of neighbourhood and childhood socio-economic environment with self-assessed health and health-related behaviours. *Health and Place* 12, 394-403.

Moon, G., Gould, M. and colleagues. (eds.) 2000. *Epidemiology: an introduction* Buckingham: Open University Press.

Moon, G., Quarendon, G., Barnard, S., Twigg, L. and Blyth, B. 2007. Fat nation: deciphering the distinctive geographies of obesity in England. *Social Science and Medicine* 65, 20-31.

Morland, K., Diez Roux, A.V. and Wing, S. 2006. Supermarkets, other food stores and obesity: the atherosclerosis risk in communities study. *American Journal of Preventive Medicine* 30, 333-9.

Morland, K.B. and Evenson, K.R. 2009. Obesity prevalence and the local food environment. *Health and Place* 15, 491-5.

Mujahid, M.S., Roux, A.V., Shen, M.W., Gowda, D., Sanchez, B., Shea, S., Jacobs, D.R. and Jackson, S.A. 2008. Relation between neighborhood environments and obesity in the multi-ethnic study of atherosclerosis. *American Journal of Epidemiology* 167, 1349-57.

Nelson, M.C., Gordon-Larsen, P., Song, Y. and Popkin, B.M. 2006. Built and social environments. Associations with adolescent overweight and activity. *American Journal of Preventive Medicine* 31, 109-17.

Oliver, L.N. and Hayes, M.V. 2008. Effects of neighbourhood income on reported body mass index: an eight year longitudinal study of Canadian children. *BMC Public Health* 8, 16.

Park, Y., Neckerman, K.M., Quinn, J., Weiss, C. and Rundle, A. 2008. Place of birth, duration of residence, neighborhood immigrant composition and body mass index in New York City. *International Journal of Behavioral Nutrition and Physical Activity* 5, 19 doi:10.1186/1479-5868-5-19.

Pearce, J., Hiscock, R., Blakely, T. and Witten, K. 2009. A national study of the association between neighbourhood access to fast-food outlets and the diet and weight of local residents. *Health and Place* 15, 193-7.

Pearce, N. 1996. Traditional epidemiology, modern epidemiology, and public health. *American Journal of Public Health* 86, 678.

Pearl, J. 2001. Causal Inference in the Health Sciences: A Conceptual Introduction. *Health Services and Outcomes Research Methodology* 2, 189-220.

Pickett, K.E., Kelly, S., Brunner, E., Lobstein, T. and Wilkinson, R.G. 2005. Wider income gaps, wider waistbands? An ecological study of obesity and income inequality. *Journal of Epidemiology and Community Health* 59, 670-4.

Poortinga, W. 2006. Perceptions of the environment, physical activity, and obesity. *Social Science and Medicine* 63, 2835-46.

Procter, K.L., Clarke, G.P., Ransley, J.K. and Cade, J. 2008. Micro-level analysis of childhood obesity, diet, physical activity, residential socioeconomic and social capital variables: where are the obesogenic environments in Leeds? *Area* 40, 323-40.

Ravussin, E. and Bogardus, C. 2000. Energy balance and weight regulation: Genetics versus environment. *British Journal of Nutrition* 83, S17-S20.

Robert, S.A. and Reither, E.N. 2004. A multilevel analysis of race, community disadvantage, and body mass index among adults in the US. *Social Science and Medicine* 59, 2421-34.

Ross, N.A., Tremblay, S., Khan, S., Crouse, D., Tremblay, M. and Berthelot, J.M. 2007. Body mass index in urban Canada: Neighborhood and metropolitan area effects. *American Journal of Public Health* 97, 500-8.

Rothman, K.J. 1976. Causes. *American Journal of Epidemiology* 104, 287-92.

Rothman, K.J. and Greenland, S. 2005. Causation and causal inference in epidemiology. *American Journal of Public Health* 95, S144-S150.

Rundle, A., Field, S., Park, Y., Freeman, L., Weiss, C.C. and Neckerman, K. 2008. Personal and neighborhood socioeconomic status and indices of neighborhood walk-ability predict body mass index in New York City. *Social Science and Medicine* 67, 1951-8.

Rychetnik, L., Frommer, M., Hawe, P. and Shiell, A. 2002. Criteria for evaluating evidence on public health interventions. *Journal of Epidemiology and Community Health* 56, 119-27.

Saelens, B.E., Sallis, J.F., Black, J.B. and Chen, D. 2003a. Neighborhood-based differences in physical activity: an environment scale evaluation. *American Journal of Public Health* 93, 1552-8.

Saelens, B.E., Sallis, J.F. and Frank, L.D. 2003b. Environmental correlates of walking and cycling: findings from the transportation, urban design, and planning literatures. *Annals of Behavioral Medicine* 25, 80-91.

Santana, P., Santos, R. and Nogueira, H. 2009. The link between local environment and obesity: a multilevel analysis in the Lisbon Metropolitan Area, Portugal. *Social Science and Medicine* 68, 601-9.

Shrewsbury, V. and Wardle, J. 2008. Socioeconomic status and adiposity in childhood: A systematic review of cross-sectional studies 1990-2005. *Obesity* 16, 275-84.

Singh, G.K., Kogan, M.D. and van Dyck, P.C. 2008. A multilevel analysis of state and regional disparities in childhood and adolescent obesity in the United States. *Journal of Community Health* 33, 90-102.

Smith, D.M. and Cummins, S. 2009. Obese cities: How our environment shapes overweight. *Geography Compass* 3, 518-35.

Smoyer-Tomic, K.E., Spence, J.C., Raine, K.D., Amrhein, C., Cameron, N., Yasenovskiy, V., Cutumisu, N., Hemphill, E. and Healy, J. 2008. The association between neighborhood socioeconomic status and exposure to supermarkets and fast food outlets. *Health and Place* 14, 740-54.

Spence, J.C., Cutumisu, N., Edwards, J. and Evans, J. 2008. Influence of neighbourhood design and access to facilities on overweight among preschool children. *International Journal of Pediatric Obesity* 3, 109-16.

Stafford, M., Bartley, M., Mitchell, R. and Marmot, M. 2001. Characteristics of individuals and characteristics of areas: investigating their influence on health in the Whitehall II study. *Health and Place* 7, 117-29.

Stafford, M., Cummins, S., Ellaway, A., Sacker, A., Wiggins, R.D. and Macintyre, S. 2007. Pathways to obesity: Identifying local, modifiable determinants of physical activity and diet. *Social Science and Medicine* 65, 1882-97.

Stafford, M., Duke-Williams, O. and Shelton, N. 2008a. Small area inequalities in health: Are we underestimating them? *Social Science and Medicine* 67, 891-9.

Stafford, M., Sacker, A., Ellaway, A., Cummins, S., Wiggins, D. and Macintyre, S. 2008b. Neighbourhood Effects on Health: A Structural Equation Modelling Approach. *Schmollers Jahrbuch*, 128, 109-20.

Sturm, R. and Datar, A. 2005. Body mass index in elementary school children, metropolitan area food prices and food outlet density. *Public Health* 119, 1059-68.

Sundquist, J., Malmstrom, M. and Johansson, S.E. 1999. Cardiovascular risk factors and the neighbourhood environment: A multilevel analysis. *International Journal of Epidemiology* 28, 841–5.

Susser, M. 1973. *Causal thinking in the health sciences.* New York: Oxford University Press.

Susser, M. 1991. What is a cause and how do we know one? A grammar for pragmatic epidemiology. *American Journal of Epidemiology* 133, 635-648.

Susser, M. 2001. Glossary: causality in public health science. *Journal of Epidemiology and Community Health* 55, 376-8.

Swinburn, B., Egger, G. and Raza, F. 1999. Dissecting obesogenic environments: The development and application of a framework for identifying and prioritizing environmental interventions for obesity. *Preventive Medicine* 29, 563-70.

Swinburn, B., Gill, T. and Kumanyika, S. 2005. Obesity prevention: a proposed framework for translating evidence into action. *Obesity Reviews* 6, 23-33.

Townshend, T. and Lake, A. 2008. Obesogenic urban form: theory, policy and practice. *Health and Place* doi:10.1016/j.healthplace.2008.12.002.

Van Lenthe, F.J. and Mackenbach, J.P. 2002. Neighbourhood deprivation and overweight: the GLOBE study. *International Journal of Obesity* 26, 234-40.

Vandegrift, D. and Yoked, T. 2004. Obesity rates, income, and suburban sprawl: an analysis of US states. *Health and Place* 10, 221-9.

Veugelers, P., Sithole, F., Zhang, S. and Muhajarine, N. 2008. Neighborhood characteristics in relation to diet, physical activity and overweight of Canadian children. *International Journal of Pediatric Obesity* 3, 152-9.

Von Kries, R., Koletzko, B., Sauerwald, T., Von Mutius, E., Barnert, D., Grunert, V. and Von Voss, H. 1999. Breast feeding and obesity: Cross sectional study. *British Medical Journal* 318, 147-50.

Wang, M.C., Cubbin, C., Ahn, D. and Winkleby, M.A. 2008. Changes in neighbourhood food store environment, food behaviour and body mass index, 1981-1990. *Public Health Nutrition* 11, 963-70.

Whitaker, R.C. and Dietz, W.H. 1998. Role of the prenatal environment in the development of obesity. *Journal of Pediatrics* 132, 768-76.

Wrigley, N. 2002. 'Food deserts' in British cities: policy context and research priorities. *Urban Studies* 39, 2029-40.

Wrigley, N., Warm, D. and Margetts, B. 2003. Deprivation, diet, and food-retail access: findings from the Leeds food deserts' study. *Environment and Planning A* 35, 151-88.

Chapter 13
Measuring Obesogenic Environments – Representing Place in Studies of Obesity

Dianna Smith, Kim Edwards, Graham Clarke and Kirk Harland

1. Introduction

Obesity is a significant social problem reaching pandemic levels in the developed world. While obesity is influenced by genetic and behavioural factors, the environmental influences have yet to be fully explored and understood. Obesity prevention and treatment have focused on pharmacological, educational and behavioural interventions, but a novel and longer term approach would be to identify and combat the environments that promote high energy intake and sedentary behaviour. Obesogenic environments have been defined as 'the sum of influences that the surroundings, opportunities, or conditions of life have on promoting obesity' (Swinburn et al. 1999: 564). These environments are perceived to be a driving force behind the escalating obesity epidemic and therefore warrant a multidisciplinary approach to investigate this concept.

Many of the chapters in this book have highlighted the growing problem of obesity around the world and the importance of a geographical perspective on understanding variations in the patterns that can be seen in cities and regions. The aim of this chapter is to review the various methods for measuring obesogenic or leptogenic environments at the small area level. Thus we do not deal in any detail with factors relating to individuals, such as childhood and family background. Individual level variation in disease is less likely to be explained by contextual effects where there is minimal spatial variation. The focus instead is more on the correlations between environmental factors and obesity in the belief that overcoming or responding to these factors would facilitate the reduction of health inequalities.

Many of the links between geography, the environment and obesity are founded on the premise that deprivation is a crucial factor in understanding spatial variations in obesity. Poverty or deprivation (family or individual) is commonly associated with obesity, although the relationship is not straightforward, depending on the timing of the outcome measure of obesity (that is, whether it is in childhood or adulthood). A thorough review in 1999 (Parsons et al. 1999) found a relationship between low socioeconomic status (SES) in childhood and subsequent adulthood obesity, which concurs with subsequent work by Hardy et al. (2000) and Okasha et al. (2003), both using father's occupation as the indicator of childhood SES. This relationship was also shown more recently using a more sophisticated indicator of deprivation (a ranking of three different factors – education level, occupation of head of household and current employment status) (Monden et al. 2006). The increased prevalence of obesity in adults and children from more deprived backgrounds could be due to a multitude of factors: dietary differences are often apparent; no safe play area for the child; lack of opportunity/funds for activities, so TV viewing is the primary leisure activity by default; food deserts (lack of accessible, affordable, healthy (low energy dense) food); constraints on calories per pound, which focuses purchases on energy dense foods. These various aspects associated with geography and deprivation in particular will be explored below.

To help explore the geography of obesity, GIS and spatial analysis (models) are often at the heart of many studies. A GIS is a computer based system for collecting, editing, integrating, visualizing and analysing spatially referenced data. The visualization of geographical information is an important features of a GIS as a map can help to strip away irrelevant or distracting information and provide a useful exploratory tool to develop hypotheses. The main strength of a GIS is that it permits manipulation of spatial data so as to derive new information or make inferences about the phenomena being studied. A GIS is an indispensable tool for managing, visualising and analysing geographically referenced data. There are many examples of this type of analysis for a number of different health outcomes (see for example Rusiecki et al. 2006, Nandakumar et al. 2005, Noonan et al. 2005, Morris et al. 2003). A GIS framework often involves spatial modelling. This will be explored below in relation to techniques or methodologies for estimating spatial patterns of obesity.

The rest of the chapter is structured as follows. In section 2 we explore the importance of what we call the food environment. This involves exploring concepts such as food access and the availability of different types of food. In section 3 we then explore the physical activity environment. This includes access to recreational spaces and the 'walkability' of the local environment. Both these sections include a brief discussion on the role of schools in promoting healthy lifestyles. In section 4 we look at the main set of estimation techniques based on these types of environmental factors. This usually involves regression or multiple regression techniques. These models are used to try and estimate small area rates of obesity and are an important component of geographical studies. If real world small area obesity rates are available the models can be calibrated and refined thus providing more faith from end users that the estimates are accurate and reliable. However, it is well known that using geographical data presents some additional problems when analysing disease patterns. These problems are outlined in section 5. New estimation techniques are introduced in section 6 that aim to address some of these concerns when using geographical data. Finally some concluding comments are offered in section 7.

2. Measuring the Food Environment

The school environment may be critical in understanding variations in obesity. The school has two main influences: food available during the day and the exercise regime (see Chapters 6 and 9). A number of studies have argued that the availability of fruit and vegetables in school lunches often forms the majority of the overall child's consumption of fruit and vegetables (Story et al. 2006). However, it is clear that the home environment can sometimes be in conflict with school practices. A UK study of children in primary schools in Leeds by Sahota et al. (2001) suggested that positive changes in school meals, tuck shops and playground activities produced only nominal overall behavioural changes in the children themselves. This is especially true in more deprived areas as parents find it difficult to sustain and support the healthier lifestyles introduced in schools, resulting in a return to unhealthy food choices for both parents and children (see also Gould et al. 2006). Case studies of individual families and individual schools can offer useful insights into causality that many quantitative analyses may fail to pick up. These techniques include interviews (with children, parents, school staff and so on), focus groups, diary recall studies and the interpretation of meaning and value from the use of observation and photographs of food consumed by children. Rawlins (2008) provides a good example (but also see Proctor et al. 2008b and Story et al. 2006 for an examination of the school environment in relation to obesity patterns). GIS methods have been utilized to characterize the environments around schools. For example, Walton et al. (2009) used GIS network tools to estimate the number of food outlets and outdoor food advertisements that children from four primary schools in New Zealand were likely to pass on their journey to and from school. The authors found that schools with a higher proportion of students passing food outlets and advertisements considered that their presence impacted on efforts within schools to improve the food environment.

In recent years there has been increasing concern over poor accessibility for certain geodemographic groups to good quality, cheaply priced fresh fruit and vegetables deemed to be so important in a healthy and balanced diet. As such products are more likely to be fresh and plentiful in supermarkets than corner shops, research has focused on geographical variations in access to supermarkets or large grocery stores. In the UK in the 1990s the *Low-income Task Force* labelled areas with seemingly poor access as 'food deserts' (see Wrigley 2002). Concern over the existence of possible food deserts is heightened when the demographic groups include the elderly, persons on low incomes, the disabled and so on. Although some studies have concluded that access to food stores shows very little differences between high and low income areas (Winkler et al. 2006, Turrell 2004, Cummins and Macintyre 1999, 2002) there is plenty of alternative evidence which suggests that food deserts do exist in many deprived areas (Guy et al. 2004, Clarke et al. 2002, Furey et al. 2002, Wrigley 2002).

A key issue in the literature is how to measure food deserts. GIS has often been used to buffer or demarcate areas around neighbourhoods and then count the number of stores or floorspace within those buffers to give a raw provision score (perhaps floorspace divided by number of persons or households). Given that the Social Exclusion Unit in the UK defined a 500 metre boundary as the threshold for measuring food deserts (the distance an average person could walk in 6-7 minutes), 500 metres is often used as the buffer size. Guy and David (2004) give a good illustration of measuring food access in this way. If small area zones are used (such as census districts) this can often lead to a rather 'lumpy' provision score – that is very high in census tracts with a large store to lots of zones with zero scores. Other studies have used retail models to estimate accessibility indicators for residential neighbourhoods. These models can allow a smoother provision or access scores as they measure access to food outlets over an entire city or region. Thus although a census tract has no stores (that is zero provision score on the GIS index) the models can measure where people actually shop and estimate the share of floorspace which is effectively enjoyed by those residents. Further details appear in Clarke et al. (2002) and Smith et al. (2006) who have built retail models based on spatial interaction models to estimate these model-based 'provision

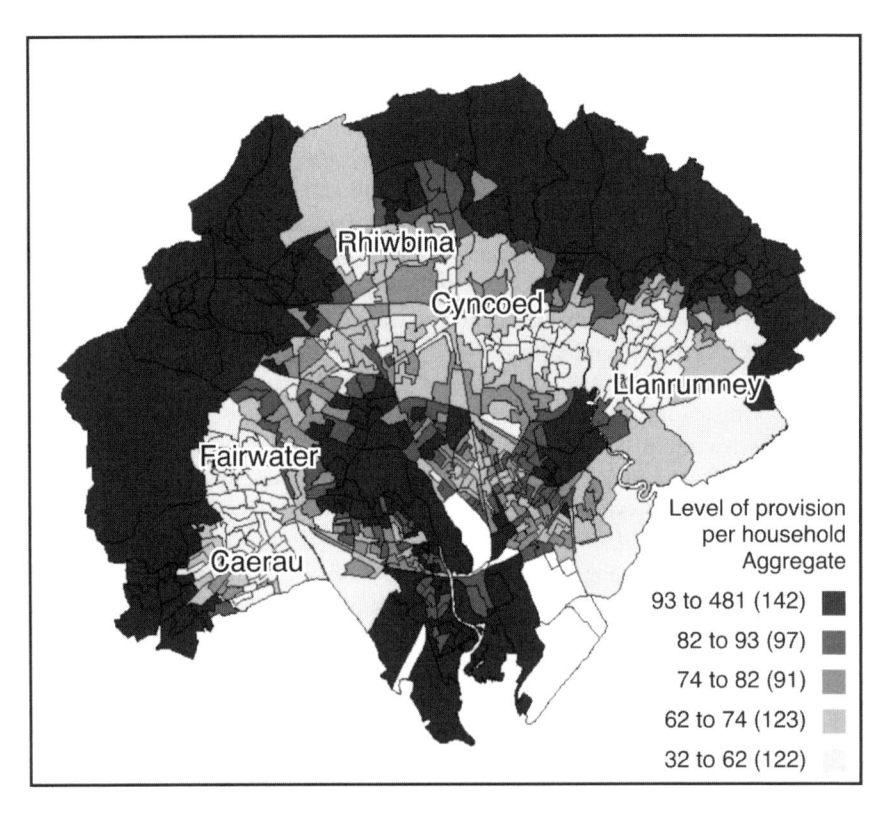

Figure 13.1 Provision scores for access to food outlets in Cardiff, 2002
Source: Clarke et al. 2002.

scores'. Areas with low scores (not just zero scores from the GIS analysis) thus also become possible food deserts. Figure 13.1 shows a map of these scores for Cardiff, UK. Provision can be seen to be high in the outer, more affluent suburban areas and lowest in the inner ring of less affluent suburbs such as the Caerau and Fairwater areas.

As well as having fewer, high quality stores in their neighbourhoods many low income residents often also find higher food prices in the stores that are present and that the healthy foods that are available are often more expensive than unhealthy options (Drewnowski et al. 2004, Furey et al. 2002). This can obviously have a great impact on diet. Although again there is mixed evidence on geographical variations in diet (see Hughes et al. 2009, Pearson et al. 2005, Wilson et al. 2004), there are a number of large scale studies which have investigated diets in low income areas before and after the intervention of opening a new large superstore. One example concerns research in the deprived suburb of Seacroft in Leeds, UK shown by Clarke et al. (2002) to have very low provision scores. Wrigley et al. (2003, 2004) show that the new superstore did have a considerable impact on increasing fruit and vegetable consumption in this area.

Research into food environments and obesity has also focused heavily on access to fast food outlets. Access to fast foods (often associated with high fat and sugar content) is one of the most commonly identified elements of an obesogenic environment (White 2007, Austin et al. 2005, Booth et al. 2005, Cummins et al. 2005). Many studies have concluded that fast food outlets are often more prevalent in deprived areas, supporting the theory that ease of access to fast food outlets contributes significantly to obesity (Pearce et al. 2007, Morland et al. 2002). However, researchers have found it harder to identify a clear causal link between fast food access and obesity. For example Pearce et al. (2009) concluded that although residents with poor access to fast food outlets were more likely to eat vegetables, they could find no clear evidence that neighbourhood access to fast food outlets was associated with a poorer diet and individuals being overweight (p. 195) (see also Pearce et al. 2007, Cummins and Macintyre 2006).

3. Measuring the Physical Activity Environment

Physical activity has often been identified as important in understanding levels of obesity (Jebb 2007). Numerous studies have focused on the idea of walkability (Frank et al. 2004, 2007, Ewing et al. 2006). In general, there seems to be strong evidence that environments which encourage walking (to all types of destination) are less obesegenic (see Chapters 7 and 8). Frank et al. (2005), for example, suggest that residents in high walkable neighbourhoods are 2.4 times more likely to reach recommended physical activity levels than residents of low-walkable neighbourhoods. As Smith et al. (2008) explain, neighbourhoods that are designed to support walking may encourage greater physical activity and thereby help prevent obesity. They suggest three factors may encourage walkability in a region:

pedestrian friendly street connections; areas with a broad mix of land uses; and older neighbourhoods designed in eras predating mass car usage. This links partly to the built form and layout of cities. Sallis and Glanz (2006) support the argument that older, more 'traditional' neighbourhoods are often more walkable than newer suburbs built for car owners. These are often associated with high levels of connectivity for pedestrians, especially to their place of work or shopping.

It has also been shown that neighbourhood characteristics such as deprivation can impact behaviours such as walking and keeping fit. It may be that proximity to (or access to) parks and green spaces, for example, has an impact on obesity in children especially by impacting their physical activity levels (for example playing on swings) or diet (for example consuming ice creams and sugary drinks), although the little research that has been undertaken in this area tends not to show a clear-cut relationship (Timperio et al. 2005). Estabrooks et al. (2004) found fewer parks, sports fields, fitness clubs and so on in lower income areas of the US compared to affluent ones, suggesting that low income children may face greater barriers to physical activity.

It is likely that perceived neighbourhood safety is an important determinant of (childhood) obesity. Without safe places to play near home for example, children may spend more time being inactive indoors (Sallis and Glanz 2006). However, once again the evidence is not clear cut (Lumeng et al. 2006, Burdette and Whitaker 2005). Crime, both perceived and actual, can affect a parent's decision whether to let the child outside to play, as can road safety issues such as safe road crossings, pavements and the speed of traffic (Timperio et al. 2005).

Again, a key issue is how to measure walkability. GIS is useful for mapping land-use patterns and for analysing connectivity in urban street networks. A mixed pattern of land-use may encourage greater walking between activities (a home might be close to shops, offices, restaurants, work and so on). Land use patterns are often available from government agencies in GIS format. Journey to work data by mode of travel and age of housing tend to be standard census variables and thus transferable into GIS packages. Standard GIS routines can then be used to help estimate 'walkability measures'. For example Smith et al. (2008) estimate pedestrian friendly design in terms of street connectivity, defined as the number of intersections within 0.25 miles of a resident's home. This distance can be easily drawn as a buffer within GIS and standard overlay facilities will estimate population and number of intersections within such a buffer.

As noted above, the school environment can also play a major role in influencing the physical activity environment (see also Chapter 9). At one level this may simply involve increasing education concerning obesity. For example Robinson (1999) found that a school-based intervention to decrease television viewing (and encourage outdoor activity) did reduce the prevalence of obesity in the children. Other studies have explored various interventions to make physical education classes more active and, in moderation, more vigorous. Datar and Sturm (2004) for example noted a considerable decrease in obesity levels for girls following the introduction of more time spent on physical activity within the school.

4. Estimation Techniques (1): Regression and Multi-Level Modelling

The literature on the geography of obesity also includes a body of research which attempts to look for explanation through the use of statistical analysis to measure the importance or significance of the sorts of multiple potential casual factors described above. These can take the form of various types of multiple regression models or more ambitious small area synthetic population/disease prevalence estimation models (and some of these techniques are also now available in powerful GIS packages such as ESRI's ARC). Studies which have looked at multiple aspects of an obesogenic environment rather than single attributes are probably more useful as environmental factors do not operate to shape our health behaviours in isolation (see discussion above). A good illustration of this multiple regression approach is the study by Nelson et al. (2006). The authors used cluster analysis procedures to identify six different environment patterns related to physical activity levels and obesity using eight different residential variables (income, race, education, SES, crime rates, road type, walkability and recreation facilities). The data showed significant differences in physical activity levels and obesity between the different neighbourhood patterns (see also Schlundt et al. 2006 and the discussion in section 6).

Many of the models currently in use (for both obesity and type 2 diabetes which is strongly related to obesity) are created using multilevel frameworks (Moon et al. 2007, Pearce et al. 2003, Twigg and Moon 2002, Twigg et al. 2002), epidemiological models (Forouhi et al. 2006) and Bayesian estimation with epidemiological models (Congdon 2006). Multilevel modelling frameworks use knowledge of individual and area level data (from national/regional surveys and/ or the Census) to create probabilities of health outcomes/behaviour in areas where data are unavailable. Multilevel models allow researchers to explore the interactions between individual- and area-level variables and the outcome of interest, first considering the proportion of cases explained by various individual-level factors (for example age, sex, ethnicity) and then introducing area characteristics (for example deprivation) with an aim to explain the remaining cases; cross-level interactions extend the utility of this analysis (Twigg and Moon 2002). Then, the model's parameter estimates can be applied to the individuals and areas where the behaviour/outcome prevalence is unknown to create new prevalence estimates (Twigg and Moon 2002). Prior to the implementation of a multilevel model, relevant predictors for the health outcome need to be identified. Each of the predictors need to be relevant to the health outcome and linked spatially to small-area population health outcome data (Twigg and Moon 2002). Logistic regression models are preferable in situations where the outcome is a dichotomous value (not diabetic/diabetic; non-smoker/smoker) and the predictors are either a continuous scale (such as age) or categorical (such as ethnic groups) (Gatrell 2002).

A slightly different methodology utilizes both regression and Bayesian estimates to create small-area disease prevalence rates. Congdon (2006) uses data on age, sex, ethnicity from the 1999 and 2003 HSE to calculate age, sex and ethnic group specific prevalence rates for both type 1 and 2 diabetes (the latter increasingly associated with obesity). These rates are then applied to the known age-sex-ethnic group population distributions in 2001 Census wards. The inclusion of some measure of social deprivation is vital in diabetes prevalence modelling, as the literature (as with obesity) shows a strong relationship between deprivation and type 2 diabetes (Evans et al. 2000 cited in Congdon 2006). Although a multilevel model is not used in this research, area-level deprivation is used as a proxy for individual socioeconomic status; sex-specific logistic regression modelling for the influence of deprivation on age-sex-ethnic diabetes risk allowed for the inclusion of deprivation. The Bayesian methods employed by Congdon introduce a 1999 diabetes risk factor in the final model to create accurate predictions of diabetes prevalence and confirm the probabilities for diabetes created from the regression of 2003 data.

5. Geographical Data Issues

The use of spatial models and GIS is not without its problems. There are many obstacles to consider when handling geographical data. First, Gatrell (2005) has recently acknowledged the difficulty of accurately modelling health outcomes using predictive models due simply to varying levels of complexity in understanding variations in disease patterns. The issues raised by Gatrell include: the inability of models to account for interactions between variables (beyond the link between layers of data in multilevel models); the often simplified, linear nature of models that are unable to account for non-linear relationships; the inherently complex nature of relationships between people and place; and the idea of epidemiology as a 'web' of inter-connected mechanisms which uniquely combine in individual lives (Gatrell 2005).

There are other problems when using spatial data and spatial analysis techniques. Confidentiality of patients is a primary concern which normally leads to the use of aggregate data with an associated loss of spatial detail. A key question then is what is the right spatial scale to use? Data is often not available for small census areas and thus it is often necessary to use larger regions. This in turn evokes other problems. Two of the most frequent problems are the ecological fallacy and the modifiable areal unit problem (MAUP). The ecological fallacy appears when researchers assume that zonal averages apply across an area or region. For example if obesity in an area is given as 25 per cent, it is important not to assume that this is constant across that area. In fact, the area to the east of the zone (for example) may have a rate of only 10 per cent whilst the area to the west has a rate of 40 per cent. Another version of the fallacy is to assume such aggregate area level statistics also apply to all individuals in the zone. MAUP also occurs when analysing data by region. As area units are modifiable (that is boundaries can be changed), results can look very different when different zoning systems are used.

If small zones such as census tracts can be used then that may bring fresh problems. How such zones actually fit into what residents or policy makers perceive as neighbourhoods or community areas is an interesting question. Our experience suggests that analysis can be easier to present and interpret for parts of cities people can label and attach community values to than census tract names (or worse numbers!!). Also, the smaller the spatial zones the increased likelihood that close spatial units are statistically interdependent: that is if the observed value of a variable at one locality is dependent of the values of the variable at neighbouring localities. Spatial autocorrelation is a label attached to this issue of similar values inhabiting adjacent positions on a map. This can lead to problems with statistical tests producing overestimates of confidence values or 'statistically significant' results when none exist (see Kulldorf et al. 2005). The good news is that most GIS provide tools to measure the level of spatial autocorrelation (e.g. Moran's I). There are also issues when using spatial data concerning small number problems. However techniques such as Empirical Bayes estimation can be used as a smoothing approach to give risk estimates based on mean risks of the area chosen and that of neighbouring zones. In this way it prevents undue attention being focused on areas with small numbers.

6. Estimation Techniques (2): Microsimulation Modelling

It is partly because of the problems discussed in section 5 that researchers have become interested recently in methods which can overcome problems associated with zonal or aggregated data by building data sets for individuals or individual households. Microsimulation is one such technique. It is a technique which enables us to effectively add a geographical location to households (or individuals) in a particular survey by matching the attributes of households contained in a survey to households within small area census tracts based on the attributes of those particular households (low income, no car, unemployed, rented accommodation and so on). The end-product is thus a database for an entire city or region containing variables (in this case potential obesogenic variables) for all persons or households.

We argue that there are at least five unique strengths of microsimulation for health geography: (1) the power to link datasets based on a common variable, providing a small area population with previously unavailable combinations of (multi) attributes; (2) the ability to fill gaps in sparse datasets by simulating individuals and households from a small (1-3 per cent) representative sample (which helps to prevent the ecological fallacy problems as patterns are based on individuals not zonal averages); (3) using the simulated population to predict the individual or household-level effect of government interventions *prior* to implementation; (4) it helps to overcome MAUP (where different zoning systems can produce different geographical patterns) as the re-aggregation is based on individuals or households not zonal boundaries. Hence model output can be plotted at whatever geographical scale is required; and (5) with microsimulation there is the ability to update, forecast and model as datasets often need to be refreshed. Forecasting can be done using static ageing or by ageing a population dynamically over time.

The preferred methodology for creating microsimulation models is reweighting (see Williamson et al. 1998). This involves starting with the joint probabilities found between variables in surveys and then adding more geography. Hence a survey (that is the Health Survey for England) can be reweighted so that it fits small area geographies (the census data we know for each census tract; in the UK the output area (OA) is the smallest scale available in the census). In effect, households in the survey are tested to see how close they match households in each OA. If there is a good match the households are effectively 'cloned' so that every household in an OA is given the attributes of the household it most resembles in the survey. Naturally, the greater the sample size the better the simulations will be. One of the main issues in reweighting is which variables to match on and the order of that matching (the choice of so-called constraint variables) and how to calibrate the model when effectively estimating 'missing data'. Clearly, the more variables that are matched the better. Proctor (2007) and Proctor et al. (2008a) discuss the general issue of constraints and calibration in detail elsewhere. Smith et al. (2009) used a novel methodology to select a different set and order of constraints for different parts of the study area. Thus in areas dominated by non-White ethnic groups ethnicity should be the main constraint used to match the survey data to the OA census geography most accurately. A clustering or geodemographic profile of areas can be made first in order to set the most important variable to be used as a constraint (see Smith et al. 2009 and Smith 2007 for more details).

For validation the microsimulation model must be able to reproduce variables used as constraints very accurately. In addition, the model can be tested to see how well it matches on variables not used as constraints (again see Proctor et al. 2008a and Proctor 2007 for a detailed discussion of calibration). The obesity models used here produced good fits for both the constraint variables and the variables not used in the matching process. Before looking at the estimates of obesity it should be noted that the microsimulation models can be used to estimate local variations for a range of activities which can lead to obesity (or type 2 diabetes). Smith (2007) for example estimates a number of such activities (see Table 13.1 which gives the mean and standard deviation values across OAs in Leeds and Bradford). Figure 13.2 plots the geography for one of these variables as an illustration.

Table 13.1 Estimating variations in factors associated with obesity

Variable	Mean % (range)	Standard deviation
Smoking	29.13 (13.22-40.61)	3.73
0>2 portions fruit and veg	36.97 (23.83-55.24)	3.57
5+ portions fruit and veg	18.68 (12.13-24.97)	2.11
Low physical activity	34.85 (22.76-63.14)	5.09
Medium-high phys activity	64.80 (15.32-45.52)	3.70
Poor store access	5.46 (0.97-14.56)	1.39

Measuring Obesogenic Environments 287

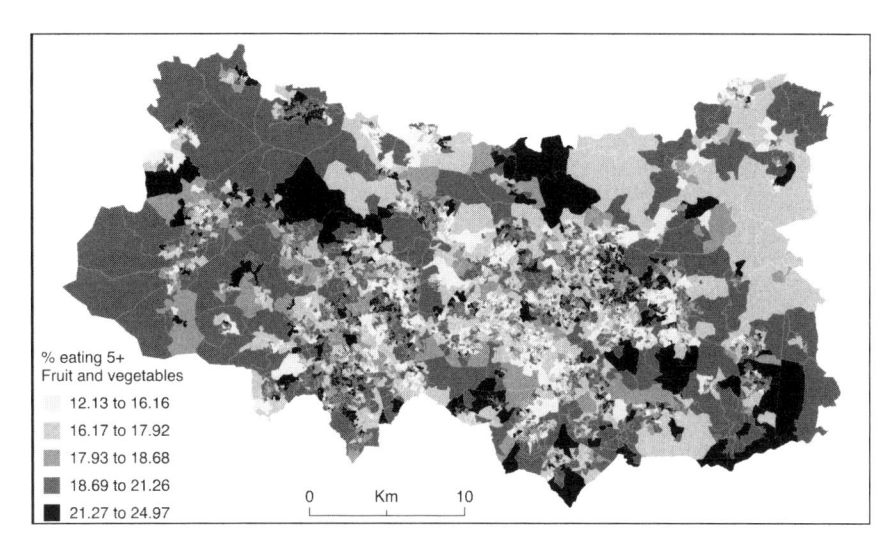

Figure 13.2 Fruit and vegetable consumption (>5 portions/day) by OA

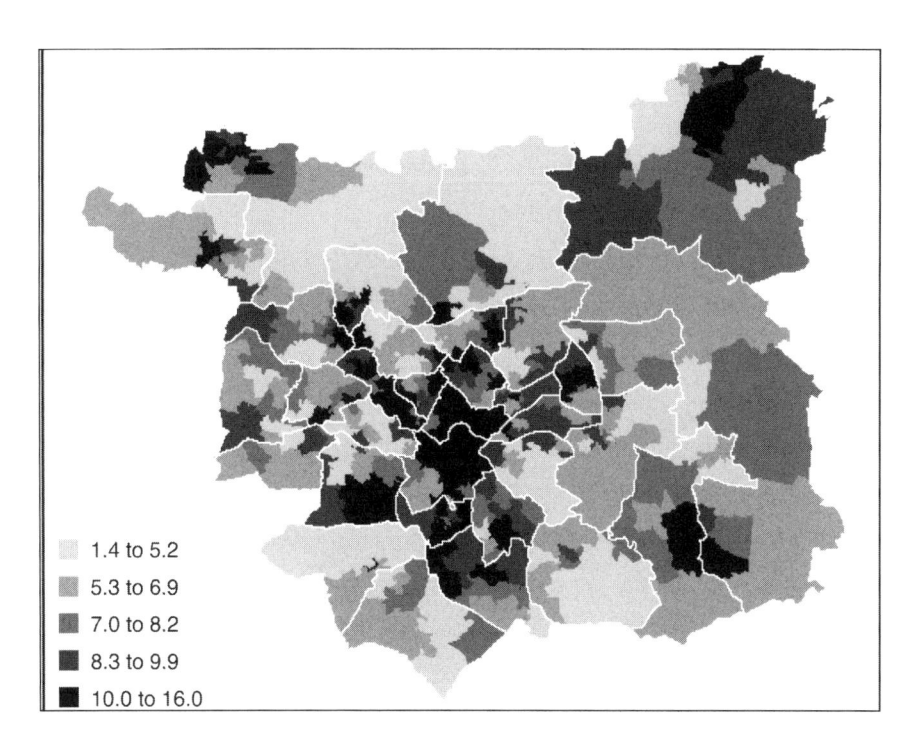

Figure 13.3 Childhood obesity across Leeds (proportion of measured children, in each area, who are obese)

288   *Geographies of Obesity*

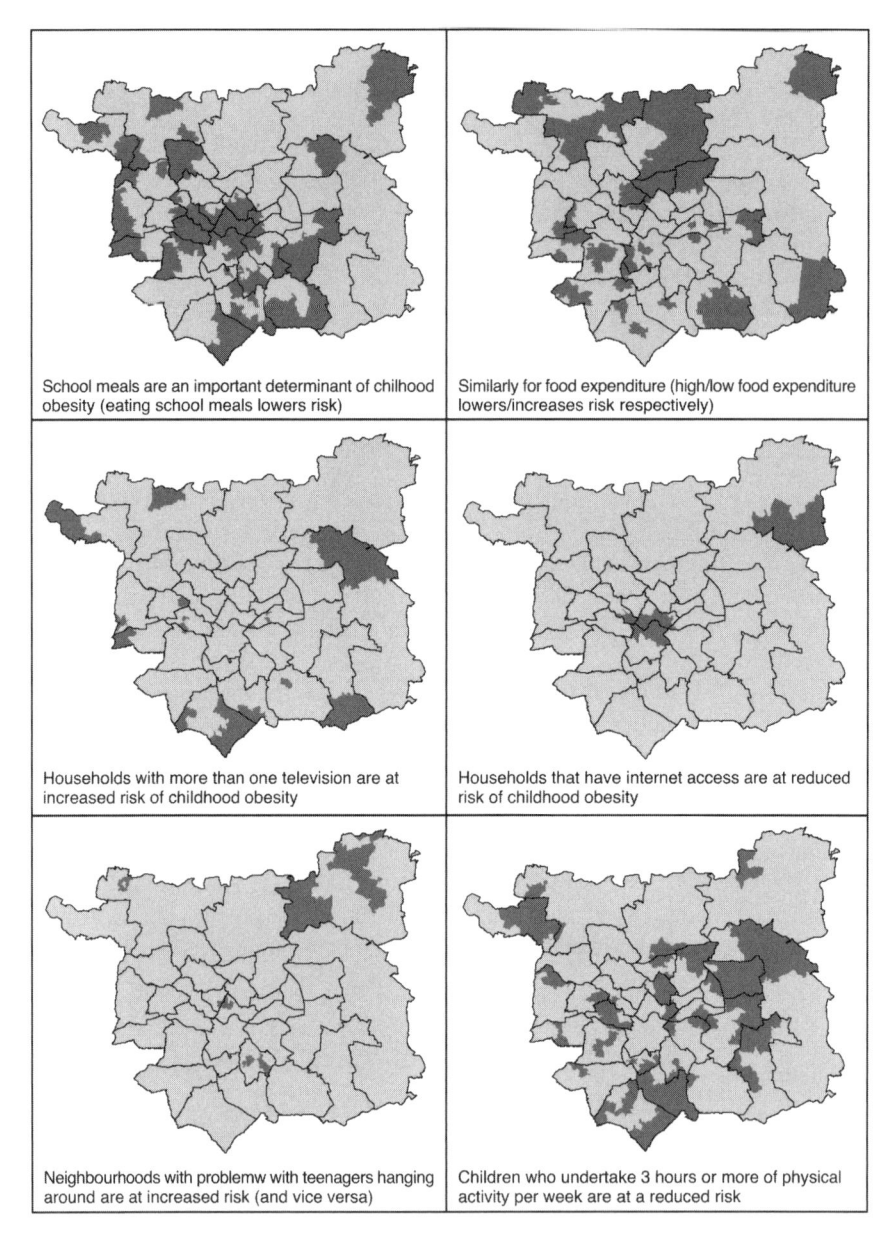

School meals are an important determinant of chilhood obesity (eating school meals lowers risk)

Similarly for food expenditure (high/low food expenditure lowers/increases risk respectively)

Households with more than one television are at increased risk of childhood obesity

Households that have internet access are at reduced risk of childhood obesity

Neighbourhoods with problemw with teenagers hanging around are at increased risk (and vice versa)

Children who undertake 3 hours or more of physical activity per week are at a reduced risk

Figure 13.4   Output areas (the dark shading) where key local determinants of childhood obesity differ from the key global determinants

Looking at fruit and vegetable consumption, a greater number of people living in or near the city centres did not eat five servings of fruit and vegetables on the day surveyed (Figure 13.2). No output area estimates exceeded 25 per cent of the population eating the recommended five servings. Patterns on the map do seem to show that the areas where five–a–day consumption is lower than average (the grey and blue areas) are more concentrated around the deprived areas of the inner city. Hughes et al. (2009) further investigate the geography of diet and nutritional eating in different regions and localities within the UK.

Using such combinations of variables Proctor et al. (2008a) then estimate the distribution of obesity across Leeds (see Figure 13.3). The map shows high concentrations in the city centre and more deprived inner city areas (as we might expect given the links in the literature between deprivation and obesity) but also significant concentrations or clusters in more affluent suburbs (to the north and east of the City). Following this, they went on to break down the global analysis of obesity in Leeds by examining the main causation factors in different parts of the City. Using geographically weighted regression (GWR – where parameter values in the model are allowed to vary geographically; see Fotheringham et al. 2002) they were able to find variations in the importance of different parameters and thus suggest that different causal factors might be important in different areas of the City.

This local analysis also determined those areas where the most important local obesogenic covariates are different to the key global obesogenic covariates – namely, expenditure on food, number of household TVs, problems with teenagers, internet access, school meals and children's levels of physical activity. The areas where these different factors are influential drivers of childhood obesity are highlighted in Figure 13.4. Overall the analysis shows that although deprivation can be shown to be associated with childhood obesity in Leeds hot spots also exist in affluent areas. Given that relationships between obesogenic covariates and childhood obesity vary across the City it is important that public health policies are tailored to find the right solution in the right location!

7. Conclusions

This chapter introduced various factors associated with the geography of obesity. We have discussed a wide ranging number of variables which have been linked to obesity. We have seen that there are factors associated with lifestyles and the food and physical activity environments which make up a particular neighbourhood. These seem equally important in understanding small-area variations in obesity. Not surprisingly, there are many alternative methodologies for studying the geography of obesity. We have looked at a number of different techniques (especially GIS and statistical models) which have been used to try and understand and explain variations in obesity.

In the second half of the chapter we explored a number of difficulties involved when using geographical data and argued that it would be useful to use a modelling methodology which focused on the individual or individual household. This would allow flexibility in terms of aggregation and allow us to explore variations within small-areas thus helping to remove the problems of ecology fallacy and MAUP. This flexibility in population estimation is the primary reason for choosing a spatial microsimulation model over other methods of statistical analysis or synthetic population estimation. There are now many more individual level characteristics available for analysis, which would not be available had we used other population estimation methods. The second set of advantages relate to the variables we can create. The final destination of the models has been to estimate small-area levels of obesity. However, along that route it has been useful to estimate other variables associated with obesogenic environments such as consumption of fruit and vegetables, smoking rates, access to high quality food outlets, degree of social capital, TV viewing and so on. These are variables which are not available from standard data sets. Finally, one of the most important results of these estimations is the ability of the models to predict health outcomes based on *local* populations. As we have shown, spatial epidemiology usually focuses on the application of a global model to small area populations to create disease estimates. Here, the use of microsimulation and GWR models shows that richer geographical patterns can be observed compared with global models.

The population estimates are subject to some limitations. The most significant limitation is the lack of data for comparison with the prevalence estimates. Although the estimates appear to follow a logical pattern based on socioeconomic status, age and ethnicity, real world data are vital to the realistic evaluation of such models; otherwise, these are estimates based purely on demographic population profiles of individuals who are obese in a national survey (Moon et al. 2007). This is certainly an area for further collaboration and research as more data on obesity (and type 2 diabetes) are released by local health authorities.

References

Austin, S.B, Melly, S.J., Sanchez, B.N., Patel, A., Buka, S. and Gortmaker, S. 2005. Clustering of fast food restaurants around schools: a novel application of spatial statistics to the study of food environments. *American Journal of Public Health* 95, 1575-81.

Booth, K.M., Pinkston, M.M. and Poston, W.S. 2005. Obesity and the built environment. *Journal of the American Dietetic Association* 105, 110-117.

Burdette, H.L. and Whitaker, R.C. 2005. A national study of neighbourhood safety, outdoor play, television viewing and obesity in preschool children. *Pedatrics* 116(3), 657-62.

Clarke, G.P., Eyre, H. and Guy, C. (2002) Deriving indicators of access to food retail provision in British cities: studies of Cardiff, Leeds and Bradford. *Urban Studies* 39, 2041-60.

Congdon, P. 2006. Estimating diabetes prevalence by small area in England. *Journal of Public Health* 28, 71-81.

Cummins, S. and Macintyre, S. 1999. The location of food stores in urban areas: a case study in Glasgow. *British Food Journal* 101, 543-53.

Cummins, S. and Macintyre, S. 2002. 'Food deserts' – evidence and assumption in health policy making. *British Medical Journal* 325, 436-8.

Cummins, S. and Macintyre, S. 2006. Food environments and obesity – neighbourhood or nation? *International Journal of Epidemiology* 35, 100-4.

Cummins, S., McKay, L. and Macintyre, S. 2005. McDonald's restaurants and neighbourhood deprivation in Scotland and England. *American Journal of Preventive Medicine* 29, 308-310.

Datar, A. and Sturm, R. 2004. Physical education in elementary school and body mass index: evidence from early childhood longitudinal study. *American Journal of Public Health* 94(9), 1501-6.

Drewnowski, A. 2004. Obesity and the food environment: dietary energy density and diet costs. *American Journal of Preventive Medicine* 27(3 Suppl), 154-62.

Estabrooks, P.A., Lee, R.E. and Gyurcsik, N.C. 2004. Resources for physical activity participation: does availability and accessibility differ by neighbourhood socioeconomic status? *Annals of Behavioural Medicine* 25, 100-4.

Evans, J.M., Newton, R.W., Ruta, D.A., MacDonald. T.M. and Morris, A.D. 2000. Socio-economic status, obesity and prevalence of type 1 and type 2 Diabetes Mellitus. *Diabetic Medicine* 17, 478-80.

Ewing, R., Brownson, R.C. and Berrigan, D. 2006. Relationship between urban sprawl and weight of United States youth. *American Journal of Preventive Medicine* 31, 464-74.

Forouhi, N.G., Merrick, D., Goyder, E., Ferguson, B.A., Abbas, J., Lachowycz, K. and Wild, S.H. 2006. Diabetes prevalence in England 2001 – estimates from an epidemiology model. *Diabetic Medicine* 23, 189-97.

Fotheringham, A.S., Charlton, M. and Brunsdon, C. 2002. *Geographically weighted regression: the analysis of spatially variable relationships*. Chichester: Wiley.

Frank, L.D., Andresen, M.A. and Scmidt, T.L. 2004. Obesity relationships with community design, physical activity and time taken in cars. *American Journal of Preventive Medicine* 27, 87-96.

Frank, L.D., Saelens, B.E., Powell, K.E. and Chapman, J.E. 2007. Stepping towards causation: do built environments or neighbourhood and travel preferences explain physical activity, driving and obesity? *Social Science and Medicine* 65, 1898-1914.

Frank, L.D., Schmid, T.L., Sallis, J.F., Chapman, J. and Saelens. B.E. 2005. Linking objectively measured physical activity with objectively measured urban form: findings from SMARTRAQ. *American Journal of Preventive Medicine* 28, 117-25.

Furey, S., Farley, H. and Strugnell, C. 2002. An investigation into the availability and economic accessibility of food items in rural and urban areas of Northern Ireland. *International Journal of Consumer Studies* 26, 313-321.

Gatrell, A.C. 2002. *Geographies of health: an introduction*. Oxford: Blackwell.

Gatrell, A.C. 2005. Complexity theory and geographies of health: a critical assessment. *Social Science and Medicine* 60, 2661-71.

Gould, R., Russel, J. and Barker, M.E. 2006. School lunch menus and 11 to 12 year old children's food choice in three secondary schools in England – are the nutritional standards being met? *Appetite* 46, 86-92.

Guy, C.M., Clarke, G.P. and Eyre, H. 2004. Food retail change and the growth of food deserts: a case study of Cardiff. *International Journal of Retail, Distribution and Management* 32, 72-88.

Guy, C.M. and David, G. 2004. Measuring physical access to 'healthy foods' in areas of social deprivation: a case study in Cardiff. *International Journal of Consumer Studies* 28, 222-34.

Hardy, R., Wadsworth, M. and Kuh, D. 2000. The influence of childhood weight and socioeconomic status on change in adult body mass index in a British national birth cohort. *International Journal of Obesity* 24(6), 725-34.

Hughes, R., Clarke, G.P., Edwards, K., Cade, J. and Ransley, J. 2009. *A Geodemographic Analysis of Diet and Nutrition in the UK*, Working Paper. Leeds: School of Geography, University of Leeds.

Jebb, S.A. 2007. Dietary determinants of obesity. *Obesity Reviews* 8, 93-97.

Kulldorff, M., Song, C., Gregorio, D., Samociuk, H. and DeChello, L. 2006. Cancer Map Patterns. Are They Random or Not? *American Journal of Preventive Medicine* 30(2 Suppl 1), S37-S49.

Lumeng, J.C., Appugliese, D., Cabral, H.J., Bradley, R.H. and Zuckerman, B. 2006. Neighbourhood safety and overweight status in children. *Archives of Pediatric and Adolescent Medicine* 160(1), 25-31.

Monden, C.W., van Lenthe, F.J. and Mackenbach, J.P. 2006. A simultaneous analysis of neighbourhood and childhood socio-economic environment with self-assessed health and health related behaviours. *Health and Place* 12(4), 394-403.

Moon, G., Quarendon, G., Barnard, S., Twigg, L. and Blyth, B. 2007. Fat nation: deciphering the distinctive geographies of obesity in England. *Social Science and Medicine* 65, 25-31.

Morland, K., Wing, S., Diez Roux, A. and Poole, C. 2002. Neighbourhood characteristics associated with the location of food stores and food service places. *American Journal of Preventive Medicine* 22, 23-9.

Morris, R.W., Whincup, P.H., Emberson, J.R., Lampe, F.C., Walker, M. and Shaper, A.G. 2003. North-south gradients in Britain for stroke and CHD are they explained by the same factors? *Stroke* 34 (11), 2604-9.

Nandakumar, A. Gupta, P.C., Gangadharan, P., Visweswara, R.N. and Parkin, D.M. 2005. Geographic pathology revisited: Development of an atlas of cancer in India. *International Journal of Cancer* 116(5), 740-54.

Nelson, M.C., Gordon-Larsen, P., Song, Y. and Popkin, B.M. 2006. Built and Social Environments: Associations with Adolescent Overweight and Activity. *American Journal of Preventive Medicine* 31(2), 109-17.

Noonan, C.W., White, M.C. and Thurman, D. 2005. Temporal and geographic variation in United States motor neuron disease mortality, 1969-1998. *Neurology* 64 (7), 1215-21.

Okasha, M., McCarron, P., McEwen, J., Durnin, J. and Davey Smith, G. 2003. Childhood social class and adulthood obesity: findings from the Glasgow Alumni Cohort. *Journal of Epidemiology and Community Health* 57, 508-9.

Parsons, T.J., Power, C., Logan, S. and Summerbell, C.D. 1999. Childhood predictors of adult obesity: a systematic review. *International Journal of Obesity* 23(Suppl 8) S1-S107.

Pearce, J., Blakely, T., Witten, K. and Bartie, P. 2007. Neighbourhood deprivation and access to fast-food retailing – a national study. *American Journal of Preventive Medicine* 32, 375-82.

Pearce, J., Boyle, P. and Flowerdew, R. 2003. Predicting smoking behaviour in census output areas across Scotland. *Health and Place* 9, 139-49.

Pearce, J., Hiscock, R., Blakely, T. and Witten, K. 2009. A national study of the association between neighbourhood access to fast-food outlets and the diet and weight of local residents. *Health and Place* 15, 193-7.

Pearson, T., Russell, J., Campbell, M.J. and Barker, M.E. 2005. Do 'food deserts' influence fruit and vegetable consumption – a cross-sectional study. *Appetite* 45, 195-7.

Proctor, K. 2007. *Measuring the obesogenic environment of childhood obesity*, PhD thesis. School of Geography, University of Leeds, Leeds, UK.

Proctor, K.L, Clarke, G.P., Ransley, J.K. and Cade, J. 2008a. Micro-level analysis of childhood obesity, diet, physical activity, residential socio-economic and social capital variables: where are the obesogenic environments in Leeds? *Area* 40(3), 323-40.

Proctor, K.L, Rudolf, M., Feltbower, R., Levine, R., Connor, A., Robinson, M. and Clarke, G.P. 2008b. Measuring the school impact on child obesity. *Social Science and Medicine* 67, 341-9.

Rawlins, E. 2008. Citizenship, health education and the obesity crisis. ACME, *International E-Journal for Critical Geographies* 7(2), 135-51.

Robinson, T.N. 1999. Reducing children's television viewing to prevent obesity: a randomised controlled trial. *Journal of the American Medical Association* 282(16), 1561-7.

Rusiecki, J.A., Kulldorff, M., Nuckols, J.R., Song, C. and Ward, M.H. 2006. Geographically based investigation of prostate cancer mortality in four U.S. Northern Plain States. *American Journal of Preventive Medicine* 2006; 30(2, Suppl 1), S101-8.

Sahota, P., Rudolf, M.C., Dixey, R., Hill, A.J., Barth, J.H. and Cade, J. 2001. Evaluation of implementation and effect of primary school based intervention to reduce risk factors for obesity. *BMJ* 323, 1027.

Sallis, J.F. and Glanz, K. 2006. The role of built environments in physical activity, eating and obesity in childhood. *Childhood Obesity* 16(1), 89-108.

Schlundt, D.G., Hargreaves, M.K. and McClellan, L. 2006. Geographic clustering of obesity, diabetes, and hypertension in Nashville, Tennessee. *Journal of Ambulatory Care Management* 29(2), 125-32.

Smith, D.M. 2007. Potential health implications of retail food access, PhD thesis, School of Geography, University of Leeds, Leeds, UK.

Smith, D.M., Clarke, G.P. and Harland, K. 2009. Improving the synthetic data generation process in spatial microsimulation models. *Environment and Planning A* 41(5), 1251-68.

Smith, D.M., Clarke, G.P., Ransley, J. and Cade, J. 2006. Food access and health: a microsimulation framework for analysis. *Studies in Regional Science* 35(4), 909-27.

Smith, K., Brown, B.R., Yamada, I. and Kowaleski-Jones, L. 2008. Walkability and body mass index. *American Journal of Preventive Medicine* 35(3), 237-43.

Story, M., Kaphingst, K.M. and French, S. 2006. The role of schools in obesity prevention. *Childhood Obesity* 16(1), 109-42.

Swinburn, B., Eggar, G. and Raza, F. 1999. Dissecting obesogenic environments: the development and application of a framework for identifying and prioritizing environmental interventions for obesity. *Preventive Medicine* 29, 563-70.

Timperio, A., Salmon, J., Telford, A. and Crawford, D. 2005. Perceptions of local neighbourhood environments and their relationship to childhood overweight and obesity. *International Journal of Obesity* 29, 170-175.

Turrell, G., Blakely, T., Patterson, C. and Oldenburg, B. 2004. A multilevel analysis of socioeconomic (small area) differences in household food purchasing behaviour. *Journal of Epidemiology and Community Health* 58, 208-215.

Twigg, L. and Moon, G. 2002. Predicting small-area health-related behaviour: a comparison of multilevel synthetic estimation and local survey data. *Social Science and Medicine* 50, 1109-20.

Twigg, L., Moon, G. and Jones, K. 2002. Predicting small-area health-related behaviour: a comparison of smoking and drinking indicators. *Social Science and Medicine* 54, 931-7.

Walton, M., Pearce, J. and Day, P. 2009. Examining the interaction between food outlets and outdoor food advertisements with primary school food environments. *Health and Place* 15(3), 811-18.

White, M. 2007. Food access and obesity. *Obesity Reviews* 8, 99-107.

Williamson, P., Birkin, M. and Rees, P.H. 1998. The estimation of population microdata by using data from small area statistics and samples of anonymised records. *Environment and Planning A* 30, 785-816.

Wilson, L.C., Alexander, A. and Lumbers, M. 2004. Food access and dietary variety among older people. *International Journal of Retail Distribution and Management* 32, 109-22.

Winkler, E., Turrell, G. and Patterson, C. 2006. Does living in a disadvantaged area mean fewer opportunities to purchase fresh fruit and vegetables in the area? Findings from the Brisbane food study. *Health and Place* 12, 306-19.

Wrigley, N. 2002. 'Food deserts' in British cities: policy context and research priorities. *Urban Studies* 39, 2029-40.

Wrigley, N., Warm, D. and Margetts, B. 2003. Deprivation, diet and food-retail access: findings from the Leeds 'food deserts' study. *Environment and Planning A* 35, 151-88.

Wrigley, N., Warm, D., Margetts, B. and Lowe, M. 2004. The Leeds 'food deserts' intervention study: what the focus groups reveal. *International Journal of Retail, Distribution and Management* 32, 123-36.

Chapter 14
Recourse to Discourse: Talk and Text as Avenues to Understand Environments of Obesity

Robin Kearns

This chapter considers ways of thinking about sociocultural contexts that can be broadly positioned within the 'big picture' of obesogenic environments. A foundational assumption is that a range of influences upon the 'landscapes' of consumption and physical activity help shape place-specific attitudes and behaviours. Further, and regardless of how such influences become inscribed in the material landscape, I argue for their construction in part through language and talk. In the chapter, a series of case studies is presented to illustrate both the range of discursive influences at work within sociocultural environments and the methodological approaches needed to identify and interpret them. Discussion focuses on 'talk' as well as other texts and symbols that help frame everyday experience and trigger contemporary debate about obesity. To this extent, the chapter is concerned with the discursive settings in which experience is produced and occurs.

The chapter surveys two discursive domains that have a bearing on obesity: language and landscape. Language can be understood as the common structures of verbal and written communication. It can be divided into formal and vernacular. Whereas policy (e.g. the New Zealand Ministry of Health's 'Healthy Eating, Healthy Action' strategy) is framed in the considered world of the formal, the latter is the linguistic world of lay people to whom obesity may not necessarily be understood in terms of cause and effect. The term 'landscape' also warrants some comment. In popular and historical understanding, it has referred to the physical or natural backdrops to human activity. In the language of contemporary human geography, however, the term has encompassed the varied elements that comprise a context for everyday life. A fundamental recognition is, according to Cosgrove and Daniels (1988) that taken-for-granted landscapes of our daily lives are replete with meaning. The purpose of this chapter is therefore to provide an account, with reference to case studies, of the power of discourse to shape individual and group food and physical activity-related practices. The chapter illustrates that varying discursive exposures are experienced in different social-cultural settings. In so doing, this account positions discourse as an aspect of social structure that constrains agency in a less visible, but no less forceful, way than does the built environment as physical structures constrains agency.

The chapter examines and contrasts language as shaping 'the problem' of obesity, in three landscapes, drawing on the narratives yielded in qualitative studies: the controversy surrounding the presence of a fast food outlet in Auckland's Starship Children's Hospital; the experiences of Pacific state house tenants in the South Auckland suburb of Otara living with market-based rents on their state houses; and the perspectives of Aucklanders on the merits of walking and driving in a car-dominated city. Table 14.1 summarizes these case studies which complement each other in building a picture of the broader discursive environment.

Table 14.1 Summary of complementary case studies

Case study	Obesity consequence	Research question	Research approach
McDonald's in Starship Hospital (Kearns and Barnett 2000)	Tacit endorsement of fast food in a health care setting.	What pressures led to its establishment, then demise?	Analysis of interviews with hospital managers, media accounts, advertising and unsolicited children's art.
Discounting health among Otara families (Cheer et al. 2002)	Constrained access to healthy food.	How are food and other health purchases affected by high housing and 'cultural costs'?	Analysis of transcripts from in-depth interviews with low-income Samoan and Cook Island state house tenants.
Walking and driving cultures in Auckland (Bean et al. 2009)	Constrained physical activity.	How is driving constructed as a normative mobility?	Analysis of transcripts from focus groups with a range of Aucklanders – students, parents of young children and seniors.

While some internationally recognized theoretical ideas concerning discourse form the chapter's foundation, case examples will be drawn from New Zealand research. This exploration of ideas and examples will tacitly provide a counter to the predominantly quantitative and necessarily reductionist work that has dominated the literature on the geography of obesity and been reported on extensively in earlier chapters. In so doing, it will provide critical reflections that are potentially useful in expanding the horizons of researchers in this field.

Language and Landscape as Text

A key interpretive framework running through all the case studies is the tension between structure and agency. This dynamic acknowledges the relative contribution of choice and constraint in human experience (Dyck and Kearns 2006) and the ways that this ability to act is invariably structured by cultural rules, laws and economic forces. Thus all human action (for instance, health seeking behaviour) is invariably structured by the opportunities and constraints presented by the society in which they are placed. Invariably such opportunities are framed in terms of language.

Rather than an objective medium, language can be regarded as a social construction particular to times and places. Language can perform a powerful role in constructing particular issues (in this case obesity) as an urgent concern as well as legitimating the norms which influence and perpetuate attitudes and behaviour. Advertising and promotional materials, for instance, help to frame the nature of places, in particular by shaping them as sites, if not commodities, to be desired. In this sense there can be a recursive link between the material (e.g. the McDonald's signage and menagerie of iconic characters) and the ideological (e.g. the implication that corporate fast food is acceptably associated with child health). In other words, the way components of the built environment such as hospitals and clinics, and their associated symbols, are represented can therefore be of considerable interest in interpreting (potentially) obesogenic landscapes.

Discourse can be less formally constituted than in advertising. Following Lupton (1992), we regard discourse as a set of ideas or a patterned way of thinking which can be discerned within texts and identified within wider social structures. Inevitably signage and the print media only convey a selective set of discourses circulating within society. On the matter of obesity, for instance, we might well uncover markedly different discourses through talking with medical specialists or members of the public.

A discourse analytic approach regards language and meaning as social constructs. Using this approach to interpret signage, media reports and spoken narrative builds on the notion that readers' perceptions are structured by dominant discourse, yet they are (paradoxically) 'active agents' co-creating meaning. A goal for the researcher is to 'denaturalize' the text through seeking to highlight the ways in which otherwise banal and taken-for-granted messages may have ideological content within elements of text. The interest is in the sociocultural and political context in which text and talk occur (Lupton 1992). Texts are produced by socially-situated speakers and writers. In other words everyone speaks (and by extension, writes) from somewhere – both in terms of geographical and social location. The researcher's concern is to produce a critical analysis of the use of language and the reproduction of dominant ideologies, or belief systems, in discourse.

Burgers and Bandages

In this section, I follow the precedent of earlier studies of health issues as portrayed in the media, by drawing on a study using newspaper reports '...to construct a narrative illuminating the distinctive threads of alternative discourse' (Joseph and Kearns 1999: 2). This case focuses on a minor moral panic that ensued in

1997 when a proposal was unveiled to establish a McDonald's restaurant within Auckland's Starship Children's Hospital. An examination of data collected from media coverage, advertising and interviews with hospital management revealed competing discourses around the issue of fast food within a health care setting (Kearns and Barnett 2000). In a hospital already replete with playful iconography strategically placed to appeal to children, this investigation argued that the introduction of a McDonald's franchise became an ambiguous symbol in the moral geography of health care consumption.

In an international context, the example of a McDonald's restaurant within a hospital is neither new nor unusual. Indeed at the time of opening, there were reportedly 50 such restaurants in hospitals elsewhere in the world (Morris 1997). For instance, Toronto's Hospital for Sick Children has, for some time, contained a ground floor food court featuring a large number of food outlets.

The implied and problematic connection between burgers and bandages began with the August 1997 announcement that a deal had been struck between Starship management and McDonald's for a franchise to be sited within the atrium of the hospital. This would complement the presence of a Ronald McDonald House adjacent to the hospital assisting families of children undergoing hospital treatment, and modelled on the concept launched in 1974 in Philadelphia. As part of the agreement, hospital visitors could choose purchases from an 'extended menu' featuring not just the regular fare of burgers and fries but also cereals, low-fat milk and seasonal fruit. In keeping with the implied bold new directions of a hospital named 'Starship', the reputed competition between McDonald's and Burger King to be the franchise at its entrance could be seen as evidence of competition 'to boldly go' where no health enterprise had gone before in New Zealand (Kearns and Barnett 1999).

As Kearns and Barnett (2000) recount, the announcement brought immediate public reaction in the columns of the major daily newspaper. Cynicism prevailed among commentators. One regular columnist identified a link between benevolent gesture and commercial foothold, asking: 'So you thought that when McDonald's set up Ronald McDonald House for kids with cancer at Auckland children's hospital, they were doing it out of the goodness of their hearts? Think again' (Welsh 1997 cited in Kearns and Barnett 1999: 85).

Writers of letters to the editor were quick to link health and place in their expectations regarding children's welfare. For example: 'Now we have our Starship children's hospital endorsing the sale of junk food on its premises, which surely must give the wrong message to parents and children that this type of food is all right after all...otherwise it would not be promoted there?' (Browning 1997).

Lobby groups weighed in with obesity concerns. The executive officer of the National Council of Women (NCW) wrote to the *New Zealand Herald* expressing concern '...that the prominent position of McDonald's restaurant at the Starship Childrens' Hospital will send a strong message of endorsement by health professionals that takeaways are an appropriate everyday food' (Morris 1997).

The correspondent went on to cite the Ministry of Health's *Guidelines for Healthy Children* which caution that '…fast foods must be eaten in moderation' in light of the fact that 'a variety of food is essential for the maintenance of growth and good health'. The NCW's concern was '…that health promotion messages will not be effective with a visibly dominant fast food outlet with limited food variety and possibly high in fat, salt and sugar, which appears to have the endorsement of the Starship…' (Morris 1997).

The critical point raised in the foregoing narrative is the apparent 'dissonance' between official messages ('eat healthy') and the implied hospital response ('if hungry go to McDonald's'). The danger in this contradiction, which we might label a 'discursive disconnect', is not only its rendering of the Starship hospital as an exceptional space (i.e. a health care setting in which regular fast food is OK) but also that in such a rendering, *other* health care settings are at risk of having their imperatives of nutritional and moderate eating undermined by precedent.

Other correspondents to the *New Zealand Herald* were less sympathetic, implicitly invoking the agency end of the structure-agency continuum. One suggested that in succumbing to the temptations of over-indulgence, we must '…blame the weak-willed fatties, particularly the parents'. With a measure of sexism thrown into the mix, the National Council of Women themselves came into the firing line: 'As for the self-appointed guardians of the nation's health, overwhelmingly women, tell them they have the cart definitely before the horse. They should be preaching to the customers, not the suppliers' (Phillips 1997).

Within the range of commentaries and reactions, we discerned a two-layered discourse (Kearns and Barnett 2000). Most clearly, talk about the McDonald's franchise opening highlighted the potential (ill)health effects of excessive fast food consumption by especially youthful users of the hospital. This was constructed as a nutritional transgression with nigh on moral dimensions. There was also a deeper and tacit concern at the corporate alliance between a commercially-oriented public hospital and a multinational fast food chain – a move described elsewhere as '…another slice off the carcass of the ailing public health system (which is) increasingly prey to the predations of private enterprise' (Welch 1997: 23). However, we also noted a countervailing discourse of pragmatism. There were presumptions of parental gratitude at the prospect of predictable and convenient food at the hospital suggested by the hospital's general manager who saw the inclusion of McDonald's as 'simply a matter of getting the best service for the hospital' (Welch 1997: 23).

After an eight year tenure, McDonald's at the Starship closed for business prior to its lease expiring. Its spokesman cited a decline in revenue as the reason, but added that 'It wasn't perhaps the best of business decisions to put it there… there's just not that many people' (*New Zealand Herald* 2005a). Others were less circumspect. For instance, Celia Murphy, executive director of the Obesity Action Coalition expressed 'delight' at the demise saying: 'Anything synonymous with chips and burgers is not appropriate in a health organization anywhere, particularly a children's hospital' (*New Zealand Herald* 2005b).

What are we to make of the tenure of a McDonald's in the major New Zealand children's hospital? Its rise and demise speaks to the shifting discursive landscape of 'health talk' in the society-at-large. Over the last decade, concerns regarding obesity and its determinants have led to a declining acceptability for fatty, high-energy foods being sold in schools and a rise of broad-based campaigns that bring together physical activity and nutritional eating imperatives. Arguably, it was not only the amount of fast food consumed in the Starship, or the number of consumers, but also the symbolic presence of the outlet that created dissonance with the goals of child health. Perhaps a lesser known brand would have been less controversial but the Golden Arches have grown to epitomize a seductive iconography of multi-nationalism in the contemporary urban landscape and have become a potent symbol of corporate reach into local places. Thus, professional distaste for questionable food quality became mixed with public disdain for globalisation and what Ritzer (1995) calls the *McDonaldization of Society*. However, while some discourses – in this case disdain for fast food in hospitals – come to dominate, they always exist with countervailing discourse. For instance, as Kearns and Barnett (2000: 91) note, a children's art area near the outlet in 1998 featured one painting on which the caption read 'I love McDonald's. McDonald's makes me happy in hospital'. What had been a discomforting presence for health promoters and some parents was, potentially, the source of comfort to at least some children in hospital. In this case the re-structuring of a health care setting had, for some, re-written the aspirations and expectations of the in-hospital experience. For others, the arrival of McDonald's in the Starship galvanized their agency in resistance to new incursions of structural influence over nutritional norms.

Automobility

For a second example of the power of discourse I turn to a study of talk about driving and walking as mobility practices among Aucklanders (Bean et al. 2008). The foundational contention of this research was that 'automobility', or the normative use of often sole-occupant private motor vehicles, has become a dominant cultural practice. With this ascendancy, other types of travel such as walking, cycling and public transport have been marginalized in both urban planning and personal practice (Sheller and Urry 2000). Automobility is regarded as linked to obesity because it has displaced walking as a dominant form of mobility with sitting, riding and driving. This displacement accelerated through the latter half of the twentieth century through walking being re-conceptualized as an inconvenient, slow mode of travel, and something to be undertaken for recreational and discretionary rather than routine reasons. Indeed, as Amato (2004) has argued, *not* having to walk became a sign of having 'made it' in socioeconomic terms. Cumulatively, these concerns relate to obesity for, at an individual level, it has been shown that there is a correlation between the time spent in a car and the risk of obesity (Frank et al. 2004).

The Bean et al. (2009) study drew on discourses expressed in talk, addressing the contention that walking is a social and public form of mobility. This, the authors argued, can be contrasted with heightened automobility (both driving and being driven) which corrodes the convenience of walking and has negative implications for public as well as private life (e.g. social exclusion, proneness to obesity). Our argument was that decisions regarding walking and driving are socially constructed through the medium of talk and that such talk has been productive to the point that resulting actions have cumulatively created fundamentally different places (see also Hinde and Dixon 2005). In other words, just as a normative discourse of fast food being OK in health care spaces is materialized in tangible symbols like the Golden Arches, so too a hegemonic discourse of car-dominance can create downstream and concrete consequences in the form of what we might call 'autoscapes' (e.g. motorways, low density strip mall development, petrol stations, auto repair centres, 'strips' of motels, and drive-in fast food outlets). Relph (1976) calls such developments 'placelessness' and the challenge for obesity prevention is the inertia in the built environment. In other words, once constructed, the scale of such development is difficult to modify away from the need to drive or be driven to access amenities such as shops, schools and parks. Agency is constrained by aspects of urban structure. The *New Urbanism* and *Smart Growth* movements have introduced a countervailing discourse to dominant planning ideologies with the promotion of 'walking places' and 'liveable communities'. These comprise high density urban developments featuring mixed land uses in close proximity, and have greenery, rest areas, and generous amounts of public space (Solnit 2000).

In this case study, everyday travel experiences were explored by means of focus groups which involved adult participants from a range of ages and residential locations. Given the social influences upon travel behaviour, these groups were deemed to offer a good vehicle for exploring decision making around mobility and the social meanings of different travel modes (Bean et al. 2009). A subsequent content analysis of the narratives collected involved searching for common themes as well as differences in participants' viewpoints.

A few indicative narratives can illustrate the power of discourse in shaping daily mobility choices. First, we have the facilitator prompting reflection among young parents on the conscious decision to walk in their home neighbourhood – an activity at once counter-cultural within Auckland at large, but one becoming increasingly normative at the local scale. Health it appears is a particular motive:

CB: So, is health a big factor in choosing to walk? Who goes for a walk just for health?

Bridget: Yeah definitely.

Nina: Yeah.

> Andrea: I do it on behalf of my children, I consider it healthy for them to walk to school and so we do…I would much prefer that they did walk than taking the car. Especially…you read all of this stuff about child obesity and you know, they have a more sedentary life compared to what we used to have and so I like to try and get them moving as much as possible.

Here we can discern a discourse of responsible parenthood informing the decision to walk, along with a nostalgia for an earlier era in which walking was more routine for children. Children, it seems, offer permission for parents to return to walking with a sense of legitimacy. This walking with, and on behalf of children, in an auto-dominated city is not available to younger adults without children for whom walking is often constructed more starkly as an 'alternative' transport mode in Auckland. By way of example, one focus group participant in the Bean et al. (2009) study perceived their choices to walk as unusual within the city's auto-dominant discursive environment:

> Peter: I used to get hassled because I walked so much when I was younger.
>
> Cecilia: Hassled?
>
> Peter: Clear as hell, they used to say, 'Why don't you just catch a taxi?' and I used to say 'I just enjoy walking'.

The dominant discourse is one of inevitability and necessity. Two other younger adult participants concurred:

> Matthew: Yeah, I cannot imagine not having a car to fall back on if I needed it.
>
> Cecilia: That's right, (it's) absolutely imperative.

For Peter, a love of walking is framed within the societal discourse that speaks to the pleasures of speed, independence and being in control:

> Peter: I really do love walking, but absolutely do love driving as well, propelling myself with petrol. I think it's the thrill…it's like being on the edge the whole time, being in control…I just enjoy driving, fast.

For planners the message is that, for most, there will not be a black-and-white choice between driving and walking. Rather support and affirmation of non-motorized transport needs to be offered. This should be tempered by the recognition that as long as it is affordable, driving is a discursive framing, with material consequences, that will remain a potent force in shaping daily mobility. In other words, while individual agents may in principle desire to walk, the various structural constraints acting upon everyday lives (i.e. ranging from time through to urban design) can re-shape and modify intentions and actions.

Discounting Health

The language deployed in the news media comprises a discursive backdrop to everyday life that, through simplification, easily and sometimes strategically diminishes the significance of the structural determinants of obesity. For instance, a report on global obesity rates was reported in the *New Zealand Herald* (2007a) with the headline 'Eight out of the ten "fattest" countries are in the Pacific', and the accompanying photo depicted a human torso, devoid of face and feet but emphasising bulging upper thighs and midriff. Two themes are at work here: first the generalization of obesity to all members of any Pacific nation, and second the depersonalization of the overweight person in favour of presenting the body as the problem. This 'sound-bite' approach in both print and spoken media can, through its attenuation of explanation, implicate otherwise sympathetic experts in the construction and maintenance of victim-blaming discourse. For instance, in the same article, Daniel Epstein of the WHO Regional Office of Americas is quoted as saying: 'Poor people have an easier time of eating junk food. People fill up on things that have a high caloric value but little nutritional value' (*New Zealand Herald* 2007a). Observation bereft of explanation can result in simplification.

Follow-up articles and commentaries need to work hard to counter discourses of blame and ill-discipline that implicitly stress agency over structure. In a follow-up article, Dr Annamarie Christiansen, of Brigham Young University, Hawaii, is reported as saying 'cultural perceptions of food and the body determine Polynesians, eating habits and management of their bodies'. This assertion takes the attribution of responsibility back to a source in material culture in which 'historically, size represented wellbeing and wealth of a person and community'. Here, an agency-based explanation is eschewed for a broader structural interpretation with the disruption to traditional foods and lifestyle being a subset of a larger scale disruption by colonialism (*New Zealand Herald* 2007b).

The cult of the expert in contemporary society privileges the formally educated and institutionally endorsed spokesperson. It is rare that the experts in coping with the challenges of daily living are cited so as to insert ground-level insight into the discursive environment. Qualitative research focussing on narrative provides some endorsement that broader discourses are at work in influencing Pacific peoples' expenditure patterns. Through their analysis of the experiences of Pacific households in the lower socioeconomic Auckland suburb of Otara, Cheer et al. (2002) explored the links between food purchasing practices and housing costs, establishing the latter to be a major constraint and structural determinant of healthy eating. Indeed, healthy food was identified as a main area of spending that was 'discounted' for the sake of other expenses (along with power, heating and telephone expenses). Three narratives are indicative:

> There is nothing more important than a roof over your head. Everything else comes after.

> I'll pay the bills first and then if there's money left over I'll buy food. Otherwise I'll just wait until food does come…The most important thing is tea and sugar. When there's no food I'll just live on cups of tea…

> The amount of food changes each week depending on what cultural occasions come up. If there's none then there's more money for food. If there's many then its back to jam and bread…last week there was only $30 for food because there was a funeral to pay for.

Thus, housing and 'cultural costs' are indirectly leading to the consumption of cheaper foods, foods that are energy-dense and nutrient-poor with high levels of sugar and saturated fats. Combined with reduced physical activity and an historic endorsement of size as signalling status, such 'discounting' can be seen as tacitly promoting obesity.

Collectively these narratives offer a different discourse of healthy eating and its consequential converse: obesity. Here the structural determinants of food insecurity are tacitly emphasized and in so doing, the speakers' perspectives are placed in a wider discursive setting that gives full cognisance to competing pressures on households' limited financial resources. Such pressures come to bear because food is not a fixed expense (as with rent) and nor does non-payment result in repossession of goods or risk of eviction. Further, the imperatives of being part of a robust web of familial relations accompanied by specific expectations of generosity and mutual support results in extra pressures on the food budget.

In addition to loan repayments because of cultural obligations such as funerals, it is not uncommon for people to make donations to families in their churches or those from the same village on the home island when someone dies or weds. As one respondent said:

> Our cultural obligations are more important than food and so when something comes up like that it will be just jam and bread that we eat. Food comes after cultural donation.

While it is, in principle, possible for householders to spend a small amount of money on health-enhancing foods such as fresh vegetables, such practices were not reported. Rather, processed high energy and bulky food purchases were reported, as in:

> It's hard to buy healthy food. Vegetables are especially expensive. I'd rather buy flour and make things from that.

> I'd rather pay $5 for the more unhealthy stuff than $10 for the healthy stuff.

In other words, the discursive setting of food purchasing for the poorer Pacific families in this study includes the influence of a complex set of discursive exhortations relating to what it is to be a 'good' householder (i.e. pay your rent at all costs, keep the family together) and family/congregation member (give at all costs). The case study highlights the multidimensional structural pressures that bear upon families in ways that mitigate against compliant agency-based responses to healthy eating- healthy action imperatives.

Conclusion

This chapter has served as both a methodological reflection and a presentation of observations on real-world settings and landscapes. While the scientific sensibility of epidemiology offers undeniable assistance in partialing out the pathways of causality in the quest to understand obesity, qualitative approaches stressing the value of talk and text help discern a wider web of influences upon attitudes and actions. For attitudes and behaviours can too easily be subjected to a reductionist examination. Rather, physical activity and food consumption – the domains of concern in the foregoing case studies – are more than simply context-dependent behaviours rationally chosen by individuals. Drawing on Bourdieu (1998), they are sets of social practices '…adopted by population subgroups to accrue different forms of capital and to position themselves socially' (Hinde and Dixon 2005: 35). In other words, this view would see the introduction of a McDonald's into the hospital to be a strategic expression of the power of corporate marketing to capture the symbolic landscape. So too, decisions to drive and choices to donate before buying healthy food are less about individual choices and more responses to structural constraints presented by the social, economic and built environments.

To be seen driving, and to be seen donating, may arguably be as strong a motive as the satisfaction in being in one's car or giving towards funeral expenses. In other words consumption must be regarded as a wide web of relations rather than moment-in-time acts of purchase or eating. To this extent, the worlds of signage, advertising and architecture all contribute to the pull and persuasion towards specific acts of, and dispositions towards, consumption (Kearns and Barnett 1997) and must be taken seriously in the broader landscape of obesity research. While undertaken and discussed separately, the foregoing case studies are therefore clearly connected within such a global perspective on consumption and its place-specific cultural economy (in this case, Auckland, New Zealand). One could, however, weave the cases together more tightly and explore connections between, say, car reliance and food consumption, or the abandonment of auto-dominance discourse and the growing unacceptability of fast food in hospitals. As Hinde and Dixon (2005) remark, the pathways are many as are the links to public health.

Those who argue that human agents are the authors of their own health status claim that individual decisions are the primary influences upon health outcomes. The problem with this view is that decisions are not made within impermeable vacuums. Rather, we are all exposed – to varying degrees – to influences that shape decisions which, even if as banal as the choice to walk or drive, or to buy food or donate to a church, cumulatively can profoundly affect the health of ourselves and that of our dependents. These discourses to which we are daily exposed may be as subtle as corporate endorsements or the tacit encouragement to drive when there are poor quality pavements (as in first two case studies) or as direct as the demands of donations to relations in need (as in the third). In sum, however, these discourses are too easily overlooked in both research and public policy. In both domains, responsibility is, I contend, too readily reduced to the individual and her/his agency when decisions, however individual in origin, are influenced by the discursive contours of a structural landscape.

References

Amato, J.A. 2004. *On Foot: A History of Walking*. New York: University Press.

Bean, C., Kearns, R.A. and Collins, D.C. 2009. Exploring social mobilities: Narratives of walking and driving in Auckland, New Zealand. *Urban Studies*, 45(13), 2829-48.

Bourdieu, P. 1998 *Practical Reason*. Cambridge: Polity Press.

Browning, R, 1997. Hospital giving the wrong food message. Letter to the editor *New Zealand Herald*, 27 August, A7.

Cheer, T. Kearns, R.A. and Murphy, L. 2002. Housing policy, poverty and culture: 'discounting' decisions among Pacific peoples in Auckland, New Zealand. *Environment and Planning C: Government and Policy*, 20(4), 497-516.

Cosgrove, D. and Daniels, S. (eds.) 1988. *The Iconography of Landscape*. Cambridge: Cambridge University Press.

Dyck, I. and Kearns, R.A. 2006. Structuration theory: Agency, structure and everyday life, in *Approaches to Human Geography* edited by G. Valentine and S. Aitken. Thousand Oaks, CA: Sage, 86-97.

Frank, L., Anderson, M. and Schmid, T., 2004. Obesity relationships with community design, physical activity and time spent in cars. *American Journal of Preventive Medicine*, 27(2), 87-96.

Hinde, S. and Dixon, J. 2005. Changing the obesogenic environment: insights from a cultural economy of car reliance. *Transportation Research Part D*, 10(1), 31-53.

Joseph, A. and Kearns, R. 1999. Unhealthy acts? Interpreting narratives of community mental health care in Waikato, New Zealand. *Health and Social Care in the Community*, 7, 1-8.

Kearns, R.A. and Barnett, J.R. 1997. Consumerist ideology and the symbolic landscapes of private medicine. *Health and Place*, 3,(3), 171-180.

Kearns, R.A. and Barnett, J.R. 1999. To boldly go? Place, metaphor and the marketing of Auckland's Starship Hospital. *Environment and Planning D: Society and Space*, 17(2), 201-226.

Kearns, R.A. and Barnett, J.R. 2000. Happy meals in the Starship enterprise: Towards and moral geography of health care consumption. *Health and Place,* 6(2), 81-93.

Lupton, D. 1992. Discourse analysis: a new methodology for understanding the ideologies of health and illness. *Australian and New Zealand Journal of Public Health*, 16(2), 145-50.

Morris, D. 1997. Starship's enterprise annoys. Letter to the editor, *New Zealand Herald,* 1 September, A7.

New Zealand Herald. 2005a. McDonald's set to close. 30 March, A3.

New Zealand Herald. 2005b. McDonald's discharges itself after eight years at Starship. 2 April, A3.

New Zealand Herald. 2007a. Pacific Islands worst in world for obesity. 21 February, A3.

New Zealand Herald. 2007b. Perception, key to obesity issue in the Pacific. 30 November A3.

Phillips, B.J. 1997. Blame the eater. Letter to the editor. *New Zealand Herald,* 3 September, A7.

Relph, E. 1976. *Place and Placelessness*. London: Pion.

Ritzer, G. 1996 *The McDonaldization of Society: Revised Edition*. Thousand Oaks,CA: Pine Forge Press.

Sheller, M. and Urry, J. 2000. The City and the Car. *International Journal of Urban and Regional Research,* 24(4) 737-757.

Solnit, R. 2000. *Wanderlust: A History of Walking*. New York: Penguin Books.

Welch, D. 1997, Starship and fries. *New Zealand Listener,* 6 September, 22-23.

PART VI
Conclusions

Chapter 15
Conclusions:
Common Themes and Emerging Questions

Jamie Pearce and Karen Witten

Introduction

The preceding chapters have documented the rapid increase in the global prevalence of obesity over a relatively short period of time, the pervasiveness of obesity promoting factors in our political, economic, physical and social environments, and the powerful influence that routine practices of daily life can have on diet and physical activity. At a population level the foods we eat and the way we move between the destinations of home, work, school and play have morphed enormously over recent decades. Foods are sweeter, more processed and energy dense and our built environments, designed to accommodate the motor vehicle, are discouraging walking and cycling and promoting car reliance. Trends in overweight and obesity first observed in industrialized countries are also emerging at an accelerating pace in many developing countries.

As the authors of the chapters in this book have shown, obesity is complex and its causes are multi-factorial across biological, psychological, economic, social, and cultural domains. In particular, we have seen some of the ways in which the environments in which we live are 'obesogenic'. In this concluding chapter we reflect on some of the key themes identified in the preceding chapters, and question how compelling the evidence is for environmental explanations of the rise in rates of obesity.

Common Themes

Given the structure of the book it is not surprising that certain observations and ideas raised in one chapter reappear to be discussed from a new perspective in subsequent chapters. We have selected three recurring themes to comment on in this concluding chapter: the positioning of the obesity epidemic within historical processes of industrialization and urbanisation; the value of conceptual frameworks to understand and respond to the epidemic; and the importance of recognizing the embedded nature of social practices around food and mobility.

Obesity as the Outcome of Historical Processes

Viewed collectively the chapter authors embrace a wide historical frame as they describe and attempt to understand the factors that have contributed to the obesity epidemic. Popkin (Chapter 2) and Kim and Kawachi (Chapter 3) use the diets and exercise patterns of our Palaeolithic ancestors as a benchmark against which to contrast contemporary food and mobility practices. These authors are also amongst those who reflect on industrialization and urbanisation as significant historical processes that have set the scene for major changes in the nature of work, leisure and domestic life, and in the production and marketing of food and private motorized transport – all domains that can be plausibly linked to a change in energy intake and energy expenditure. The accelerating rate of overweight and obesity, particularly as it is unfolding in the developing world, has also been linked by several of the chapter authors to more recent phenomenon associated with globalisation. The developing world has become a marketplace for products, behaviours and lifestyle practices that are well established in the developed world. The impacts on the diet, physical activity levels and body size of citizens in these countries are following a remarkably similar course to those observed in the developed countries.

Popkin (Chapter 2) traces the gradual emergence of overweight and obesity in industrialized countries through the 1900s and the rapid increases from the 1970s and 80s. He contrasts this with the far sharper rise of obesity in middle and lower income countries since the 1990s. To illustrate the extent of the problem he lists six countries, three lower income – Egypt, Mexico and South Africa – and three higher income – United States, Australia and the United Kingdom – in which almost two-thirds or more of adults are overweight or obese. While more urbanized countries have higher rates of overweight and obesity, evidence of increasing obesity in rural settings is also presented. Compelling data on childhood obesity as an emerging worldwide phenomenon are also reported. The unequal burden of obesity is not restricted to geographical differences between countries but also has an important socio-spatial dimension. Many authors in this book highlight that rates of obesity and obesity-related health outcomes are often higher among more socially disadvantaged groups. For example both Thornton and Kavanagh (Chapter 5) and Turrell (Chapter 8) note that obesity, weight and related health outcomes as well as the environmental driving forces of these phenomena are unequally distributed across small areas. More socially disadvantaged areas often tend to have poorer obesity-related health outcomes as well as environments that are less conducive to a nutritious diet and being physically active.

Hoek and McLean (Chapter 4) describe the process of industrialization in the agricultural sector and how this has led to increased efficiencies in food production and the commercialisation of the food supply. Political and economic drivers of changes over time in food composition and consumption are discussed with particular reference to the rapid rise of high fructose corn syrup as a food sweetener, a rise that is attributed to US government subsidies to support corn production. Overproduction of food in developed countries in combination with the higher profits that can be made from 'value-added' processed and refined foods are positioned as key macro level political and economic factors behind changing dietary patterns. In addition to these major trends the authors note the emergence of multi national food companies and supermarket chains as substantial factors that have reconfigured the foods available to consumers. The increase in food preparation, preserving, packaging and distribution undertaken at the industrial scale is identified by Kim and Kawachi (Chapter 3) and Hoek and McLean (Chapter 4) as a factor that is likely to have contributed to a decline in household level food preparation and to an increase in the quantity and types of foods consumed.

With regard to the physical activity environment Giles-Corti and colleagues (Chapter 7) reflect on the post-World War II technological, environmental and societal changes that have profoundly altered the built environment, and many of the mundane routines of daily life. Industrial and labour saving devices such as washing machines, motorized lawn mowers and television remote controls, as well as various forms of computer technology, designed for convenience, have reduced the time required to complete everyday tasks diminishing the need for physical effort and encouraging more sedentary behaviours. The car and car reliance are the supreme example. The pervasiveness of the motor vehicle across the globe has not only altered our behaviours but transformed the design of our cities. As car ownership levels increase, urban environments become increasingly attuned to enhancing vehicle mobility. Autocentric planning, in combination with land use zoning regimes designed to separate housing from polluting land uses, has increased the distances between the activities of daily living. Commuting distances for both adults (see Chapter 7) and children (see Chapter 9) have risen. Zoning ordinances have inadvertently made our cities more sprawling, less walkable and less connected.

The Value of Conceptual Frameworks

Robust research requires a sound conceptual model. The ANGELO (Swinburn 1999) and International Obesity Task Force Working Group (OTFWG) conceptual frameworks are introduced by Kim and Kawachi (Chapter 3) as useful tools for theorising and representing our emerging understanding of the multiple scales, sectors and settings that interact to create obesogenic and leptogenic environments. The book's structure has borrowed from the macro to micro, international to local dimensions of these frameworks. The models have also been used by the authors of several chapters as a point of departure for conceptualising relationships between environmental factors and obesity-related outcomes. Notwithstanding the immense value of these tools for conceptualising possible explanatory pathways it is also noted that to retain their value, models need to have the flexibility to facilitate new theorising and to reflect an emerging evidence base. In this light, Kim and Kawachi (Chapter 3) suggest the models would be enhanced as explanatory heuristics by adding a temporal dimension. This would enable the shifting and dynamic nature of economic, political, social and physical environments to be incorporated into a model and, they suggest, may help to explain the 'occurrence and trajectory of the obesity epidemic'. More comprehensive models would also reflect emerging evidence that environmental factors in varying domains, and at different spatial scales, may have differential effects on the obesity-related outcomes of specific socio-demographic groups. Effects of environmental exposures may vary according to age, life stage, gender, socioeconomic status and many other factors. Dimensions that will always be difficult to capture in a static model are effects of environmental exposures which accumulate over the life course.

Changing Social Practices

Kearns (Chapter 14) stretches the notion of obesogenic environments beyond the sectors, scales and settings of the ANGELO model and the societal policies and process of OTFWG model. He contends that the different discourses that circulate within sociocultural environments are another potent dimension of the obesogenic environment. To illustrate the diverse imperatives that can shape the dietary practices of different sociocultural groups he describes the impacts financial constraints arising from cultural obligations can have on the food choices and consumption patterns of Pacific families in Auckland, New Zealand. It is acknowledged that sociocultural practices such as those described are layered on the many other contextual enablers and constraints that shape daily life in particular neighbourhoods. It is not assumed that local foodscapes and physical environments will have the same meaning for different groups or provide similar opportunities for making healthy choices around diet and mobility. Culturally specific foods and familiar or favoured opportunities for physical activity may or may not be available in particular localities.

Giles Corti and colleagues (Chapter 7) also reflect on the embedded nature of social change with respect to the workforce participation of women, family life and children's mobility. A temporal mode shift from foot to car for the trip to school has been tracked with consistent findings across many developed countries. Giles-Corti et al. (Chapter 7) report on studies that have linked children's mode use for the school journey to the employment status and the hours worked by their mothers. Parents' concern for their children's safety, particularly their exposure to traffic, is another social factor that has placed constraints on children's independent mobility. Oliver and Schofield (Chapter 9) talk of parents' overriding concern for children's safety contributing to the paradoxical 'social trap' (Tranter and Pawson 2001): as more parents chauffeur their children the number of cars on the road increases which in turn heightens the danger to remaining child pedestrians. As fewer children walk the demand for pedestrian infrastructure drops, the auto landscape is further entrenched and so too is the social practice of chaperoning children.

Interactions between the physical and sociocultural aspects of place as they influence dietary and physical activity patterns are also touched on by Kim and Kawachi (Chapter 3). Their discussion of Japanese and Pima Indian migrant studies demonstrate the interactive and adaptive processes that occur over time between people and place. For both groups of people – the Japanese and Pima Indian migrants – moving countries was associated with changing food practices and an increase in dietary fats. In Chapter 4, Hoek and McLean offer another example when they mention that an increase in the number of meals eaten away from the home corresponded to a period during which adult workforce participation and household disposable incomes increased. As social practices changed in response to economic prosperity, the number and range of out of home food outlets increased thereby modifying the food landscape and enticing further changes in food–related social practices. Over time, what were once novel behaviours become taken-for-granted practices of daily life and their banal nature can make the downstream consequences of such social practices for body size hard to detect.

Kearns (Chapter 14) argues for the utility of social theorising, and in particular the sociological concepts of structure and agency as an interpretive framework, as we try to understand the relationship between environments, diet, physical activity and body size. The discursive environments of home, work, school and play are conceptualized as exposures from which messages and meanings are absorbed that form part of the 'social structure that constrains agency in a less visible, but not less forceful way than the built environment'. Integral to theorising on the relationship between structure and agency is the idea of recursivity: just as the environment (structure) shapes agency, the agency of individuals and groups also shapes the environments in which they live. With respect to childhood obesity, Walton and Signal (Chapter 6) comment on the potential influences of the environments children pass through – home, school and community – each with different physical, social and cultural contexts, with competing social norms and underpinning discourses. To understand children's exposures to obesogenic environments requires an understanding of the characteristics and influences of the various places in which they spend time.

Neighbourhood Environments, Diet, Physical Activity and Obesity: The Evidence?

The overwhelming conclusion to be drawn from the evidence reviews that comprise Chapters 5 (Thornton and Kavanagh) and 6 (Walton and Signal) relating to the food environment, and Chapters 8 (Turrell) and 9 (Oliver and Schofield) with regard to the physical activity environment is that it is early days for research in these fields. In both areas findings are described as mixed, equivocal or inconsistent more often than they are definitive. This observation echoes those recorded in a recent supplement to the *American Journal of Preventive Medicine* reporting on the proceedings of a workshop on 'Measures of the Food and Built Environments'. The field is described as 'in its infancy' (Story et al. 2009: S187) and the measures in use as 'first generation' (McKinnon et al. 2009: S81) – an eclectic mix of tools drawn from a range of disciplines.

Turrell (Chapter 8) examines the evidence for associations between a number of physical and social environment characteristics and body size. Although he explains that the evidence does not indicate with any certainty that the neighbourhood environment influences the likelihood of being overweight or obese he cautiously states that '[W]hen statistically significant difference were found, they were usually (although not always) in the expected direction: in short, a 'better' neighbourhood tended to be associated with a lower likelihood of being overweight or obese'. He goes on to list the attributes of a 'better' neighbourhood such as good street connectivity, a variety of land use, high density, low levels of crime and visible incivilities; as he observes, many of the same characteristics that have been shown to be associated with higher rates of physical activity.

Likewise Oliver and Schofield (Chapter 9) conclude from their review of the evidence that support for an association between built environment factors, and the physical activity and body size in children is equivocal. Evidence is accumulating that shows children who walk or cycle to school are more physically active overall than those who do not, and that trip distance contributes to car reliance for the journey to school. But with regard to children's activity levels and urban design attributes other than trip distance, they emphasize that the relationships are complex and outcomes appear to differ by factors like the age and sex of the child, and community and cultural norms. Walton and Signal (Chapter 6) come to equally tentative conclusions with respect to available evidence on factors in children's food environments that contribute to the prevalence of overweight and obesity.

As Moon (Chapter 12) outlines, much of the uncertainty in the evidence base regarding the environmental drivers of obesity, relates to limitations in study design and data collection. Many of the limitations of research in these fields as noted in the preceding chapters and in the supplement noted above, are those that are commonly identified across the neighbourhoods and health field: inadequate theorising of explanatory pathways; a lack of robust environmental measures; no consistent use of context or outcome measures across studies; reliance on self reports of health status and related behaviours; a lack of measures sensitive to specific population groups and neighbourhood types; measures made at varying spatial scales; and static definitions of neighbourhood that may not reflect residents experiences of neighbourhood boundaries (Story et al. 2009, MacIntyre et al. 2002, Turrell (Chapter 8), Thornton and Kavanagh (Chapter 5), Oliver and Schofield (Chapter 9)). Moon (Chapter 12) draws particular attention to the issue of endogeneity which has been an impediment to many studies in the neighbourhoods and health field, and studies seeking to understand the environmental explanations for obesity are no exception. In an obesity context, an example of endogeneity would be self selection by less obese people into environments that are conducive to a nutritious diet and/or physical activity. Hence any association between area measures of the food or physical activity environments that are identified in cross-sectional studies cannot be assumed to be causal: the causal pathway may be in the opposite direction. Moon advocates for a greater use of longitudinal studies and natural experiments to address this issue. As Smith and colleagues (Chapter 13) detail, recent methodological developments in the field of spatial epidemiology also offer considerable potential to assist in overcoming some of the limitations that have restricted our understanding of the environmental drivers of obesity. For instance, Geographical Information Systems (GIS) approaches provide a framework for integrating spatial data from a range of sources and more precisely characterising features of the local food and physical activity environments. Similarly geographical techniques such as multilevel modelling, cluster analysis and microsimulation are serving to overcome important concerns including the ecological fallacy and criticisms relating to the aggregation of data into geographical zones.

The priorities for research in the area identified by a working group at the 'Measures of the Food and Built Environment' workshop noted above were succinctly reported as: 'factors (identified); mechanisms (understood); measures (developed); natural experiments (evaluated); surveillance (established); and standards (developed)' (Story et al. 2009: S183). The conclusions reached by many of the chapter authors reflect on the limitations of the current evidence base. Where future research directions are discussed a number of the recommendations echo those outlined by Story et al. (2009): frameworks that acknowledge the range and complexity of environmental influences and take account of changing exposures over time; investment in the development of valid environmental and health outcome measures and their consistent use across studies; international comparison studies that apply common environmental and outcome measures; studies that enable the determinants of specific outcomes to be examined for different sub-population groups (age, gender, socioeconomic, sociocultural) in various settings (school, workplace, rural/urban); the use of longitudinal designs to determine causality; methodologies that address problems of endogeneity; greater use of natural experiments; the systematic evaluation of intervention programmes designed to reduce obesogenic aspects of the foodscape and built environment; and an exploration of sociocultural factors that sustain social practices such as car reliance and food consumption patterns that undermine population health.

Research into the environmental drivers of obesity is clearly in its infancy with evidence of the obesogenic potency of many environmental attributes far from conclusive. However the magnitude of the obesity problem has demanded a policy response. The rigorous evaluation of the implementation and outcome of policies designed to increase physical activity, improve food consumption patterns and reduce the prevalence of overweight and obesity will make an important contribution to the accumulating evidence base.

Health Promoting Public Policies

Giskes (Chapter 10) and Riva and Curtis (Chapter 11) review the policy responses that have arisen at the international, national and local levels that aim to modify obesogenic aspects of the food and physical environments. The authors argue that the formative nature of the evidence base on the environmental determinants of obesity should not be interpreted as a justification for policy inaction. 'Policy' in this context incorporates laws, regulatory measures, and funding priorities for intervention programmes and other actions. Whilst the authors of both chapters recommend caution in some areas of policy development and acknowledge the need for further high quality research in many areas, they nonetheless provide compelling evidence that a number of environmental interventions have significant potential for addressing the increase in obesity. The authors suggest that policy responses operate within a complex web of determinants, in which the causes of obesity, and in all probability the solutions, are multifactorial. They advocate for policies that target interventions at multiple governance levels: national, regional, local, and organizational with effective linkages between the various levels. With regard to the food environment Giskes identifies policy opportunities at various points in the food supply chain including primary production (for example agricultural policy), food processing (for example food standards and labelling regulations), food distribution (for example subsidies and taxes), food marketing (e.g. product displays and sponsorship), retail outlets (for example number, type and location of food outlets) and the catering and food service industry (for example school food policy).

Both Giskes (Chapter 10) and Riva and Curtis (Chapter 11) argue for the importance of coordinated policies across sectors and at various levels of governance to create health promoting environments. These assertions are supported by key policy documents such as the World Health Organization's *Global Strategy on Diet, Physical Activity and Health* (World Health Assembly 2004) and the Foresight (2007) report in the UK which advocate for the development of enabling environments to improve nutrition and increase physical activity. Giskes contends that policy developments around healthy eating are more likely to be successful where healthy food choices are supported by coordinated polices at other levels in the food supply chain and Riva and Curtis (Chapter 11) make a similar case with respect to creating supportive environments for physical activity, emphasising the importance of policy integration across sectors as well as at varying spatial scales. A challenge for public health practitioners tasked with implementing environmental interventions to address the obesity epidemic will be to identify viable legal frameworks and regulations which can be used to encourage positive nutritional choices and greater participation in physical activity (see Dietz et al. (2009) for an overview of the legal framework as applied to obesity). As suggested by a number of authors in this book, examining the legal frameworks and approaches aimed at reducing the consumption of tobacco may prove fruitful.

Concluding Comments

The global rise of obesity is overshadowing many other public health concerns with the downstream human and financial costs of obesity-related diseases looming in many countries. Its reach is extensive, affecting children as well as adults and increasingly those residing in rural and urban areas in developing and industrially developed countries. The prevalence of obesity-related diseases does not fall evenly across population groups and in many countries these diseases are contributing to growing social inequalities in health.

Obesity has proven impervious to public health messages and clinical treatment targeting individual-level behavioural change. This lack of success at changing the course of the epidemic through individual-level strategies has turned attention to alternative approaches including the identification and modification of obesogenic aspects of our everyday worlds: the foodscapes and urban design of neighbourhoods and the political and economic drivers of these landscapes.

The chapters in this book provide an update on current evidence on an array of environmental determinants of obesity, the limitations of the evidence base, and an indication of what are likely to be promising strategies for curbing obesity's rising toll. Social practices around food consumption and mobility are embedded within the culture of households and communities, a likely contributory factor to the elusiveness of effective change strategies. Eating and moving are amongst the most taken-for-granted aspects of everyday life. And, as the evidence in preceding chapters has shown, they are also practices which are influenced by societal processes that stretch across political, economic, social and cultural arenas. The mundane nature of food and mobility practices and the pervasiveness of the societal level processes that influence them are likely, at least in part, to explain why alarm over our changing food consumption and mobility patterns was slow to build. In addition, many of the changes that have occurred in the transport and food domains have had positive as well as negative impacts on wellbeing further clouding our collective view and making it harder to spot and tease out the devastating longer term health impacts of incremental changes in policies and practices. Looking forward, these factors are also likely to compromise our combined efforts to identify and to implement effective environmental changes to promote healthier eating and mobility practices.

References

Dietz, W., Benken, D. and Hunter, A. 2009. Public Health Law and the Prevention and Control of Obesity. *Milbank Quarterly* 87(1), 215-28.

Foresight. 2007. *Tackling Obesities: Future Choices*. London: Government Office for Science.

Macintyre, S., Ellaway, A. and Cummins, S. 2002. Place effects on health: how can we conceptualise, operationalise and measure them? *Social Science and Medicine* 55, 125-39.

McKinnon, R.A., Reedy, J., Handy, S. and Brown Rodgers, A. 2009. Measuring the Food and Physical Environments: Shaping the Research Agenda. *American Journal of Preventive Medicine* 36(4 Suppl 1), S81-S85.

Story, M., Giles-Corti, B., Lazarus Story, M., Giles-Corti, B., Lazarus Yaroch, A., Cummins, S., Frank, L.D., Huang,T. and Blair Lewis, L. 2009. Work Group IV: Future Directions for Measures of the Food and Physical Activity Environments. *American Journal of Preventive Medicine* 36(4 Suppl 1), S182-S188.

Swinburn, B., Egger, G. and Raza, F.1999. Dissecting obesogenic environments: the development and application of a framework for identifying and prioritizing environmental interventions for obesity. *Preventive Medicine* 29, 563-70.

Tranter, P. and Pawson, E. 2001. Children's Access to Local Environments: a case-study of Christchurch, New Zealand. *Local Environment* 6, 27-48.

World Health Assembly. 2004. *Global Strategy on Diet, Physical Activity and Health (WHA Resolution 57.17)* [Online: World Health Organization]. Available at: http://apps.who.int/gb/ebwha/pdf_files/WHA57/A57_R17-en.pdf [accessed: 9 June 2009].

Index

Page numbers in *italics* refer to figures and tables.

abdominal obesity 30
Ackerson, L.K. 264
active living research 138, 140
active transport behaviour *see* transport behaviour
advertising *see* marketing
aesthetics, neighbourhood 160–61
Alderman, J. 57, 69
Alter, D. 219
Amato, J.A. 302
Amish/Mennonite studies 139–40
ANGELO conceptual framework 39, *42*, *42*–43, 315
animal source foods 25–26, 113
asthma 24
Australia 6, 18
    away-from-home intake 62
    child obesity 21, 118, 218
    fast food restaurants 93, 118, 218, 219, 261
    food availability and cost 97–98, 216
    fruit and vegetable consumption 84
    school food environments 221
    urban sprawl 263
    walkability 48, 136
automobility case study 302–5
away-from-home intake 62–64, 115–16
Ayala, G.X. 260

Ball, K. 82
Barnett, J.R. 300
Bassett, D. 140
Bayesian estimates 283–84
Bean, C. 303
Bell, A.C. 139, 160
Bell, J.F. 263
Berke, E.M. 158, 262
beverages 26, 48–49, 59

Beydoun, M.A. 112
Bhopal, R. 254
Bjork, J. 263
Black, J.L. 261, 265
Block, J.P. 219
Body Mass Index (BMI) 30, 111
Boehmer, T.K. 155, 158, 159, 160, 160–61, 162, 264
Bradford Hill, Austin 252
Bradford Hill criteria 252–53, *253*, 265
Bray, G.A. 59
Brazil 21, 23, 217
Britain *see* United Kingdom
Brown, A.L. 262
Brownell, K.D. 63, 66, 67, 68
Brownson, R.C. 65, 72
built environment and physical activity 151, *154*, *228*, 228–29, 262, 263–65
    active living research 138–40
    green spaces 160, 240, 263, 282
    and motorized travel 133–36
    and social trends 136–38
    solutions, possible 140–41, 231–34, *233*, *234*, 240–41, 243–44
        local action 237–39, *238*
        national targets 234–37, *235*, *236*
        promoting active transport 239–41
        at schools and workplaces 241–42
        socioeconomic differences, dealing with 242–43
    urban sprawl 47–48, 134–35, 262
    see also transport behaviour; walkability
Bull, F.C. 230
Burdette, A.M. 161, 264
Burdette, H.L. 219, 261, 262

Caballero, B. 6
Canada 24, 219, 263
    food availability and cost 216, 218

physical activity policies *235*, 235, 236–37
cancers 24
cardiovascular disease 102, 112, 176
case-control studies 256
Casey, A.A. 260, 262
Catlin, T.K. 155, 159
causality 49–50, 252, 259, 265–66
　Bradford Hill criteria 252–53, *253*
　dietary environment and obesity 259–61
　direction, importance of 254–55
　exercise environment and obesity 261–63
　future research needs 266–67
　randomization 255
　and research design 255
　　endogeneity 258
　　hierarchy *255*, 255–56
　　multilevel modelling 257–58
　　structural equation modelling 256–57
　Rothman's criteria 253–54
　socioeconomic status and obesity 264–65
Chambers, E.C. 264
Chang, V.W. 260, 265
Cheadle, A. 98
Cheer, T. 305
Chester, J. 66
child obesity 24, 111–12, 175, *287*, *288*
　causes of 113
　and food accessibility 260
　health consequences 24
　interventions 120
　research needs, future 196
　and transport behaviour 135, 191–93, 241–42
　trends in 3, 15–16, 21, 175
　see also children's food environments; physical activity, of children
children's food environments 114, *114*, 119–21
　community environment 117–19
　home environment 114–16
　school environment 48–49, 63, 116–17, 220–21, 279
China 15, 21, 23, 26, 28
　motor vehicles, impact of 29, 139, 160

Chopra, M. 62
Christakis, N.A. 47, 266
Christiansen, Annmarie 305
Clarke, G.P. 280, 281
Closing the Gap in a Generation (WHO report) 7
Cohen, D.A. 163
cohort studies 255, 256, 258
Coleman, K.J. 155, 156, 157, 159
comfort eating 137
communities *see* neighbourhoods
community nutrition environments 80, 82–84, 93–96, 101, 117–19
commuting *see* transport behaviour
conceptual frameworks 50, *81*, 101, 315–16
　ANGELO framework 39, *42*, 42–43, 315
　International Obesity Task Force framework 39, 43–44, *44*, 315
Congdon, P. 283–84
consumer nutrition environments 80–82, 97–100, 101
contextual influences 40, 79–80, 166, 230
convenience stores 93, 117, 120, 260
Cosgrove, D. 297
Crawford, D.A. 118, 261
crime and safety 45, 161–62, 178, 282
cross-sectional studies 255, 256, 265
　limitations of 138, 165, 244, 256, 258, 318
Cummins, S. 97, 117, 165, 219, 258, 259, 266–67
Cummins, S.K. 175
Curtis, S. 319, 320
cycling *see* transport behaviour

Daniels, S. 297
Darmon, N. 115
Datar, A. 117–18, 259, 282
Delgado, C.L. 25
Denmark 15–16
deprivation *see* disadvantaged areas
developing world
　built environment and physical activity 141, 237
　dietary changes 26, 28, 215–16
　obesity trends 3, *19*, *20*, 313, 314
　and trade policies 60, 215–16

diabetes 23–24, 31, 112, 176, 284
diet *25, 27*, 62–64
    changes in 16, 24–27, 28, 41, 215–16
dietary behaviours 79, 115, 316–17
    contextual forces on 80–82, *81, 85–92*, 101–2
        community nutrition environments 82–84, 93–96
        consumer nutrition environments 97–100
        other nutrition environments 100
    and household income 64–65, 115, 119, 305–7
    improving 220, 221
    see also marketing
dietary environments 259–61
    children's food environments 114, *114*, 119–21
        community 117–19
        home 114–16
        school 116–17
    macro-level influences 45, 68–69, 80
    micro-level influences 45–49, 79, *81*, 93–96, 99–100, 101
        community nutrition environments 80, 101
        consumer nutrition environments 80–82, 97–100, 101
        and dietary behaviours and health outcomes 84, *85–92*, 93
        organizational nutrition environments 82, 100, 101
        other nutrition environments 82
        research, future 101–2
    see also fast food restaurants; food accessibility
Diez-Roux, A.V. 264
disadvantaged areas 79, 153, 262, 264, 277–78, 314
    and fast food restaurants 46, 83, 96, 219–20
    and food accessibility 61–62, 83, 97–98, 117, 120, 281
discourse studies *298*
    case studies
        Starship Children's Hospital 299–302

        transport behaviours in Auckland 302–5
    discourse defined 299
    discursive disconnect 301
    importance of 307–8
    landscapes 297, 299
    language 297, 299
    prioritizing 305–7
    structure and agency 298–99
    victim-blaming discourse 305
Dixon, J. 307
Doyle, S. 158, 161
Drewnowski, A. 59, 62, 64, 115, 260, 264

eating practices *see* dietary behaviours
ecological studies 255, 261–62, 263
economic resources, household 64–65, 115, 119, 305–7
education, consumer 71, 99, 215, 221
Egypt 18
Eid, J. 262
Ellaway, A. 160–61, 162, 262, 264
Elliott, S.J. 264
endogeneity 96, 258, 318
energy balance equation 4, 251
energy-dense foods 24–26, *27*, 63
    price, influence of 64–65, 115, 214, 215–16, 306
England 3–4, 178, 219, 236, 241, 257, 286
environment
    changes to, recent 4
    defined 5
    influences of *5*, 5–6, 57–58
    macro-level forces in 45
    micro-level forces in 45–49
    personal behaviour, affect on 69
    see also built environment and physical activity
environmental conceptual frameworks 42–45
Eny, K. 219
Epstein, Daniel 305
Estabrooks, P.A. 282
Evenson, K.R. 259

fast food restaurants 27–28, 63, 95, 98, 118
    accessibility of 93, 100, 219–20, 259, 260–61, 281

siting of 46, 49, 83, 96, 258
Starship Children's Hospital case study 299–302
within-store influences 97, 98–99
fat intake 8, *25*, 40, 41, 207, 211
   policies to restrict 214–15, 221
   see also HFSS (high fat, salt and sugar) foods; LFSS (low fat, salt and sugar) foods
foetal development 46, 47
fibre intake 27
food accessibility 61–64, 82–83, 117, 207, 216, 259–61, 279
Food and Agricultural Organization (FAO) 25
food choices
   and accessibility 61–64, 82–83, 117, 207, 216, 259–61, 279
   and price 64–65, 98, 115, 119, 305–7
   see also dietary behaviours
food consumption *see* diet
food deserts 61, 83, 261, 279, 279–81, *280*
food disappearance data 25
food manufacturing industry 45
   economic goals of 58, 59, 66
   hard policies, resistance to 208–9
   regulation of 67–72
   responses to concerns 66–67, 70
   structure of 60–62
   see also marketing
food preparation 28, 45, 62, 115, 136, 315
food pricing 59, 64–65, 98, 117–18
food supply
   changes in 31, 58–60, 72
   regulation of 69–70, 211–14
food supply chain 61, 209, *210*
Ford, P.B. 260
Fowler, J.H. 47, 266
Framingham Heart Study 47
Frank, L. 155, 156, 157, 158, 160, 262, 281
Franzini, L. 262
free trade policies 60
Friedman, R. 67, 68
fruits and vegetables
   availability of 59, 61, 279
   consumption of 27, 84, 93, 279, 281, *287*, 289

cost of 64, 64–65, 98, 118, 214, 216, 259
subsidies 59, 216

Galster, G. 257–58
Garden, F.L. 263
Gatrell, A.C. 284
Gen, S. 157, 160
gene-environment interaction 29–30, 41, 57, 267
genetics 4, 29–30, 39, 57, 267
genomics 267
Giles-Corti, B. 155, 156, 159, 160, 315, 316
GIS (Geographic Information Systems) 278
Giskes, K. 319, 320
Glanz, K. 80, *81*, 282
Glass, T.A. 161, 264
Global Strategy on Diet, Physical Activity and Health (WHO) 231
Goran, M.I. 261
Gordon-Larsen, P. 262
Grafova, I.B. 155, 263, 264
green spaces 160, 240, 263, 282
grocery stores, small 93, 97, 98
Grotz, V.L. 68
Guide for Population-based Approaches to Increasing Levels of Physical Activity (WHO) 232

Hardy, R. 277
Harrington, D.W. 264
Hawkes, C. 98
Hayes, M.V. 265
Hayne, C. 72
health consequences 3, 23–24
Healthy Cities network 232, 237
Healthy City is an Active City, A (WHO) 232
Heinrich, K.M. 262
HFSS (high fat, salt and sugar) foods 9, 28
   and government policies 59, 60, 65
   marketing of 45, 66, 70, 217
   pricing of 64–65
   reformulation of 66–67
high fructose corn syrup (HFCS) 58 59, 314–15
Hill, T.D. 161, 264

Hinde, S. 307
Hoek, J. 113, 314, 315, 317
home food environments 95, 114–16, 119
Hynes, H.P. 166
hypertension 24, 112

incivilities, neighbourhood 162
income and food consumption 64–65, 115, 119, 305–7
India 15, 24, 26
individuals
    policies aimed at 6, 7, 12, 120, 208, 321
    vs. populations 39–40
    responsibility of 68–70, 307–8
industrialization 58, 314
International Obesity Task Force conceptual framework 39, 43–44, *44*, 315

Jackson, R.J. 175
Jacobsen, M. 61, 62, 66
Jalaludin, B.B. 263
Jeffery, R. 66
Jeffery, R.W. 261
Joshu, C.E. 155, 159, 160–61, 162, 263

Kahn, H.S. 264
Kavanagh, A.M. 314
Kawachi, I. 63, 67, 258, 314, 315, 316
Kearns, R. 316, 317
Kearns, R.A. 300
Kelly, C.H. 243
Kersh, R. 68
Kim, D. 63, 65, 67, 314, 315, 316
Kubik, M.Y. 116
Kwate, N.O. 260

Lake, A. 261
land use 6, 133–34, 135, 156, 282
    and transport behaviour 135–36, 157, 159, 193, 240, 263
land zoning 134
landscape 297
Lang, T. 113
language 297, 299
Larson, N.I. 260
Latin America 28

learned behaviour 266
Lee, Harper 131
leptogenic environments 8, 41
LFSS (low fat, salt and sugar) foods 64, 65
Li, F.Z. 156, 260, 263
Lobstein, T. 112
Lopez, R.P. 155, 157, 166
Lopez-Zetina, J. 262
Lovasi, G.S. 263
Ludwig, D. 63, 70
Lumeng, J.C. 264

Macdonald, L. 219
Macinko, J. 261, 265
Macintyre, S. 46, 97, 117, 165, 219, 259
Mackenbach, J.P. 264
macro-level environmental forces 45, 68–69, 80
Maddock, J. 219, 260
Maher, A. 63, 118
marketing 45, 61, 66–67
    to children 63, 65–66, 118–19, 217–18
    outdoor advertising 84, 118
    regulation of 67–68, 216–18
Marone, J. 68
Matheson, F.I. 264
Maziak, W. 64, 66
McCormack, G.R. 48
McDonald's in children's hospital 299–302
McGinnis, J. 66
McKeown-Eyesen, G. 257
McLean, R. 113, 314, 315, 317
Mehta, N.K. 260
Mexico 18, 26
micro-level environmental forces 45–49, 79, *81*, 93–96, 99–100
    children's food environments 114, *114*, 119–21
    community 117–19
    home 114–16
    school 116–17
    community nutrition environments 80, 101
    consumer nutrition environments 80–82, 97–100, 101
    and dietary behaviours and health outcomes 84, *85–92*, 93
    influences of 101

organizational nutrition environments 82, 100, 101
other influences 82
research needs, future 101–2
microsimulation modelling 285–89, *286*, *287*, *288*
migrant studies 40, 265
Miles, R. 262
Mobley, L.R. 156, 159, 161, 264
Monden, C.W. 264
Montgomery, K. 66
Moon, G. 257, 264, 318
Morland, K. 61, 219, 259
motor vehicles, impact of 29, 133–36, 139, 141, 160, 315
Auckland case study 302–5
Mujahid, M.S. 151, 161–62, 263, 264
multilevel modelling 257–58, 283–85
Murphy, Celia 301

natural environments 227–28, 240
neighbourhoods 163–67, 317–19
aesthetics 160–61
crime and safety 45, 161–62, 178, 282
incivilities 162
land use *see* land use
opportunity structures 158–59
physical activity, influence on *see* built environment and physical activity
planning strategies 240–41
recreational facilities 137, 158–59, 194
research review 163–65
methodological considerations 152–53
overview of studies 153–55
research needs, future 165–67
results 153–63
residential density 134, 157, 193
social capital 46–47, 49, 163, 264
socioeconomic environments 45–46, 82, 115, 218, 242–43
street characteristics 155–56, 192, 193, 240–41
street connectivity 154–55, 192, 240–41, 282
transport *see* transport behaviour
walkability *see* walkability

see also community nutrition environments
Nelson, M.C. 264, 283
Nestle, M. 45, 59, 61, 62, 66, 70
New Urbanism movement 232, 244, 262, 303
New Zealand 4, 46
automobility 302–5
fast food restaurants 46, 219, 261
food accessibility 61–62, 84, 118
green spaces 46
marketing 65, 118–19, 279
physical activity policies *235*
prioritizing household income 305–7
school nutrition environments 63, 115–16, 117
Starship Children's Hospital, Auckland 299–302
Nollen, N.L. 116
Nutrition Environment Measures Surveys (NEMS) 99
nutrition environments *see* dietary environments
nutrition labelling 71, 99, 215
nutrition transition 16

Oakes, J.M. 48
obesity, defined 111
obesogenic environments 41, 113, 259, 277
estimation techniques for measuring
limitations of 290
microsimulation modelling 285–89
regression and multi-level modelling 283–85
Okasha, M. 277
Oliver, L.N. 265
Oliver, M. 316, 318
opportunity structures 158–59
organizational nutrition environments 82, 100, 101, 220–21
Ottawa Charter for Health Promotion (1986) 6, 69
overweight, defined 30, 111
Owen, N. 138

Park, Y. 265
Pearce, J. 46, 61–62, 219, 261, 281
Pendola, R. 157, 160

Peters, J. 57
Philippines 28
physical activity 29
   and built environment *see* built environment and physical activity
   of children *176*, 176, 178, 262–63
      active transport behaviour 191–93, 241–42
      environmental influences *177*, *179–90*, 191–94
      parental safety concerns 136, 178, 282, 316
      at school 194–95, 241–42, 282
   community settings and 228–29
   global statistics 132
   influences on *229*
   measuring PA environment 281–84
   promotion of *see* policies, physical activity promotion
   psycho-social mechanisms for 137
   safety concerns 45, 156, 161–62, 241, 243
   transport behaviour 29, 135–36, 159–60, 239, 261–62
      Auckland case study 302–5
   and urban sprawl 47–48
   and workplace 242
Pickett, K.E. 264
policies 5, 6–7, 10, 67–72, 319–20
   child obesity, tackling 113
   concerted action, need for 221–22, 244
   food industry, resistance of 208–9
   food supply, influence on 59–60
   industrialization, support for 58
   levels of governance 209–11, *212–13*
   physical activity promotion 229–30, 231
      barriers to, possible 243
      built environment, changing 237–39, *238*
      community planning 240–41
      information, provision of 243
      international projects 231–33, *233*
      intersectoral collaboration 244
      national targets and policy guidance 234–37, *235*
      public health sector, role of 244
      research needs 244

      social inequalities, dealing with 242–43
      transport planning 239
   policy action framework 211
   policy instruments available *212–13*
      catering/food service regulations 220–21
      food distribution regulations 215–16
      food manufacturing regulations 214–15
      marketing regulations 216–18
      primary production subsidies and taxes 211, 214
      retail regulations 218–20
   soft policies, failure of 208
   suggestions for 71–72
Poortinga, W. 159, 162, 163, 264
Popkin, B.M. 64, 113, 314
populations vs. individuals 39–40
portion size 64, 98
Pothukuchi, K. 62
poverty *see* disadvantaged areas
Powell, L.M. 117
pricing of food 59, 64–65, 98, 117–18
Proctor, K. 286
Proctor, K.L. 286, 289
public health sector, role of 244
public transport 134, 135, 157, 239
Puhl, R. 66

Rayner, G. 113
reciprocity 113, 120
recreational facilities 137, 158–59, 194
reformulation of food products 66–67
regression models 278, 283–85
regulation of food industry *see* policies
Reither, E.N. 264
Relph, E. 303
residential density 134, 157, 193
restaurants 97, 99, 100, 260
   see also fast food restaurants
retail models 280
Rigby, N. 59, 72
Riva, M. 319, 320
Robert, S.A. 264
Robinson, T.N. 282
Rose, Geoffrey 39

Rosenkranz, R. 62, 64
Ross, N. 157, 263, 264
Rothman, Kenneth 253
Rothman's criteria 253–54
Rothschild, M. 71
Rundle, A. 155, 156, 157, 160, 263
rural vs. urban areas 18–21, *19*, *20*, 191–92
Russia 21, 28
Rutt, C.D. 155, 156, 157, 159

Sacks, G. 209, 211
Saelens, B. 158, 262
safety concerns and physical activity 45, 137, 156, 161–62, 241, 243
  parental concerns 136, 178, 282, 316
Sahota, P. 279
Sallis, J. 138
Sallis, J.F. 282
salt *see* HFSS (high fat, salt and sugar) foods; LFSS (low fat, salt and sugar) foods
Santana, P. 263
Schofield, G. 316, 318
schools
  food environments 48–49, 63, 116–17, 220–21, 279
  and physical activity 194–95, 241–42, 282
Schwartz, M. 63, 66, 67, 68
Scotland 46, 97, 219, 262, 266–67
sedentariness 132–33, 194
Shrewsbury, V. 264
Signal, L. 317, 318
Singh, G.K. 263
Smart Growth movement 232, *234*, 244, 303
Smith, D. 319
Smith, D.M. 280, 286
Smith, K. 281
Smith, K.R. 155, 157
Smoyer-Tomic, K. 61
social capital 46–47, 49, 163, 264
social influences 47, 62, 118, 302–3
social marketing campaigns 71–72
social trends 136–38
socioeconomic environments 45–46, 82, 115, 218, 242–43
socioeconomic status 261, 264–65

soft drinks 26, 48–49, 59
South Africa 18, 28
spatial analysis models 277
  estimation techniques
    microsimulation modelling 285–89
    regression and multi-level modelling 283–84
  geographical data issues 284–85
  measuring food environments 279–81
  measuring physical activity environments 281–84
Spence, J.C. 262–63
Stafford, M. 257, 263, 264
Stanton, R. 62
Starship Children's Hospital, Auckland 299–302
statistics, global 3, *17*, 131
  prevalence patterns 18, *18*
  data for 16
  rates of increase 21, *22*
  urban vs. rural areas 18–21, *19*, *20*
Story, M. 58, 59, 61, 63, 64, 65, 319
strategies to tackle obesity *see* policies
street characteristics 155–56, 192, 193, 240–41
street connectivity 154–55, 192, 240–41, 282
stress 47, 137, 264
structural equation modelling 256–57
Sturm, R. 117–18, 259, 282
Subramanian, S.V. 258
sugar 8, 26, 41, 63, 211, 218
  policies to restrict 214–15, 221
  see also HFSS (high fat, salt and sugar) foods; LFSS (low fat, salt and sugar) foods
Sundquist, J. 264
supermarkets 28, 61, 281
  accessibility to 61–62, 79, 83, 218, 279
  and dietary behaviours and health outcomes 84, 93, 117
  within-store influences 97, 98, 99
Susser, Mevyn 254
sweeteners 24, 26 *27*, 58–59
Swinburn, B. 70, 71, 113

television watching 29, 115
theoretical frameworks 49–50
Thornton, L.E. 93, 314
Tillotson, J. 58, 60, 61, 63, 71
Tilt, J.H. 160–61
Timperio, A. 192
To Kill a Mockingbird (Lee) 131
tobacco controls 67, 71, 72
Tolley, Anne 63
Townshend, T. 261
transport behaviour 29, 135–36, 159–60, 239, 261–62
    active transport in children 191–93, 241–42
    Auckland case study 302–5
Tremblay, M. 139, 140
Turrell, G. 314, 318

United Kingdom 6, 18, 21, 175
    physical activity policies 235, *235*, 236, *236*, 241
United States 6, 18, 45–46, 60, 62–63, 97, 215
    child obesity 112, 175, 263
    fast food restaurants 63, 93, 98–99, 100, 219, 260
    food availability 83, 84, 97, 98, 117, 218, 260
    health consequences 23–24
    marketing 45, 84
    obesity trends 3, 23
    physical activity and environment 191, *235*, 262, 263
    school nutrition environments 48–49, 116, 117, 221
    sweeteners, use of 26, 59
    walkability 48, 178, 243
urban sprawl and physical activity 47–48, 134–35, 262
urban vs. rural areas 18–21, *19*, *20*, 191–92
US Institute of Medicine 40
Utter, J. 66

Van Lenthe, F.J. 264
Vandegrift, D. 262

vegetables *see* fruits and vegetables
vending machines 48, 100, 117
Veugelers, P. 263

walkability 48, 135, 157–58, 240–41, 263, 281–82
    measuring 282
    and safety concerns 136, 137–38
    and street characteristics/connectivity 192
    and time/distance involved 192–93
    see also street characteristics; street connectivity
walking *see* transport behaviour; walkability
Walton, M. 279, 317, 318
Wang, M.C. 259–60
Wang, Y. 112
Wardle, J. 264
Whitaker, R. 219, 261, 262
WHO (World Health Organization) *see* World Health Organization (WHO)
Wilson, D.K. 155, 159, 161, 163
Winson, A. 59, 61, 68, 99
within-store influences 97–98, 220
Witten, K. 46, 159
workplaces
    food environments 79, 100, 221
    and physical activity 242
World Health Organization (WHO) 6, 209, 231–32
    Closing the Gap in a Generation 7
    Global Strategy on Diet, Physical Activity and Health 231, 320
    Guide for Population-based Approaches to Increasing Levels of Physical Activity 232
World Trade Organization (WTO) 216
Wrigley, N. 281

Yamamoto, J.A. 99
Yoked, T. 262

zoning of land 134